Kai-Uwe Schrogl
Spyros Pagkratis
Blandina Baranes (eds.)

Yearbook on Space Policy 2009/2010

Space for Society

SpringerWienNewYork

Kai-Uwe Schrogl
Spyros Pagkratis
Blandina Baranes
European Space Policy Institute, Vienna, Austria

Softcover reprint of the hardcover 1st edition 2011

SpringerWienNewYork is a part of
Springer Science + Business Media
springer.at

Cover illustration: ESA-ADES Medialab (ESA's Earth explorer CryoSat 2 mission, launched on 8 April 2010, dedicated to the study of polar ice)

Typesetting: Thomson Press (India) Ltd., Chennai

Printed on acid-free and chlorine-free bleached paper

SPIN: 80084405

With 23 Figures

CIP data applied for

ISSN 1866-8305
ISBN 978-3-7091-1122-2 SpringerWienNewYork

SpringerWienNewYork

Yearbook on Space Policy

Edited by the
European Space Policy Institute
Director: Kai-Uwe Schrogl

Preface

The financial and economic crisis is still an issue of great concern for the global space sector. While space activities fared quite well during the first year of the crisis, effects on public programmes and commercial activities might still become more visible in the future, when public budgets have to be confirmed and when investment cycles in the private sector are completed. So far, however, the governments as well as the companies in the sector have kept to their promises and have been able to even modestly increase their business. This shows that space is regarded on the governmental level as a strategic asset and that it has generated a robust market, which through its services in telecommunications, direct broadcasting, navigation and Earth observation still has a huge potential that can be even further tapped during a situation like the global crisis. Europe is well positioned in this context, but the largest dynamic can be seen in emerging countries, which are partners, markets and competitors at the same time. But so far, growth in the space sector has allowed for beneficial international cooperation and joint economic growth. Europe is taking strong efforts in further developing its internal structures for governing space activities efficiently and seeing a competitive industrial base with manufacturers, operators and service providers grow.

One remarkable event in the timeframe that to is covered by the Yearbook – the period from July 2009 to June 2010 – has been the issuing of a new U.S. Space Policy. A rare expression of a comprehensive approach to all space activities, this document has become the point for extended analysis. While it contains remarkable statements and also changes from the last document of its kind, its impact will have to be seen only in the future. More immediate impacts and concrete effects had a number of policy discussions and events, which all are related to one of the largest issue area for space applications: natural disasters, where space plays a crucial role in their mitigation and related global discussions, as the Summit in Copenhagen, epitomising the problem of climate change. Through this, space received a large visibility and demonstrated its impact. It is for this reason that the thematic title of this Yearbook reflects on "Space for Society", since the application issues – not only for disaster management but also for other areas such as telecommunications, navigation and Earth observation – are highlighted throughout this volume.

As usual, the Yearbook on Space Policy comprises three parts. The first part shows an overview on the global space endeavours. It is prepared in-house in ESPI and it contains the whole spectrum of actors, issues, policies and economic developments. While its perspective is European, is provides an analytical whole of space around the world. The second part again contains contributions from highly distinguished experts in the field. We have been able to assemble personalities mainly from the academic sector, adding also views from agencies and users. Issues which are covered have been highlights during the period of mid 2009 to mid 2010, of course reflecting on the new U.S. Space Policy and the Copenhagen summit, but also highlighting important European issues, like Galileo or the Lisbon Treaty, and in addition looking into international relations and benefits from space activities for societies world-wide. For this purpose, we have again invited contributors from within and outside Europe, thus showing that the network established by ESPI, the European Space Policy Research and Academic Network (ESPRAN) is getting more and more global. The third part of the Yearbook maintains the additional character of the Yearbook as an archive for space activities. Again prepared in-house in ESPI, a chronology, a bibliography and data about institutions is provided, where readers of the now four volumes of the Yearbook can identify statistical developments and trends.

An important milestone in the preparation of the Yearbook was again ESPI's Autumn Conference, where the authors met for an exchange on drafts of their contributions. Having taken place in Vienna in September 2010 and sponsored by the German Aerospace Center DLR, it provided the forum for a constructive exchange and coordination of the contributions. We appreciated very much the excellent discussion culture at that meeting, which lead to new insights and shared analyses. The discussions at the Autumn Conference were additionally supported by members of ESPI's Advisory Council (its Chairman Herbert Allgeier and its member Alfredo Roma), which also acts as the Editorial Advisory Board to ESPI's book series and the Chairman of its General Assembly (Harald Posch). Thanks also go to Johannes Pseiner, Conor Francois and Renaud Abram.

Kai-Uwe Schrogl, Spyros Pagkratis, Blandina Baranes
ESPI Editorial Team

Table of contents

List of acronyms xv

PART 1
The Year in Space 2009/2010

Chapter 1. European space activities in the global context. *Spyros Pagkratis*

1. Global political and economic trends 2
   1.1. Global economic outlook 2
   1.2. Political developments 7
      1.2.1. Security 7
      1.2.2. Environment 9
      1.2.3. Energy 11
      1.2.4. Resources 14
      1.2.5. Knowledge 15
      1.2.6. Mobility 17
      1.2.7. The financial crisis and its consequences for the space sector 18
   1.3. Main science and technology indicators relevant for space activities 20
      1.3.1. Science and Technology inputs 21
      1.3.2. Science and Technology outputs 23
2. Worldwide space policies and strategies 24
   2.1. The United Nations system 25
      2.1.1. United Nations General Assembly (UNGA) 25
      2.1.2. UNGA Committees 25
         2.1.2.1. Disarmament and International Security Committee 26
         2.1.2.2. Committee on the Peaceful Uses of Outer Space (COPUOS) 26
      2.1.3. Other UN bodies and organs monitoring outer space activities 26

2.1.3.1. International Telecommunication Union (ITU) . . . 27
2.1.3.2. UN-SPIDER . . . 27
2.1.3.3. UN Programme on Space Applications (SAP) . . . 27
2.1.3.4. International Committee on Global Navigation Satellite Systems (ICG) . . . 27
2.1.3.5. United Nations Spatial Data Infrastructure (UNSDI) . . . 28
2.1.3.6. Conference on Disarmament (CD) . . . 28
2.1.3.7. United Nations Institute for Disarmament Research (UNIDIR) . . . 28
2.2. The Group on Earth Observation . . . 29
2.3. Europe . . . 29
2.3.1. European Space Agency . . . 29
2.3.2. European Union . . . 32
2.3.3. Eumetsat . . . 36
2.3.4. National governments . . . 37
2.3.4.1. France . . . 37
2.3.4.2. Germany . . . 39
2.3.4.3. Italy . . . 42
2.3.4.4. The United Kingdom . . . 43
2.4. The United States . . . 45
2.5. Russia . . . 55
2.6. Japan . . . 56
2.7. China . . . 58
2.8. India . . . 59
2.9. Emerging space powers . . . 60
3. Worldwide space budgets and revenues . . . 61
3.1. Overview of institutional space budgets . . . 62
3.2. Overview of commercial space markets . . . 66
3.2.1. Direct Broadcast Services . . . 67
3.2.2. Fixed Satellite Services . . . 68
3.3. Evolution of the space industry . . . 74
3.3.1. Industrial evolutions in Europe . . . 74
3.3.2. Industrial evolutions in the United States . . . 79
3.3.3. Industrial evolutions in Russia . . . 85
3.3.4. Industrial evolutions in Japan . . . 86
3.3.5. Industrial evolutions in China . . . 87

3.4. Industrial overview 87
3.4.1. Launch sector 88
3.4.2. Satellite manufacturing sector 92
3.4.3. Satellite operators sector 96
4. The security dimension 97
4.1. The global space military context 97
4.2. Europe 98
4.2.1. National initiatives 98
4.2.2. European Union level 100
4.2.3. European Space Agency 101
4.3. The United States 102
4.4. Russia 111
4.5. Japan 113
4.6. China 114
4.7. India 115
4.8. Other selected space actors 116
4.9. Threats to the space environment 117

Chapter 2. Developments in space policies, programmes and technologies throughout the world and in Europe.
Spyros Pagkratis

1. Space policies and programmes 134
2. Space transportation 134
2.1. Europe 134
2.2. United States 135
2.3. Russia 139
2.4. Japan 140
2.5. China 141
2.6. India 142
2.7. Emerging actors 142
3. Space sciences and exploration 143
3.1. Human spaceflight activities 143
3.2. Lunar exploration 146
3.3. Mars exploration 147
3.4. Saturn exploration 150
3.5. Venus exploration 150
3.6. Neptune exploration 152

3.7. Jupiter exploration 152
3.8. Solar observation 152
3.9. Outer solar system exploration and observation 154
3.10. International cooperation in space exploration. 157
4. Satellite applications 159
4.1. Space-based communications. 159
4.2. Space-based positioning, navigation and timing systems 159
4.3. Space-based Earth observation. 162
5. Technology developments 163
5.1. Propulsion 163
5.2. Information technology. 166
5.3. Spacecraft operations and design 168
5.4. Suborbital activities. 169
5.5. Other technologies 171
5.6. Innovation policy 174

PART 2
Views and Insights

1. Space in the financial and economic crisis. *Christophe Venet* 184
1.1. Introduction. 184
1.2. The financial and economic crisis 185
1.2.1. Basic facts and figures 185
1.2.2. National and international reactions to the crisis 185
1.2.3. The uneven impact of the crisis 186
1.3. Space in the crisis. 187
1.3.1. The space economy 187
1.3.2. The commercial space sector in the crisis. 189
1.3.3. The institutional space sector in the crisis 190
1.4. The political implications of the crisis for the space sector . . . 192
1.4.1. The strategic relevance of space. 192
1.4.2. Space in the framework of a European policy against the crisis 194
1.4.3. The need for a real paradigm change. 195
2. The legal personality of the European Union and its effects on the development of space activities in Europe. *Lesley Jane Smith*. 199
2.1. Introduction. 199
2.1.1. General. 199
2.1.2. European space organisations and activities 199

2.1.3. Tackling EU space activities 200
2.1.4. The national and EU paradigm 201
2.2. Legal personality in international law 202
2.2.1. General 202
2.2.2. Legal personality and the EU 203
2.3. Personality, capacity and competence distinguished 204
2.3.1. Constitutionality 204
2.3.2. Jurisdiction of the Court of Justice 204
2.4. Legal personality, the European legal order 205
2.4.1. Ambit of the new legal order 205
2.4.2. Constraints on legal personality. 206
2.4.3. Competence creep and sovereignty 206
2.4.4. Legal personality, European governance and integration 207
2.5. Demarcation and conferral 208
2.5.1. Principles 208
2.5.2. Conferral defined 208
2.6. Structure of regulating space activities in Europe 209
2.6.1. Challenges 209
2.6.2. Civilian and military aspects of space. 209
2.6.3. Common foreign and security policy and agencification 210
2.7. Conclusion 210
3. Institutional development of satellite navigation in Europe. *An interview of Heike Wieland to ESPI Resident Fellow Spyros Pagkratis* 217
4. The new 2010 U.S. space policy. *Michael Sheehan* 230
4.1. Introduction 230
4.2. Key features of the Obama space policy 231
4.3. The gap 233
4.4. The next generation spacecraft. 234
4.5. U.S. space policy governance 235
4.6. The national security dimension 237
4.7. Conclusions 240
5. The potential for transatlantic cooperation in the International Space Station programme and space exploration. *Ian Pryke* 242
5.1. Introduction 242
5.2. International cooperation in a broader space policy context. . . 242
5.2.1. The Obama administration's civil space policy 243
5.2.2. European civil space policy 246

5.3. Future prospects for transatlantic cooperation 248
5.3.1. The International Space Station 248
5.3.1.1. Potential international cooperation within the context of the existing ISS partnership 250
5.3.1.2. Potential transatlantic cooperation involving nations that are not ISS programme Partners 251
5.3.2. Space exploration 252
5.3.2.1. Implementing an "exploration vision" in the U.S. 253
5.3.2.2. Implementing a European exploration strategy 255
5.3.2.3. The U.S. and Europe working together . . . 255
5.3.2.4. The future 257
5.4. Conclusion 258
6. Trends in shaping space policies around the world. *Deganit Paikowsky and Isaac Ben Israel* 260
6.1. Introduction 260
6.1.1. Space as a symbol and a means of power 260
6.1.2. Greater international cooperation 262
6.1.3. Expansion of national security space missions 263
6.1.4. Commercialisation and industrial scale 264
6.2. American space policy and future space activity 265
6.3. United Kingdom 267
6.4. Australia 268
6.5. Israel 268
6.6. Singapore 269
7. Space applications after Copenhagen. *Simonetta Cheli* 272
7.1. The challenges of a changing world 272
7.2. Copenhagen's accomplishments 273
7.3. The contribution of space to climate 275
7.4. The role of the European Union 278
7.5. Towards Mexico 279
8. Satellite Earth observation and disaster management – lessons and needs after the Indian Ocean tsunami and the Haiti earthquake. *Ray Harris* 281
8.1. Introduction 281
8.2. UN principles on remote sensing 282

8.3. International Charter 283
8.3.1. The Charter 283
8.3.2. Indian Ocean tsunami 285
8.3.3. Haiti earthquake 287
8.3.4. Access and accuracy 288
8.3.5. Other disaster information systems 289
8.4. Data policy trends for science and research 290
8.5. Lessons 292
8.6. Future needs 293
8.7. Conclusion 294
9. Implications of new trends in small satellite development.
Rainer Sandau 296
9.1. Introduction 296
9.2. Small satellite missions: facts and trends 297
9.2.1. General facts 297
9.2.2. General trends 300
9.3. Status and prospects 300
9.3.1. Capabilities of micro-satellites – optical payloads 300
9.3.1.1. Spatial resolution 300
9.3.1.2. Spectral resolution and range 302
9.3.1.3. Temporal resolution 303
9.3.2. Limitations 303
9.3.3. Distributed space systems 303
9.3.3.1. Swarms 305
9.3.3.2. Inspection and docking systems 306
9.3.3.3. Constellations 306
9.3.3.4. Satellite formations 306
9.4. Implications 307
9.4.1. Technical implications 308
9.4.1.1 Data rate and volume 308
9.4.1.2. Access to space 308
9.4.1.3. Ground systems 310
9.4.2. Regulatory implications 310
9.4.2.1. Space system registration 311
9.4.2.2. Transmission frequency allocation 311
9.4.2.3. Space debris management 311

PART 3
Facts and Figures
Blandina Baranes and Spyros Pagkratis

1. Chronology: July 2009–May 2010 314
   1.1. Access to space 314
   1.2. Space science and exploration 317
   1.3. Applications 318
   1.4. Policy and international cooperation 322
2. Country profiles 324
3. Bibliography of space policy publications July 2009–June 2010 336
   3.1. Monographs 336
   3.2. Articles 338
List of figures and tables 343
About the authors 345
Index 351

List of acronyms

A

AATSR: Advanced Along Track Scanning Radiometer
ACI: Airports Council International
ADF: Australian Defence Force
ADM: Atmospheric Dynamics Mission
AEHS: Advanced Extremely High Frequency Satellite
AFRL: Air Force Research Laboratory
ALOS: Advanced Land Observing Satellite
APRSAF: Asia-Pacific Regional Space Agency Forum
APSCO: Asia-Pacific Space Cooperation Organisation
ARATS: Association for Relations Across the Taiwan Straits
ARMC: African Resource Management and Environmental Constellation
ARTA: Ariane 5 Research and Technology Accompaniment Programme
ARTES: Advanced Research in Telecommunications Systems
AR5: 5th Assessment Report
ASAT: Anti Satellite
ASI: Agenzia Spaziale Italiana (Italian Space Agency)
ATM: Air Traffic Management
ATV: Automated Transfer Vehicle
AVIC: Aviation Industries of China

B

BAE: British Aerospace
BGAN: Broadband Global Area Network
BRIC: Brazil Russia India China

C

CALT: China Academy of Launch Vehicle Technology
CASC: China Aerospace Corporation
CASTC: China Aerospace Science and Technology Corporation
CEO: Chief Executive Officer
CEOS: Committee on Earth Observation Satellites
CFSP: Common Foreign and Security Policy
CGWIC: China Great Wall Industry Corporation

CMA: China Meteorological Administration
CMSEO: China Manned Space Engineering Office
CNES: Centre National d'Etudes Spatiales (French Space Agency)
CNNC: China National Nuclear Corporation
CNSA: China National Space Administration
COF: Columbus Orbital Facility
COFUR: Cost Of Fulfilling User Requests
COPUOS: Committee on the Peaceful Uses of Outer Space
COSMO-Skymed: Constellation of small Satellites for the Mediterranean basin Observation
COTS: Commercial Orbital Transportation Services
CSA: Canadian Space Agency

D

DARPA: Defence Advanced Research Projects Agency
DBS: Direct Broadcast Services
DLR: Deutsches Zentrum für Luft- und Raumfahrt (German Space Agency)
DMSP: Defence Meteorological Satellite Program
DOC: Department of Commerce
DoD: Department of Defence
DSTO: Defence Science and Technology Organisation
DTH: Direct-to-Home

E

EADS: European Aeronautic Defence and Space Company
EarthCARE: Earth Clouds, Aerosol and Radiation Explorer
EC: European Commission
ECB: European Central Bank
e-CORCE: e-Constellation of Observation by Recurrent Cellular Environment
EDA: European Defence Agency
EDRS: European Data Relay Satellite
EELV: Evolved Expandable Launch Vehicle
EERP: European Economic Recovery Plan
EGNOS: European Geostationary Navigation Overlay Service
EISC: European Interparliamentary Space Conference
EJSM: Europa Jupiter System Mission
ELINT: Electronic signals Intelligence
ELV: Expandable Launch Vehicle
EMS: Electromagnetic Sciences
EO: Earth Observation

EPS: EUMETSAT Polar System
ERA: European Research Area
ERC: European Research Council
ERS: European Remote Sensing Satellite
ESA: European Space Agency
ESDP: European Security and Defence Policy
ESP: European Space Policy
ESPI: European Space Policy Institute
EU: European Union
EUMETSAT: European Organisation for the Exploitation of Meteorological Satellites
EUSC: European Union Satellite Centre
EVA: Extravehicular Activity

F
FAA: Federal Aviation Administration
FAO: Food and Agricultural Organisation
FCC: Federal Communications Commission
FLPP: Future Launcher Preparatory Programme
FOC: Full Operational Capability
FP7: Framework Programme for research and technological development 7
FSS: Fixed Satellite Services
FY: Fiscal Year

G
GAC: GMES Advisory Council
GAD: General Armaments Department
GAGAN: GPS-Aided Geosynchronous Augmented Navigation System
GAO: Government Accountability Office
GCM: GMES Contributing Missions
GDP: Gross Domestic Product
GEO: Geostationary Orbit
GEO: Group on Earth Observations
GEOSS: Global Earth Observation System of Systems
GERD: Gross Domestic Expenditure on R&D
GES: Global Exploration Strategy
GIANUS: Global Integrated Architecture for iNnovative Utilisation of space for Security
GIO: GMES Initial Operations
GIOVE: Galileo In-Orbit Validation Element

GIP: Galileo Inter-institutional Panel
GIS: Geographic Information System
GJU: Galileo Joint Undertaking
GLONASS: Global Navigation Satellite System
GMES: Global Monitoring for Environment and Security
G-MOSAIC: GMES services for Management of Operations, Situation Awareness and Intelligence for regional Crises
GOCE: Gravity field and steady-state Ocean Circulation Explorer
GOES: Geostationary Operational Environmental Satellite
GOSAT: Greenhouse Gases Observing Satellite
GPS: Global Positioning System
GSA: GNSS Supervisory Authority
GSC: GMES Space Component
GSC: Guyana Space Centre
GSLV: Geosynchronous Satellite Launch Vehicle
GTO: Geostationary Transfer Orbit
G8: Group of Eight
G20: Group of Twenty

H
HDTV: High Definition Television
HR: High Resolution
HSPG: High-Level Space Policy Group
HTV: H-2 Transfer Vehicle

I
IAEA: International Atomic Energy Agency
IBEX: Interstellar Boundary Explorer
ICAO: International Civil Aviation Organization
ICBM: Intercontinental Ballistic Missile
ICG: International Committee on Global Navigation Satellite Systems
ICT: Information and Communication Technologies
IEA: International Energy Agency
IFAD: International Fund for Agricultural Development
IGS: Integrated Geo Systems
IGT: Innovation Growth Team for Space
IGY: International Geophysical Year
IHY: International Heliophysical Year
ILS: International Launch Services
IMF: International Monetary Fund

IMINT: Imagery Intelligence
IMO: International Maritime Organisation
INMARSAT: International Maritime Satellite Organisation
INSPIRE: Infrastructure for Spatial Information in Europe
IOV: In-Orbit Validation
IP: Internet Protocol
IPCC: Intergovernmental Panel on Climate Change
IRIS: Interface Region Imaging Spectrograph
ISA: Israeli Space Agency
ISAF: International Security Assistance Force
ISC: International Space Company
ISECG: International Space Exploration Coordination Group
ISRO: Indian Space Research Organisation
ISS: International Space Station
ITAR: International Traffic in Arms Regulations
ITU: International Telecommunication Union
IXO: International X Ray Observatory

J
JAXA: Japan Aerospace Exploration Agency
JEM: Japanese Experiment Module

K
KSLV: Korea Space Launch Vehicle

L
LEO: Low Earth Orbit
LM: Long March
LMCLS: Lockheed Martin Commercial Launch Services
LRO: Lunar Reconnaissance Orbiter

M
MDA: Missile Defence Agency
MDG: Millennium Development Goals
MEJI: Mars Exploration Joint Initiative
MEO: Medium Earth Orbit
MERIS: Medium Resolution Imaging Spectrometer
MHI: Mitsubishi Heavy Industries
MoD: Ministry of Defence
MoU: Memorandum of Understanding

MPLM: Multipurpose Laboratory Module
MR: Medium Resolution
MSG: Meteosat Second Generation
MSI: Multi-Spectral Imager
MSL: Mars Science Laboratory
MSS: Mobile Satellite Services
MSV: Mobile Satellite Venture
MTCR: Missile Technology Control Regime
MTG: Meteosat Third Generation
MUOS: Mobile User Objective System
MUSIS: Multinational Satellite-based Imagery System

N
NASA: National Aeronautics and Space Administration
NATO: North Atlantic Treaty Organisation
NEO: Near-Earth Objects
NGO: Non-governmental Organisation
NOAA: National Oceanic and Atmospheric Administration
NORAD: North American Aerospace Defence Command
NPOESS: National Polar-orbiting Operational Environmental Satellite System
NRO: National Reconnaissance Office
NSSA: National Security Space Authority

O
OECD: Organisation for Economic Co-operation and Development
OHB: Orbitale Hochtechnologie Bremen
OPEC: Organisation of Petroleum Exporting Countries
ORFEO: Optical and Radar Federated Earth Observation
ORS: Operationally Responsive Space
OSTM: Ocean Surface Topography Mission

P
PBEO: Programme Board for Earth Observation
PLA: People's Liberation Army
PNT: Positioning, Navigation and Timing
POES: Polar Operational Environment Satellites
PPP: Public Private Partnership
PRS: Public-Regulated Service
PSA: Programme on Space Applications
PSLV: Polar Satellite Launch Vehicle

Q

QDR: Quadrennial Defence Review

R

R&D: Research & Development
RISAT: Radar Imaging Satellite
RLV-TD: Reusable Launch Vehicle Technology Demonstrator
RSCC: Russian Satellite Communications Company
RTD: Research and Technology Development

S

SA: Société Anonyme
SAFER: Services and Applications for Emergency Responses
SAR: Synthetic Aperture Radar
SBSS: Space Based Surveillance System
SDA: Satellite Data Association
SDI: Strategic Defence Initiative
SDO: Solar Dynamics Observatory
SELENE: SELenological and ENgineering Explorer
SES: Single European Sky
SES: Société Européenne des Satellites
SHF: Super High Frequency
SHSP: Strategic Headquarters for Space Policy
SIA: Satellite Industry Association
SICRAL: Sistema Italiano per Comunicazioni Riservate ed Allarmi
SIGINT: Signal Intelligence
SME: Small and Medium Enterprise
SMOS: Soil Moisture and Ocean Salinity
SOHO: Solar and Heliospheric Observatory
SPOT: Satellite pour l'Observation de la Terre (Earth Observation Satellite)
SS2: Space Ship 2
SSA: Space Situational Awareness
SSC: Swedish Space Corporation
SSL: Space Systems/Loral
SSN: Space Surveillance Network
SSOT: Sistema Satelital para Observacion de la Tierra (Satellite System for EO)
SSTL: Surrey Satellite Technology Ltd.
S&T: Science and Technology
START: Strategic Arms Reduction Treaty
STSS: Space Tracking Surveillance System

T

TCBM: Transparency and Confidence Building Measures
TSAT: Transformation Communications Satellite

U

UAE: United Arab Emirates
UHF: Ultra High Frequency
ULA: United Launch Alliances
UN: United Nations
UNCCC: United Nations Climate Change Conference
UNEP: United Nations Environment Programme
UNESCO: United Nations Educational, Scientific and Cultural Organization
UNFCCC: United Nations Framework Convention on Climate Change
UNGA: United Nations General Assembly
UNGIWG: United Nations Geographic Information Working Group
UNIDIR: United Nations Institute for Disarmament Research
UNISPACE: United Nations Conference on the Exploration and Peaceful Uses of Outer Space
UNOOSA: United Nations Office for Outer Space Affairs
UNSC: United Nations Security Council
UNSDI: United Nations Spatial Data Infrastructure
UN-SPIDER: UN Platform for Space-based Information for Disaster Management and Emergency Response
USAF: United States Air Force
USGS: United States Geological Survey
USSTRATCOM: United States Strategic Command
UV: Ultraviolet

V

VC: Venture Capital
VHR: Very High Resolution
VNIR: Visible and Near Infrared

W

WEU: Western European Union
WFP: World Food Programme
WGS: Wideband Global Satcom
WHO: World Health Organisation
WTSA: World Telecommunication Standardisation Assembly

PART 1

THE YEAR IN SPACE 2009/2010

European space activities in the global context

Spyros Pagkratis

1. Global political and economic trends

In 2009 the global financial crisis entered a new stage, in which the adverse effects of last year's credit crisis started to weigh on worldwide economic activity. The year was marked by a fall in global industrial production and trade activity and a consequent steep rise in unemployment. However, the first signs of improvement also made their appearance, as bank earnings and capital levels began to rise again and GDP growth started to return, although it is not expected to reach pre-crisis levels for several years. In 2010 this trend is expected to continue, but economic recovery will be slow and precarious. This year's economic policies are expected to focus on continuing the reform of the financial and banking system, rebalancing the patterns of global trade, boosting private consumption, enhancing international cooperation and restraining unemployment rates before they change from cyclical to structural. The pace of economic recovery is expected to be slow and very different from country to country. Emerging economies will exit the crisis at a quicker pace than advanced ones, but the whole process will remain fragile and extremely vulnerable to adverse events such as rising commodity prices, geopolitical events, or a resurge of protectionism.

1.1. Global economic outlook

In 2009 the global economy appears to be expanding again and this trend is expected to continue in 2010. At present, Asian economies seem to be the driving force behind global economic recovery, whereas stabilisation and modest improvement is the case elsewhere. Apart from Asia however, recovery is projected to be weak and slow by historical standards and GDP growth will remain well below pre-crisis levels until 2014 at least.[1] For 2010 global activity is expected to expand by approximately 3%, after a 1% contraction in 2009. Growth in emerging economies will be significantly higher.[2] This sluggish recovery will be marked by long lasting post-crisis characteristics such as low inflation, a drop in private consumption and investment, and a steep rise in unemployment which may become structural. Markets and financial institutions have been stabilising and will continue to do so in 2010. Nevertheless, market financial stress and risk aversion will remain elevated for the foreseeable future, which will put considerable stress on households and medium-size enterprises, and will consequently continue to increase bank loan delinquencies. On the upside, international capital flows are on the way to recovering.[3]

In the financial sector the year has been marked by a slow return of risk appetite that has led to considerable currency fluctuations, with the Euro strengthening its position against both the Dollar and the Yen on the second half of 2009, before falling again in 2010. Bank loans to the private sector however are still stagnating, especially in advanced economies. In fact, credit risks remain elevated and the sustainability of bank earnings is still precarious at best: in October 2009 global bank write-downs were estimated to reach $2.8 trillion and more than half of this amount has not yet been recognised. The bulk of these losses are attributed to U.S., UK and Euro zone banks. In addition to this, a further $1.5 trillion wall of maturing dept will have to be met by 2012.[4] By comparison to European banks, U.S. banks have deleveraged faster and this may help credit conditions in that country to ease sooner. Nonetheless, financing conditions for consumers and medium-size companies in developed countries are expected to remain difficult.

In the second half of 2009 global markets continued to stabilise and this is expected to continue in 2010. Even though investment will not attain pre-crisis levels in the foreseeable future, a certain risk appetite has returned. For the moment, however, market recovery seems fragile, a number of financial stress indicators remains high and the fear of a possible reversal weighs heavily on investors. In the context of the credit conditions described above, global markets are thought to remain extremely sensitive to external factors such as geopolitical events or real-estate-related shocks. Real-estate in particular will continue to put pressure on bank balance sheets, whereas subsequent low construction activity is expected to create additional risks for the financial sector in general.[5]

On a global scale inflation moderated to 1% in mid 2009 down from 6% a year earlier and is expected to remain low in 2010 as well. Inflation rates in emerging economies varied considerably from region to region, dropping in Asian countries and rising in East European ones. Advanced economies are still facing mild deflation risks as the pace of economic recovery remains slow, even though inflation rates are expected to rise above zero in 2010. Deflationary dangers in these countries are aggravated by the fact that interest rates have been brought close to zero and there is little room left for additional financial stimulus from monetary policy measures.[6]

Unemployment rose throughout 2009 and is anticipated to continue rising in advanced economies throughout 2010. Both in the U.S. and the Euro zone, unemployment rates are anticipated to exceed 10% in 2010. Non-financial corporations and medium-size companies will continue to lay off workers due to the aforementioned difficult financial conditions. Countries with proportionately greater construction sectors will suffer even greater job losses. Euro zone countries are projected to face higher unemployment rates than the U.S. (up to 12% in 2010) due to a more sluggish recovery and a less adjustable job market. In the medium-term, historical evidence suggests that in the aftermath of major economic crises and the protracted recovering period that succeeds them, unemployment can become structural and difficult to deal with. This might be the case in the Euro zone, where unemployment rates are not expected to fall bellow 10% before 2014 at the earliest.[7] In any event, rising unemployment will pose a major challenge to all advanced economies throughout 2010.[8]

As a result of the above-mentioned factors, governments worldwide will continue to implement extraordinary public support measures for financial institutions well into 2010. These measures however will have to face the challenge of transforming from short-term financial stimulus schemes to medium-term comprehensive reform policies. Formulating these policies faces three major challenges: rallying the necessary public support, choosing the right timing, and respecting as much as possible macroeconomic budgetary and fiscal constraints. Indeed in 2009 and the first half of 2010, public support for the recapitalisation of financial institutions diminished considerably, especially in advanced economies. Public opinion is becoming more and more sceptical on measures that are perceived as generous government bailouts for firms that were largely responsible for the credit crisis in the first place.[9] This development, in conjunction with increasing unemployment, will make governments reluctant to increase recapitalisation measures in the face of mounting political pressure to do the opposite.

In 2010, political considerations together with an improving financial environment will push governments to consider lifting the extraordinary monetary accommodation that they offered to financial institutions in 2008. It seems that the most difficult task ahead will be to carefully choose the timing of this decision. If the unwinding of public intervention comes too soon, it will place the progress made in 2009 in jeopardy. If it is protracted for a longer period than necessary, it will distort market incentives and create fiscal problems for national budgets.[10] Although monetary accommodation measures are likely to stay in force throughout 2010, governments will probably have to decide on this matter before the end of the year.

Finally, lifting recapitalisation measures will have to be accompanied by medium term policy decisions on reforming the financial sector framework, while restructuring fiscal policies to accommodate the large public dept that the crisis generated in many countries. Prudent macroeconomic decisions will have to be made on both issues in 2010 and this development is already under debate both on a national and an international level. In fact in 2009, there has been an unprecedented level of international cooperation in tackling the credit crisis aftermath. In 2010, this cooperation is expected to expand into taking specific regulatory decisions on reforming the financial sector operating framework, stabilising the economic circle, and avoiding financial protectionism. Indeed, protecting public finances and especially central banks' balance sheets already became a key plank of economic measures in the second half of 2009, and this is expected to continue. In conclusion, the main challenge that advanced economies are facing in 2010 is the need to find room for adequate macroeconomic countercyclical policies in the face of fiscal problems caused by accumulated public dept during the crisis period.[11]

One of the key trends in 2009 and 2010 has been that emerging economies have entered recovery much faster and easier than advanced ones. This is particularly the case for China and India, which escaped a severe recession. With considerable help from its robust fiscal position and the overall health of its banking sector, China has initiated large policy stimuli (up to 5% of its GDP in 2009) and successfully managed to overcome the fall of its exports, which in 2009 were reduced by 30% compared to 2008. This was mainly achieved through boosting domestic demand (private credit rose by 25% in the first half of 2009) and undertaking major infrastructure and industrial retooling projects. This led to an 8.4% GDP growth in 2009 and a continued expansion in 2010.[12]

In fact, China has been the driving force behind the recovery of the entire SE Asia region, where capital flows resumed in 2009 and markets rose sharply. Nevertheless, given the slow pace of recovery in advanced economies, it remains unclear whether Chinese growth will be able to sustain itself beyond 2010 without an adequate increase in exports. At the same time, boosting domestic demand by prolonged credit growth may increase inflationary pressure in the medium term. The Indian economy grew at a somewhat slower pace in 2009 and 2010 as well, at an annualised rate a little above 6%. Growth has been facilitated by adequate monetary policies and a relatively smaller dependence of the Indian economy on exports.[13]

In 2009, Russia experienced an estimated 8.7% contraction of its GDP.[14] This development was the result not only of the world credit crisis, but also of the fall of the oil price that occurred. Low oil prices caused a considerable surge in capital flows in the first half of 2009, which led to an important 5.9% depreciation of the ruble, but this trend was reversed in the 4th Quarter, following a rise in oil prices and a considerable increase in exchange and gold reserves.[15] Domestic demand in the country fell sharply, followed by production (−12.6% in tradable goods in

2009) and investment. Unemployment adjusted to 7.6% at the third Quarter of 2009, from 9.2% at the beginning of the year and it is projected to remain stable in 2010 as well. From the fourth Quarter of 2009 industrial output has been improving slowly and consumption has been regaining lost ground, but real wage reductions and tight credits have caused non-tradable goods production to continue stagnating. At the same time the credit market is not expected to ease throughout 2010.[16]

In economically advanced countries the pace of economic recovery has been considerably slower. In the U.S. the financial situation has been stabilising throughout 2009 and the first Quarter of 2010. GDP contraction has been slowing down from −6.4% at the beginning of 2009 to a 2.2% increase in the third Quarter.[17] On an annual basis, the U.S. economy is expected to contract by 2.45% in 2009, but a modest growth of 1.5% is expected for 2010. Although economic stabilisation is likely to continue, growth will probably not exceed the rate of 2% in the medium turn. In the mean time, credit conditions remain uncertain and unemployment has risen to the highest rates since the early 1980s (in 2009 it is expected to reach 10% on an annual basis). The greatest challenge for the U.S. economy in 2010 is to prevent high cyclical unemployment rates from becoming structural, as well as addressing long-term imbalances in public, corporate and household expenditures.[18]

In Europe, recovery seems to be more sluggish than in the U.S. The Euro zone did not emerge from recession before the end of 2009, and it is predicted to attain growth rates less than 1% in 2010. Further growth will only be attained gradually and in the medium-term. Unemployment reached 10% in 2009 and might reach 12% in 2010. Credit in the Euro zone remains tight due to the greater role of banks in the financing system, as well as major exposures to cross-border risks regarding banking activity in Eastern Europe. Emerging EU economies, such as those of the Baltic States, Bulgaria and Romania, have been hit particularly hard by the crisis, whereas countries with moderate current account deficits or surpluses have shown more resilience.[19] In 2010, public expenses in most EU countries are deteriorating sharply, and addressing this problem will be of great importance. Containing the rise of unemployment and supporting demand under strict budgetary restrictions will prove a major challenge in 2010 for most European countries.

In Japan, stabilisation started in the second half of 2009 and continues in 2010.[20] After a steep GDP drop (−11.9%) in the first Quarter of 2009, modest growth (2.7–1.3%) returned during the rest of the year and continued in 2010.[21] Unemployment rates throughout the aforementioned period remained high by Japanese standards, hovering above 5% on an annual basis in 2009, while at the same time real wages continued to decline. Corporate and bank profits were substantially reduced and mild deflationary pressures appeared on prices. Business investment continued falling and uncertainty about the future of the economic outlook remained high among both investors and consumers. Nevertheless, industrial output has been increasing since the third Quarter of 2009, profiting mostly from the rise of regional commercial activity, and consumption has been increasingly showing signs of improvement.[22] In general terms, recovery in Japan is following the slow and gradual path witnessed in the rest of the advanced economies, with the addition of a relevantly elevated deflation risk.[23]

1.2. Political developments

1.2.1. Security

Security is a field in which space systems are vital. For the purposes of this report, security is defined in its traditional narrow definition related to defence and the ability to effectively engage in military operations. A broader definition of security is briefly discussed in section 1.2.5. Satellite systems are identified as key enablers of military capabilities. These space applications include image and electronic surveillance gathering, communications, meteorological and navigation/positioning data, among others.

A major development in 2009 and 2010 was the rapid deterioration of the security situation in Afghanistan. Taliban insurgents considerably improved their operational and logistics capabilities in the aforementioned period, resulting in a record high number of casualties for the ISAF coalition forces in the country. These amounted to 520 dead in 2009, a significant increase from 295 in 2008. During the same period, U.S. forces casualties marked a 100% increase, to 316.[24] The bulk of fatalities was attributed to improvised explosive device attacks, which were up by 60% from the year before. Civilian casualties also increased by 12%.[25] The total number of such incidents exceeded 7,200 from 4,169 in 2008, whereas their average explosive charges and destructive capability doubled.[26]

For the first time since August 2009, Taliban insurgents launched a series of suicide attacks inside Kabul. On 28 October 2009 a United Nations personnel residence came under an attack that resulted in the loss of 5 U.N. staff members. As a direct result of this incident, more than 340 U.N. personnel members were relocated outside the country, seriously downgrading the U.N. assistance mission's performance in the area.[27] Taliban forces also resumed their intimidation tactics against the local population with a series of targeted assassination attempts. The overall deterioration in security conditions crippled the United Nation's humanitarian aid and reconstruction programmes.[28]

Security conditions in the country were also affected by mounting political instability. On 19 November 2009, Afghanistan's President H. Karzai was finally inaugurated for a second term. This development ended two months political turmoil between the President and his principal political opponent Dr. Abdullah over the latter's accusations of electoral fraud in the 20 August presidential ballot. President Karzai was proclaimed the winner of the electoral process only because his opponent refused to participate in the second round. However, the run-up to the finalisation of the result increased civilian unrest and paralysed the government. Consequently, public confidence in the country's reconstruction and future also waned.[29]

In the midst of these negative developments, the U.S. President announced on 1 December 2009 a new strategy for Afghanistan. He announced the dispatch of an additional 30,000 troops reinforcement to the country. At the same time, President Obama reiterated his plan to begin the gradual withdrawal of U.S. forces from the country by July 2011. The additional forces proposed would increase annual war costs by $30 billion, or almost by 50% in comparison to the current budget.[30]

The new U.S. policy in Afghanistan followed from a comprehensive strategy document released on 27 March 2009. The new strategy widened the scope of U.S. objectives in the region by including neighbouring Pakistan in its scope of operations. It also recognised that the Taliban principal logistics and command posts were concentrated in Pakistan's border regions with Afghanistan. The proposed action plan included disrupting terrorist operations inside Pakistan, while at the same time increasing military and political assistance to that country. Supporting Pakistan would also involve increased financial cooperation and government building measures to promote democratic rule in that country. The new U.S. policy also called for state building actions in Afghanistan itself, including a new strategic communications and joint civilian-military counterinsurgency strategy.[31]

Another issue that continued to provoke tensions on the international scene was the negotiations regarding the Iranian nuclear programme. On 18 February 2010, the International Atomic Energy Agency (IAEA) published its latest regular two month revue of Iran's atomic energy related activities, in the framework of the relevant U.N. Security Council resolutions. In this document, the IAEA clearly stated that Iran was not cooperating in the verification of the peaceful purposes of its nuclear programme. Furthermore, the Agency found that Iran had failed to meet the requirements set by the U.N. Security Council in order to provide assurances for the nature of its programme. Finally, it particularly took notice of the continued operation of the enrichment facilities in Natanz.[32]

In a considerable departure from its past reports, the IAEA explicitly claimed for the first time that Iran could have possibly started the development of a nuclear payload for a missile. It also went on to give specific technical details of Iran's possible nuclear weapons development capabilities, based on its information.[33]

1.2.2. Environment

Space applications have an important role in environment monitoring and protection. Space assets are uniquely positioned to offer a global perspective on climate change. They often also represent a common multinational platform for collecting relevant meteorological and environmental data. This characteristic makes them ideal promoters of international understanding and cooperation in this field.

Climate change and the concerted international effort to control it continued to be the main issue in environmental policy in 2009 and 2010. Global warming remains a major threat not only to the environment, but also to long-term economic growth and prosperity worldwide. It can potentially disrupt food supply, cause major humanitarian catastrophes, destabilise developing countries and consequently endanger their population.[34] From a political point of view the most important development was the UN Climate Change Conference held in Copenhagen from 7 to 18 December 2009. Its proceedings included the 15th conference of the 193 parties to the UN Framework Convention on Climate Change (UNFCCC/COP) and the fifth meeting of the 189 parties that have adhered to the 1997 Kyoto Protocol (CMP).[35] The Conference was attended by more than 15,000 participants, including 119 Heads of State on its final day, and it attracted unprecedented public attention and press coverage.[36] The conference's principal aim was to discuss appropriate measures against global climate change that will have to be implemented before the Kyoto Protocol's provisions expire in 2012.[37]

Despite the great expectations nourished before the conference and the fact that all participants acknowledged the urgent nature of the measures that had to be taken, progress during the conference was modest and decisions did not arrive until its very last day. The conference's main declared objectives were: to set new long-term emission reduction rates for 2020; to adopt appropriate mitigation actions for developing countries; to initiate a long-term funding commitment from developed countries to sustain these actions; and to set up an appropriate institutional framework for addressing the needs of developing countries.[38] The key objective was to cut down emissions to 25–40% lower than 1990 levels by the year 2020.[39]

Reaching an agreement proved difficult due to the different views between developed and developing countries. The final result of the deliberations was the "Copenhagen Accord", an agreement for industrialised countries to voluntarily limit their emissions by 2020 and for developing countries to muster their efforts to reduce emissions and to communicate their results every two years. All voluntary pledges to limit emissions were listed in the accord by the end of January.[40] It was also agreed that the accord would be reviewed before 2015. Raising funds among developed countries for appropriate actions also proved more difficult than expected. However, a dedicated fund (the "Copenhagen Green Climate Fund") was established to support immediate action against climate change over the next three years, with a total budget of $30 billion. At the same time, the long-term goal of raising $100 billion by 2020 was also reiterated. The next UNFCCC conference is scheduled to take place by the end of 2010 in Mexico City, after two preparatory negotiating sessions in Bonn in 31 May and 11 June.[41]

In addition to this, the 3rd World Climate Conference (WCC-3) was held by the World Meteorological Organisation (WMO) in Geneva between 31 August and 4 September 2009, almost 20 years after the last one in 1990. Participation included an expert panel as well as high-level government representatives. Its main scope was to enhance international cooperation and global capabilities in the field of climate information and weather prediction. In a way, it gave scientists the opportunity to review climate related scientific practices ahead of the Copenhagen Conference. WWC-3 concluded its work with a high-level political declaration and a separate conference statement.[42] Participants agreed on establishing a Global Framework for Climate Services, in order to coordinate and strengthen production and availability of climate prediction services worldwide. They also decided to set up an independent task force of experts that will deliver recommendations on the structure of this Framework within 12 months. These recommendations will then be presented for adoption at the next WCC Conference in 2011.

At the EU level several initiatives were taken by the Swedish Presidency in the first half of 2009. The key subject was preparing the EU's participation in the Copenhagen conference. EU policy objectives for the conference were ambitious. They focused on making considerable progress towards a new comprehensive and binding global treaty to replace the Kyoto Protocol after 2012. In order to achieve this, EU members looked forward to obtaining tangible emission reduction commitments from all conference participants; assuring public finance to implement these reductions until 2020; adopting a new institutional framework for international cooperation on climate change; and implementing a strict follow-up process to monitor the progress made.[43]

In spite of the fact that the EU offered to contribute €2.4 billion by 2012 to meet its funding obligations, its expectations for the conference were not met. Although the conference's decisions were considered a positive first step, emission reductions rates were considered insufficient, long-term financing was not secured and the overarching objective of negotiating a new legally binding treaty was not attained.[44] However, EU Member States associated themselves with the Accord's provision for a 20% emission reduction and offered to increase cuts to 30%, if developing countries would agree to contribute to gas emissions as well.[45] Other key environment policy objectives during the Swedish presidency included protecting biodiversity and promoting the EU's transition into an eco-efficient economy based on renewable energy sources, energy-efficient systems and adequate community planning. Eco-efficient economy in particular is now seen as a possible competitive edge for EU that will enable it to develop new technologies and to become more self-sufficient in terms of energy and natural resources supply.[46]

1.2.3. Energy

Space systems can contribute from orbit to the exploitation of Earth's energy resources. Imaging satellites help determine surface resources and underground deposits alike. Communication and space observation satellites help operate and monitor fossil energy transport corridors. Space applications provided motivation for solar panel technology improvements that are now at the forefront of renewable energy technologies. With recent developments in solar energy gathering satellites, space could become a source of energy for terrestrial use itself.

From mid 2009 to mid 2010 energy demand began to rise again. After a turbulent 2008, when oil prices fluctuated violently, oil prices in 2009 and the first half of 2010 have stabilised to roughly $75–82 bbl.[47] This price level represents a 15 month high and a significant rise from December 2008's $33bbl.[48] Rising oil demand in the developing countries and a particularly cold winter in the northern hemisphere drove the prices up by the end of 2009. This trend continued in 2010, fuelled mostly by economic recovery in SE Asia, as well as by increasing investment flows to commodity assets. Oil supply has been rising mildly throughout this period (roughly by 0.4 billion barrels) while demand was declining. As a result, commercial oil inventories remain significantly high, to approximately 60 days of forward cover. This development, in conjunction with rising demand in 2010, has driven spot and freight prices up.[49]

Although oil demand is increasing in developing countries and it has stopped decreasing in developed ones, global oil consumption remains considerably lower than pre-crisis levels (app. −2.3 mb/d) and this trend is expected to continue throughout 2010. The continued upward trend of the oil market price in spite of sufficient production and increasing inventory volumes is mainly attributed to increasing investment flows in commodities markets. In fact, the financial sector's exposure to energy commodity assets rose by 70% in the period between December 2009 and January 2010. Since oil production fundamentals do not support these prices, we might be facing the possibility of a downward price spiral in the second half of 2010.[50]

In the medium-term these trends are not expected to change. Oil consumption will not attain pre-crisis levels before 2012, provided that recovery continues. In OECD countries it will remain particularly sluggish and global consumption will mainly be sustained by SE Asia's emerging economies. Almost 80% of the projected increase in oil consumption until 2030 is expected to come from these countries, and the transportation sector will be the driving force behind this demand. Supply is expected to rise modestly, mostly thanks to increasing output by non-OPEC countries. This increase will be the result of exploitation of non-conventional oil sources (e.g. Canadian oil sands). Although OECD countries' supply is expected to decline, natural gas and renewable sources' exploitation will compensate for this loss and overall spare oil supply capacity will remain adequate. Nevertheless, since current price levels do not encourage investment decisions in oil supply, OPEC countries are projected to slightly increase their share of the market and a new price boom cannot be excluded in the medium-term.[51]

Gas prices declined considerably in 2009, due to limited industrial demand, which dropped by up to 10%. Gas-generated power demand, in particular, fell by up to 8% because of its position in the merit order. However, cold weather has kept domestic and commercial heating demand strong and this fact has partially compensated for the decline in industrial use.[52] Price fluctuations varied from market to market. U.S. prices were relatively stable in 2009, because they had already adjusted to the crisis in late 2008. Oil-based prices in Continental Europe and Japan dropped sharply in 2009 as a result of the 2008 crisis, due to the fact that inbuilt time lags in supply contracts did not allow them to adjust earlier. Gas prices in Europe, in particular, remained considerably higher (up to 100%) than those in the U.S. where production actually increased in 2009, mostly thanks to unconventional gas production growth. This development actually allowed U.S. liquid natural gas shipment to be diverted to the Pacific market and to fuel booming industrial demand there. Nevertheless, U.S. future gas output remains one of the main uncertainties concerning future market behaviour.[53]

Another major development in gas markets in 2009 and 2010 was the dramatic increase in LNG supply capacity. Many LNG production development plans started production in 2009, leading to an unprecedented increase in output capacity (over 370 billion cubic metres). Excess supply capacity will probably test the market's flexibility and resilience in 2010.[54] As with oil, the main issue will be inadequate investment for future projects, under the adverse current financial conditions. With the pace of economic recovery remaining very slow in advanced economies, any prediction for future LNG demand is risky. As a result, most supply development projects are likely to be put on hold for a while, thus creating a shortage of new output capacity after 2012. Obviously, the longer the recovery takes, the greater the LNG supply shortage will be in the medium-term and the higher its price.[55] As gas use in advanced economies is stagnating, developing economies like China or India are emerging as major gas users. In the medium-term, both countries are expected to exceed 100 bcm in annual consumption rate. At the same time, new gas suppliers from the Middle East such as Qatar and Iran are appearing, although the latter is not expected to become a significant exporter before 2015.[56]

In Europe, the early 2009 gas supply crisis left its mark on the entire year and well into 2010 as well, making paramount the issue of strategic gas supply security. The crisis underlined chronic interconnectivity, reverse flow and storage capacity deficiencies in many European countries, especially in Central and Eastern Europe. Better and time-efficient cooperation among European countries in this area has become a key subject of discussion in 2009 and 2010. The objective for the EU in particular is to enhance energy security through varying gas sources and routes, increasing storage capacity and diversifying electrical power sources. However, implementing these policies will require considerable funding in the short-term that is not guarantied in the current financial conditions. Furthermore, diversifying power supply by embracing renewable energy sources might actually increase gas consumption in the medium-term, as environmentally poor power sources will be abandoned and renewable ones will not yet be able to entirely substitute for them .[57]

In conclusion, global and especially European energy policies from mid 2009 to mid 2010 are facing multiple challenges with often contradictory solutions. Improving environmental efficiency will necessitate adequate funding that is difficult in times of financial insecurity. Limited investment will result in greater dependency on imported gas supply in the medium-term and limited energy security. Advanced economies will have to improve energy efficiency, while at the same time coping with the price fluctuations caused by booming demand in emerging economies. Increasing gas demand in SE Asia combined with stable or declining demand in the U.S. (and possibly Europe) will raise international pressure to disassociate their price index.[58] In general, long-term policies designed to improve the environmental impact and efficiency of energy resources will decrease energy security in the medium-term by consolidating the market power of traditional resource suppliers, such as Russia and the Middle East countries. Consequently, an unprecedented level of international cooperation, prolonged investments and strict market transparency measures will be indispensable, in order to tackle all of these issues simultaneously.

1.2.4. Resources

Space applications can also be useful for accommodating international trade and improving the exploitation efficiency of other natural resources as well. Imaging and meteorological satellites can make agricultural output bigger and more reliable. Furthermore, communication satellites are indispensable for making international business transactions and payments, which continue to grow in today's globalised commercial environment.

After a sharp drop in 2008 and in the first half of 2009, international trade and commodity prices began to rise again from the third Quarter of 2009. This development was principally due to industrial output recovery in the emerging economies of Asia, and especially China. However, the slow pace of economic recovery in developed countries is still limiting commodity demand and pushing prices down. Overall global trade in 2009 contracted by approximately 17.6%.[59] As a result of weak recovery and weak base effects, commodity prices are not expected to reach pre-crisis levels before 2011 at the earliest.[60] For 2010, global trade growth is not expected to exceed 4.2%.[61] Low industrial production levels caused metal prices to plummet in 2009 (aluminium: −11%, copper: −9%). Chinese demand supported price levels considerably, mostly thanks to extensive restocking. If Chinese demand were excluded, metals price decline would have exceeded 20%.[62] In 2010, the modest price raise is expected to continue. In general, demand from China had a rather stabilising effect on commodity prices throughout 2009.[63]

Although agricultural products declined by 22% in 2009 compared to their 2008 peak, they still remain almost twice as high as the lows recorded earlier in the decade. Higher oil prices and bio fuel demand, together with an increase in stockpiling, contributed to an upward trend in prices in 2009 and 2010. Lower production costs, however, helped counterweight this tendency and keep prices stable. At the same time, most countries have eliminated export restrictions in agricultural products that were put in place at the peak of the crisis.[64] Demand for food commodities is generally insensitive to the cycles of economic activity. As a result, agriculture has been more resilient than other sectors of the economy to the effects of the global economic crisis. Agricultural commodity prices are therefore expected to rise modestly but steadily throughout 2010.[65]

In spite of a relatively stable price environment, price volatility in the commodities market remained high in 2009 and the first months of 2010. This has proven to be a major trend in the aftermath of the financial crisis and it is mainly attributed to two factors: the volatility of the dollar exchange rate through most of this period and the increasing "financialisation" of commodity markets. As commodity prices are typically denominated in dollars, its exchange rate has a bearing on their prices by intensifying pro-cyclical price movements. For example, dollar depreciation in the second half of 2009 was accompanied by a steady increase in prices.[66] For this reason, real commodity prices fluctuation has been much more moderate than nominal prices.

The second trend is related to the increasing participation of financial investors in commodity futures exchanges, in order to diversify their portfolios and hedge against possible inflation risks. Recent statistical evidence indicates that financial investors accelerated and even amplified pro-cyclical price movements, especially in food commodities.[67] This was particularly evident in cases where placements were purely speculative and tended to ignore the commodity market fundamental values. Consequently, their involvement may be considered partially responsible for both the boom in commodity prices before the financial crisis and their rapid decline afterwards, at least to the extent that it cannot be attributed to changes in demand and supply. With financial market movement remaining unpredictable and recovery slow, investor participation in commodity markets in 2010 is increasing again and with it price volatility as well. In addition to this, investor involvement in the commodities market complicates price hedging for traditional commercial users, with all the negative effects that this could have on future commodities supply.[68]

In conclusion, market speculation in the second half of 2009 and the first half of 2010 has increased price volatility and pro-cyclical effects in commodity prices, bringing an element of cyclicality even in markets where it traditionally did not exist, such as the food market. This development, in combination with the fact that agricultural productivity in poor countries is still short of keeping pace with increasing population, might create further food emergencies in the short-term.

1.2.5. Knowledge

Space systems play a key role in promoting scientific research and development in three ways. First, they are the means for taking scientific discovery beyond the boundaries of our planet, expanding our knowledge of astronomy and physics through space exploration. Second, space assets themselves are very demanding engineering inventions, the development of which motivates scientific innovation across the board in multiple disciplinary fields. Third, by offering worldwide communication services space systems contribute to the global flow of information and the free exchange of scientific knowledge. Consequently, they help promote scientific cooperation and they distribute its benefits to all of mankind.

An important step in developing a European scientific and innovation policy occurred in December 2009, when the European Security Research and Innovation Forum delivered its final report. Its plenary council of 65 members from 32 countries was mandated by the European Commission and the 27 EU Member States in September 2007 to propose a future European Security Research and Innovation Agenda (ESRIA) for the next 20 years. During its two years deliberations, the panel was supported by more than 600 experts from various government and industry sectors, making it the only high-level initiative of its kind in Europe.[69]

In conducting its research, ESRIF set up different working groups to investigate future technologies that could have an impact on European security, including one on Situational Awareness and the Role of Space in it. Its report identified key capability areas where space systems would be indispensable. These included Integrated Communication Networks, Information Management and Decision Support Systems, Command and Control etc. The report also listed a number of required space based systems that would be essential to the future EU security capabilities and prioritised them according to European needs.[70]

The panel's recommendations proposed better coordination in the use of existing space assets through collaborative and multiple uses of space services, information and data. It also stressed the importance of interoperability and the creation of common European operational picture and information distribution platforms. It asserted the crucial role of space based communications, Earth observation and satellite navigation, timing and positioning for European security. In this perspective the report praised the importance of the Global Monitoring for Environment and Security (GMES), Galileo and the European Geostationary Navigation Overlay System (EGNOS) programmes.[71]

In addition to these, the panel devoted particular attention to the creation of a European Space Situational Awareness (SSA) system. It identified the SSA programme as a major driving force for technology innovation in relative fields, such as automated satellite operation, formation flying architectures, multi-sensor fusion, protection of critical infrastructure and in-orbit networking, among others. Finally, ESRIF recognised that building the SSA infrastructure would require a cooperative approach from all stakeholders, including the European Space Agency (ESA), the European Commission and the European Defence Agency (EDA).[72]

In conclusion ESRIF adopted a holistic approach to security, calling for making security related innovation an EU priority, developing common European rules and procedures and exploiting knowledge synergies for security purposes whenever possible. Its key recommendation was that technological innovation and security planning should interact systematically, with the latter being an integral part of the former.

1.2.6. Mobility

Mobility is another activity area revolutionised by space technologies and their applications. Maritime commerce accounts for the bulk of global trade, whereas airplanes carry most of the world's passenger traffic. Space assets are indispensable to both, as they provide meteorological, navigation and communication services that make sea and air transport safer and cheaper.

The transport sector continued to suffer from the effects of the global financial crisis in 2009 and 2010. As the crisis has proved, the global economy works in a completely interdependent and concerted fashion, to the point where a crisis in any place can affect the entire system. As the transport sector is the epitome of this global trade interconnectivity and interdependence, it was hit particularly hard by the current economic crisis. As supply and demand fell sharply, the transport of materials and goods followed suit.[73] Furthermore, the financial crisis put a strain on the credit flow that is essential to international commerce transactions, with several banks refusing even to issue letters of credit. According to some sources, unmet demand for trade financing in developing economies is estimated between $100 and $300 billion.[74]

Maritime transport that represents the bulk of global transport (90%) suffered the greatest blow. The financial crisis put an end to a constant growth in maritime trade since 1993, one of the longest in recorded maritime history. The timing of the crisis was particularly adverse, as ship owners had enjoyed the most profitable financial results of all time before the crisis, and had an unprecedented number of vessels under order, accompanied by an equal increase in shipyard capacity.[75] During the last 12 months decreased maritime activity has led to a wave of cancellations of ship orders, an unprecedented level of distress demolitions (projected to reach 15–18% of world fleet capacity in 2010) and an almost six-fold contraction in shipping revenues. If these estimates materialise, all sectors of the maritime industry will suffer from considerable unemployment. A further medium-term consequence of the financial crisis for sea trade could be the appearance of protectionist measures that would further hinder world trade.[76]

Another challenge for the maritime industry in the past 12 months has been the increased number of piracy incidents, especially off the Somalia coast. Although international military presence in the region has somewhat increased security, piracy incidents have persisted. This surge in piracy acts at the Gulf of Aden has generated considerable costs, especially for Europe as 80% of shipments that pass through the area are coming from or to this continent. Re-routing shipments around the Cape of Good Hope alone is estimated to generate over $7.5 billion of additional shipping costs annually. At the same time, insurance costs for ships passing through Suez currently stand at 40 times their normal price, because of the war risk coverage included.[77] The problem has taken such proportions that on 2 December 2009 the International Maritime Organisation Assembly in London adopted a resolution calling for more international cooperation in the fight against piracy, quicker adoption of the Djibouti Code of Conduct by all states in the region, and an enhanced role for the United Nations.[78]

Air transport was equally struck by the crisis' repercussions. According to ICAO, 2009 saw the worst performance of airline traffic in history. International passenger traffic declined by approximately 3.9% and domestic traffic by 1.8%. However, domestic flights traffic decreased primarily in advanced economies, with emerging economies maintaining a positive albeit modest growth rate, especially in SE Asia and the Middle East.[79] In Europe and North America, low cost carriers performed somewhat better than more traditionally operating airlines. A modest increase of 3.3% in passenger traffic is expected for 2010 according to some observers, but full recovery will have to wait until 2011 at the earliest. Furthermore, cargo traffic also contracted by 15% in 2009, including in developing regions of the world. It is noteworthy that air traffic activity contraction in 2009 even exceeded that of the 9/11 aftermath.[80]

This drop in airline traffic translated into approximately $9 billion revenue loses in 2009. As with the shipping industry, bank financing became scarce and customer confidence waned. Most analysts agree that the picture will remain unchanged in 2010 as well, on a yearly basis. The biggest challenge ahead for airline companies is to manage excessive passenger capacity, which might lead to an increasing liberalisation of the market worldwide and possibly to major job cuts. Cutting costs throughout the chain of supply, increasing capital flow, abolishing ownership restrictions and encouraging international regulatory convergence will probably be the keys of this liberalisation concept that is becoming known under the name of "Open Aviation".[81]

1.2.7. The financial crisis and its consequences for the space sector

As demonstrated by the global economic and political outlook presented above, the time period under consideration has been marked by the consequences of the financial crisis that started in 2008. 2009 saw the spreading of these consequences from the financial sector to the entire economy. In 2010, we are witnessing the geopolitical repercussions of the economic downturn.

One resilient characteristic of the 12 months in question has been the quicker recovery of emerging economies, by comparison to developed ones. The main cause behind this fact seems to be the emerging economies' greater adaptability to the crisis conditions. In particular, when faced with rapidly decreasing demand for their manufactured products, developing economies were fast in directing a greater part of their output to regional and domestic demand. Developed economies apparently did not react as quickly. One possible explanation could be that in their case regional and domestic demand was already saturated before the crisis, because of their greater degree of economic/social development and regional economic integration (as in the case of Europe, for example). In other words, the current economic crisis may be interpreted as a slowdown of global economic integration in favour of a more regional one, especially in the case of regions where principal emerging countries cluster, such as SE Asia and the Middle East. Consequently, the different pace in which developed and developing countries exit the crisis may accelerate the already evident slow migration of the global economic centre from the western edge of the Eurasian Continent to the eastern one.

The above conclusion implies that the economic crisis will certainly have global geopolitical consequences of a currently unpredictable nature and magnitude. No country has been left untouched by the crisis and they have all entered it at approximately the same time. However, each one of them seems to be exiting the crisis at a different pace. The relative pace at which countries will recover will also determine their power and influence on the international scene. Economies that have some kind of "edge" seem to respond better to the crisis conditions. The first in line for recovery seem to be countries with rich natural resources, especially in the Middle East. Russia could also potentially fall under this category. Second are countries with huge internal markets and a relatively cheap working force, like China, India, or Brazil. Third are countries with strong industrial output and accumulated profits from positive commercial balances. Finally, the last to recover would be countries that have relied heavily on the financial and services sector for their development.

In addition to this, the different pace in which recovery comes in different parts of the world will probably exacerbate global and regional antagonisms and increase the centrifugal forces in the international relations system. In regions where recovery is more or less homogenous, as in SE Asia, regional economic cooperation and commercial relations will develop further. In regions where the pace of recovery differs from country to country, as in Europe, the resulting economic disparity could impede further regional economic integration and it might even encourage protectionist measures. In general, the crisis has created a very fluid and unpredictable situation on the international scene, where the relative economic as well as geopolitical value of countries and even entire regions can change rapidly.

As far as space activities are concerned, the financial crisis seems to have two distinct effects. On the up side, the commercial space sector could profit from the various financial stimulus funding that governments distribute to the private sector to boost recovery. Some of these measures for example include communication infrastructure development, which could certainly involve satellite communication services operators. On the down side, the credit crunch has made banks more risk wary and investing in space systems still entails a great deal of development and operational risks.

The greatest difficulties, however, might be encountered in the public funding of space activities. After last year's generous recapitalisation aid to financial institutions failed to keep the crisis from spreading to the "real" economy, public opinion in developed countries may become increasingly sceptical of indiscriminate government spending. In the face of rising unemployment, people could demand that more government funding is directed to creating jobs and mitigating the effects of the crisis to the "real" economy. In this context, space programmes that have high development costs, a slow technological maturity process and long-term benefits, could be considered as superfluous in the face of other, more urgent fiscal needs. In this sense, expenses that have no immediate effect on economic recovery could come under public scrutiny, and space budgets could fall under this category.

In order for the space sector stakeholders to successfully avert such a development, they would have to engage public opinion more than ever before. Explaining to people how space activities produce concrete financial and social benefits that are worthwhile should become a principal task for all actors involved. Furthermore, future space programmes should demonstrate their capacity to produce such positive results for society even from their conceptual phase of development.

1.3. Main science and technology indicators relevant for space activities

The space sector demands generally cutting-edge technologies which are the input of a global network within the society. The investments necessitate an effort not only carried out by the private entities but also by the states. Economy, space activities and states sustain thus complicated relations which ask a great involvement and a form of synergy between the different entities concerned. This is particularly true for the European Union's economy allegedly based mainly on knowledge and innovation. A lot of measures have recently symbolised this determination like the ERA[82] (European Research Area). The space sector is at the forefront of this reality. Input are defined as 'investments in the resources necessary to conduct scientific activities, like money and technical personnel while outputs are what comes out of theses activities, namely knowledge and invention.[83] Since the 1960s, the input-output are commonly used as tool to gauge sciences activities closely related to the Space sector. New technologies are therefore a key issue in competitive market and developed countries, it becomes even the determinant element in the race to Space.

1.3.1. Science and technology inputs

Those last years have been particularly eventful concerning the R&D due to the financial crisis. After having known a steady increase in most countries, it has been particularly expected to assess the consequences on the R&D of the world economic turmoil. The different effects concerning the crucial period of 2008–2009 will be measured by two statistics tool, namely the GERT (gross domestic R&D expenditure), the R&D intensity and the Government budget appropriations or outlays for R&D (GBAORD) related to the GDP. The GERD shows the nominal evolution of the expenditures in R&D. It is noticeable that the spending in R&D increased in Russia of 12.7%[84] during this period while in the EU during the same period the amount decreased of 1.3%. Without constituting a sharp drop it is all the same an important phenomenon given that the European economy policy is due to be mainly based on knowledge and innovation. However, it is worthy to note that the distribution is not equivalent among the European members.

The most noticeable exceptions concern Poland which increases its spending of 17.7%, Norway 4, and France 2.5 while Romania falls of 20.9% and Sweden 5.5. The GERD considers a nominal amount particularly affected by the financial crisis which has entailed a fall of the GDP. In comparison the investments of Japan have dropped from 8.3% in the same period. However these figures must be put into perspective with the R&D intensity which measures more efficiently the effort provided by a country in R&D. This indicator reveals that Europe is rather constant in the intensity of its expenditures related to the GDP, which is a good indicator of the strategy chosen by the states to bypass the crisis. Germany, Ireland and Finland particularly increase the part of GDP spent in this area. Japan that is much more suffered from the economic turmoil has seen a brutal drop of its R&D intensity.[85] The consequences have not been therefore so dramatic for the EU and its member countries which have even in general slightly augmented their

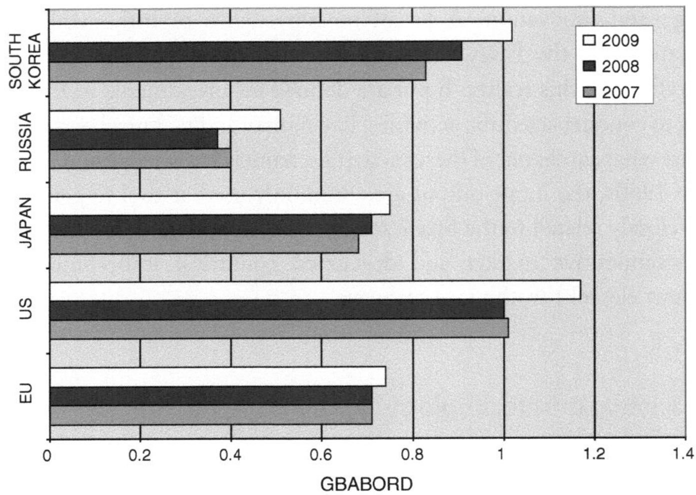

Fig. 1: *Global shares in GERD.*

investment in R&D in spite of the context. Another indication of the will of government to invest in R&D is illustrated by the Government budget appropriations or outlays for R&D (GBAORD).

The public investments in the EU are still stable especially between 2007 and 2008 while the world economic crisis just began.[86] Except in the EU public investments in R&D as a part of GDP globally increase in the other areas illustrating a strong governmental involvement to use this tool to get over the turmoil. This is particularly true concerning Russia that augments the share in 2008 from 0.37% to 0.51 in 2009. As for the U.S. we observe a rise of 0.17% between the two years whereas during the same period it is only 0.03% of the EU. Japan severely affected experienced a similar increase. The crucial role of the European Union and especially the European Commission could be in the future to take the lead in this domain which is at the core of its economic policy and counterbalances the limited investments provided by the member states by developing join programmes. This is already partially the case with a budget in 2010 devoted to improve the competitiveness of the EU around €14 Billion.[87] The EU is still overtaken in its effort by South Korea and the U.S. while it becomes very closed to Japan.

The public investments are crucial to understand the evolution of the R&D but the private sector is a major actor of it as well. A comparative study concerning the R&D intensity among the world's top 1400 companies between the U.S. and the E.U shows that the firms created before 1975 have roughly the same percentage of investments in R&D (2.8% for the EU and 3.6% for the U.S.) but the gap is constituted with the companies created after 1975 and more likely to be connected to new technologies.[88] The informatics sector illustrates perfectly this reality with a strong domination of the U.S. (Microsoft, Appel etc...) that has a lot of innovative companies in this field. This is also a result of a strong involvement of the government by military and civil researches programmes which ultimately yielded by the U.S companies enhance their competitiveness. This effect is all the more significant as the U.S. structure is made of 54.4% of companies created after 1975 against 17.8 in the EU.[89] In a more international perspective we can observe therefore that U.S. invests more by both the public and private sector. Japanese companies tend to even more spend than their American counterparts what is not astonishing given their speciality in new technologies champion like Sony dependant on innovation. EU companies are also overtaken by China's one. Within the EU concerning the most important economies, the firms from Finland, Sweden and Germany take the lead.[90]

The European effort towards innovation and a more knowledge based economy is thus to put into perspective. The public effort is still rather timid and quite similar to Japan which have been more sharply hit by the financial crisis. The real weakness of Europe would be more constituted from its private structure much less innovative than its international counterparts.

1.3.2. Science and technology outputs

The output of R&D is gaugeable by two sides, namely the scientific publications and secondly the inventions patented. In spite of the economic crisis the EU keeps the leadership in 2009 concerning the scientific publications constituting 33.4% of the worldwide ones.[91] An important part that besides experiences a decrease of 4.3% from 2000. The evolution is quite similar while the participation of China soared during the two period from 6.4 to 18.5%. However the EU is still a dominant actor in this field. An important success which is less obvious in patent deposed by European countries to the great displeasure of the European Commission.

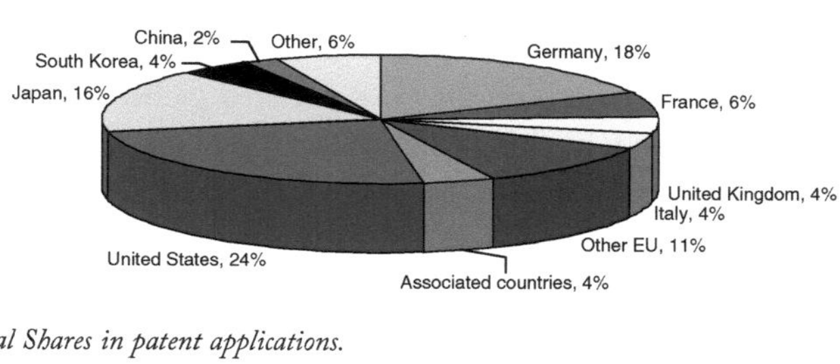

Fig. 2: *Global Shares in patent applications.*

Unfortunately it is harder to judge the effect of the economic crisis on this field given that the data available are limited to 2007. However we can already underline some trends which are not likely to change in the next years. In 2007 47% of all EPO patent were invented in Europe, followed by the U.S. 24% and Japan 16%. An analysis within the union shows quickly that Germany is the European champion of patent application representing almost haft of the all European patents deposed.[92] These figures confirm what has been studied herein with a strong correlation between business R&D expenditures and patent deposed for countries such as Switzerland, Germany and the Netherlands with a very efficient rate contrary to Central and Eastern European countries are those which invent the fewest EPO patents per euro of business R&D expenditure. An EPO patent applications by inventor's country of residence per billion GDP provides another overview of the patent policy efficiency at the international scale. Between 2000 and 2007 the EU experienced an important decrease of from 8 to 6 which is slightly got ahead by Japan which is experienced a contrary movement from 4.5 to 7.

However, this phenomenon can be in part explained by the integration of new countries with a less efficient innovation policy. As for the U.S., it is still around 3. However, the most striking observation in the figure below is the outstanding progress observed in South Korea from 2 to 7 and to a lesser extent in Japan. These two countries have by far overtaken the United States in inventing EPO patents, relative to the size of their economy.[93] Some examples which could inspire the EU in the future. It will be particularly interesting to asses the effect of the crisis on the production of patterns as soon as the relevant data are available.

2. Worldwide space policies and strategies

An interesting trend between mid 2009 and mid 2010 has been the steady rise in the number of space agencies worldwide. In spite of the global economic crisis' impact, an increasing number of governments have seen fit to create a central administration body for their space activities. This trend has begun in the late 1990's and it has continuer uninterrupted ever since. From 2000 to 2009 the number of space agencies worldwide has risen from 40 to 55, according to a study conducted by Paris-based Euroconsult. Space related global spending has also continued its upward trend, reaching $36 billion for civil (up 9%) and $32 billion (up 12%) for military programmes. In 2009 U.S. space related expenditures amounted to $48.8 billion (or 72% of total), Europe's ESA members to $7.9 billion, Japan's to $3 billion, Russia's to $2.8 billion, China's to $2 billion and India's to $900 million.[94]

2.1. The United Nations system

Various institutions within or associated with the United Nations are relevant for space policy. In this subchapter, the UN General Assembly (UNGA), UNGA Committees and other UN bodies and organs are discussed regarding space activities.

2.1.1. United Nations General Assembly (UNGA)

In December, the 64th session of the United Nations General Assembly (UNGA) was held. On 2 December 2009, it adopted the Resolution 64/28 "Prevention of an arms race in outer space". In the resolution, the GA put emphasis on transparency and confidence building measures (TCBM) to avoid an arms race in space. TCBMs were seen to possibly form an integral part of broader agreements on the prevention of an arms race. The GA recalled that the existing legal framework for outer space does not guarantee the prevention of an arms race and asked the states, especially the major space faring nations, to negotiate further. The Conference on Disarmament (CD) was seen as the sole multilateral disarmament forum. The Resolution also called for establishing and Ad Hoc Committee on the Prevention of an Arms Race in outer space within the CD. In general, it also acknowledged the complementary nature of multilateral and bilateral efforts in this issue area.[95]

Also on 2 December 2009, the GA adopted the Resolution 64/49 "Transparency and confidence-building measures in outer space activities". The Resolution was identical to the one tabled in 2008. It stated that an arms race in space would constitute a significant danger to peace and security and it invited the Member States to continue submitting proposals on TCBM to the Secretary General. In addition, the GA decided to include the issue in the agenda of the 65th session.[96]

A Resolution on "International cooperation in the peaceful uses of outer space" (64/86) was adopted on 10 December 2009 by consensus without a vote. The resolution reminded of all central aspects and challenges of the peaceful use of outer space. It also recalled the crucial importance of international cooperation to tackle the corresponding issues and it reviewed some of the steps that have been taken in this regard, like conferences, sessions of relevant entities and progress in implementation of corresponding programmes.[97]

2.1.2. UNGA Committees

The UNGA disposes of several committees that are involved in space policy and associated matters. Some of them are discussed here.

2.1.2.1. Disarmament and International Security Committee

The resolutions on the prevention of an arms race in outer space and on transparency and confidence-building measures had been introduced in the Disarmament and International Security Committee, also referred to as the First Committee, beforehand. The debates were marked by enduring differences between the U.S. on the one hand and Russia and China on the other hand.

2.1.2.2. Committee on the Peaceful Uses of Outer Space (COPUOS)

The activities of COPUOS were marked by its plenary session and the sessions of its subcommittees, along with various workshops and conferences. The Scientific and Technical Subcommittee held its 47th session from 8 to 19 February 2010. Topics discussed included the use of nuclear power sources in outer space, possible dangers from near-Earth objects, space debris, space-based disaster management support and developments in global navigation satellite systems. The Subcommittee received and considered information provided by the Member States on their activities in all these fields. Moreover, the implementation of the recommendations of the Third United Nations Conference on the Exploration and Peaceful Uses of Outer Space (UNISPACE III) were reviewed.[98]

A very important part of this session was the topic "Long-term sustainablitity of outer space activities". The Subcommitte discussed space situational awareness and agreed to establish a working group on the long-term sustainability of outer space activties, preparing a report and proposing measures and guidelines. This should include its contribution to the achievements of the Millenium Development Goals and should be consistent with the peaceful use of outer space.[99]

Also, a Symposium on National space legislation was held in Vienna during the 49th Session of the Legal Subcommittee. The topics discussed on this event were: needs for national space legislation, elements of national space legislation as well as their consequences.

2.1.3. Other UN bodies and organs monitoring outer space activities

Beyond the UN General Assembly and its Committees, there are other UN bodies, programmes and organs related to space activities. In the following, ITU (being a specialised agency of the UN), UN-SPIDER; the UN Programme on Space Applications, the International Committee on Global Navigation Satellite Systems (ICG), the United Nations Spatial Data Infrastructure (UNSDI), the Conference on Disarmament (CD) and UNIDIR are discussed.

2.1.3.1. International Telecommunication Union (ITU)

The International Telecommunication Union (ITU) held its World Radiocommunication Seminar (WRS) 2008 on 8 to 12 December 2008 in Geneva. Among other things, it discussed the application of the ITU Radio Regulations that had been changed in the course of the ITU World Radiocommunication Conference (WRC) 2007. The meeting provided a forum to exchange views on the associated technical, procedural and operational aspects. One of the relevant issues is given by the revisions made to the Fixed-satellite service plan that draws upon new technical developments and facilitates satellite system to access the frequency spectrum. The next World Radiocommunication Conference is scheduled for 6 to 10 December 2010.[100]

2.1.3.2. UN-SPIDER

Several workshops and regional meetings were organised in the framework of the United Nations Platform for Space-based Information for Disaster Management and Emergency Response (UN-SPIDER). This platform was set up by the UNGA in 2006 with the aim of providing universal access to all types of space-based information and services relevant to disaster management support.

The International Charter on Space and Major Disasters was activated several times by the Office of Outer Space Affairs (OOSA) at the request of other UN entities. The concept model for a UN-SPIDER knowledge portal was developed further in cooperation with German institutional partners.

2.1.3.3. UN Programme on Space Applications (SAP)

The UN Programme on Space Applications (SAP) is concerned with cooperation in space science and technology. Several activities were carried out under its auspices in the reporting period which dealt with, for instance, Technology Contribution to Infection Surveillance and to the Health-related MDG Goals, Basic Space Science and the International Heliophysical Year 2007, Integrated Space Technologies and Space-based information for Analysis and Prediction of Climate Change, Space Law, Integrated Applications of Global Navigation Satellite Systems, and Integrated Space Technology Applications for Socioeconomic Benefits.[101]

2.1.3.4. International Committee on Global Navigation Satellite Systems (ICG)

The aim of the International Committee on Global Navigation Satellite Systems (ICG) is to promote cooperation in matters of satellite navigation. OOSA serves as the Executive Secretariat of the ICG and the associated Providers' Forum. The fourth meeting of the ICG took place in Saint Petersburg, Russia, on 14–18 September 2009. It saw attendance from industry, governments, non-governmental organisiations and academia and it reviewed and discussed developments in global navigation systems.

The ICG work plan was organised in four working groups: on compatibility and interoperability, on enhancement of performances of GNSS services, on information dissemination, capacity building, and on interaction with national and regional authorities and relevant international organisations. In the joint statement it was noted "that substantive progress had been made in furthering the workplans of ICG and the Providers' Forum that had been approved at the previous meetings of ICG".[102] The next ICG meeting will take place in Turin in October 2010.

2.1.3.5. United Nations Spatial Data Infrastructure (UNSDI)

The United Nations Geographic Information Working Group (UNGIWG) held its tenth annual meeting in Bonn, Germany, on 19-21 October 2009. The UNSDI is understood as a comprehensive, decentralised geospatial information network to facilitate decision-making.[103]

2.1.3.6. Conference on Disarmament (CD)

The Conference on Disarmament (CD) is the only multilateral disarmament and arms control negotiating forum within the international community. It was in session from 11 June to 2 September 2009 and from 19 January to 23 March 2010. The stalemate in its work regarding space security has been ongoing. In the course of the 2009 session, the prevention of an arms race in outer space was again a central topic on the agenda.[104]

2.1.3.7. United Nations Institute for Disarmament Research (UNIDIR)

Several projects of the United Nations Institute for Disarmament Research (UNIDIR) deal with space security, directly or indirectly. Among other things, UNIDIR intends to review former proposals and to propose new options for breaking the deadlock in space weaponisation matters at the Conference on Disarmament (CD).

2.2. The Group on Earth Observation

The Group of Earth Observation (GEO) is a voluntary partnership of governments and international organisations whose task is to coordinate effort to build a GEOSS. In 2009 and 2010 GEO had a busy activity mainly composed of workshop and symposia. The main interest of such an organization is to share the best practises and experiences while preparing the future challenges. The topics treated in such event were various, that concerned among others, environmental aspects (especially forest and ocean monitoring), support agricultural monitoring, natural disaster management, water researches, help Africa to take advantage of Earth Observation, atmospheric observation and climate change. An important meeting concerned the Forest Carbon Tracking (FCT) Task Information meeting which is a worldwide issue. The event was an occasion to procure an overview of the situation and solution available to address it.

In 2010, certain workshops were especially dedicated to definite area such as Black Sea on 4 May 2010 or sector like bio energy. Some multilateral projects ongoing were also tackled concerning EnviroGRIDS Project, EuroGEOSS. 2010 finally ended with the 20th executive committee meetings and GEOSS Monitoring and Evaluation Meeting followed of 4th International Meningitis Environmental Risk Information Technologies 'MERIT' Technical Meeting. A strong and diversified activity which has been mainly focused on environmental aspects given the emergency of the situation.[105]

2.3. Europe

2.3.1. European Space Agency

After reviewing the EU space activities programmes from July 2009 to June 2010, the crucial role of the European Space Agency in all of them becomes obvious. In fact, from a space policy standpoint, the most significant development has been the de facto transformation of the agency into the implementing arm of the EU space policy. This fact becomes apparent given ESA's increased involvement in shaping and building most of the necessary infrastructure for EU space projects. This gradual process however has not yet acquired a more institutionalised or de jure form. It therefore still remains a more or less empirical and result-driven cooperation process between the EU and ESA, guided by the space policy aspirations of one and the unique capacity to materialise them of the other. This cooperation therefore remains a step-by-step process that still operates under conditions of constantly evolving tasks and the need to produce tangible results for Europe's space activities.

On a policy level, the period in question witnessed the implementation of most of the key orientations given by ESA's Ministerial Council in November 2008. These focused on expanding ESA activities into financing space applications programmes while at the same time increasing space exploration activities. ESA's increased space applications-related workload can be depicted in its participation in several EU programmes described above. Increased funding for the above-mentioned projects also accounts for an 18.6% in the agency's budget in 2009, compared to 2008. Total payment appropriations have been increasing at an annual rate of 10% over the last four years, reaching €3.35 billion in 2009. This increase brought the agency's budget to the desirable levels approved by the Ministerial Council.

On the other hand, in 2009 the continuing global economic crisis began to weigh on the ESA Member States' space budgets and contributions to the agency. Acknowledging the new financial realities, ESA Director General J.J. Dordain announced on 14 January 2010 that the agency's budget spending would remain at these levels for the next two years, 2010 and 2011. This decision was not expected to seriously affect ESA's project schedule, as the agency's budget had already attained an adequate level in 2009. Nevertheless, it was a form of recognition of the difficult new financial realities, and a message that ESA would not overstretch itself financially without previous approval by its Ministerial Council, set to convene again in 2011. In order to avoid any programme cancelations due to this freeze, Mr. Dordain explained that from now on ESA would be stretching the payment periods of any new contracts.[106]

In general, however, current ESA operations have been relatively little touched by the economic crisis and they have continued as expected. The agency's operations suffered only indirectly from the crisis in November, when it announced it would be freezing payments for contracts valued over €10 million for a month (until 2010) due to a cash-flow shortage. This temporary stop was attributed to a €400 million cash reserves deficit, created by the accelerated pace of contract payments to its industrial partners that was part of a deliberate attempt to counter the effects of the global financial crisis on the space industrial sector. It is unclear however, whether the financial stresses of some of its 18 Member States also contributed to the problem. Instead of delaying €400 million worth of contract payments, ESA officials preferred the possibility of taking out a bank loan in order to cover the gap.[107]

Nevertheless, on 18 December 2009 ESA reinitiated payments after only a three weeks self-imposed spending moratorium. Financial auditing revealed that the deficit's figure was exaggerated and that it would not exceed €200 million, which

ESA officials described as “manageable”. The first programmes to be funded (for over €500 million) were the construction of the first three Sentinel Earth observation satellites for the European Commission and preliminary work on a new upper stage for the Ariane 5 rocket that would enable it to lift a 12 tons payload to the GEO transfer orbit.[108]

On the other hand, in spite of the financial crisis Galileo’s budget overrun of €376 million was approved by the European Commission in June 2009, after three months of audits. The auditors found that additional charges were reasonable, given the contract modifications ordered by the customer. Their report concurred with a similar investigation performed by ESA earlier that year. After this development, additional payments were approved for Galileo’s in-orbit Validation phase, which includes building four satellites as well as most of the required ground control infrastructure by 2010. However these budget and schedule overrides increased fears that the system might exceed its projected €3.4 billion budget and might not be fully operational by 2013 as planned.[109]

Among the ESA’s key policy related activities, one can distinguish the joint EU-ESA International Conference on Human Space Exploration, held in Prague on 23 October 2009. During this conference, it was decided that a space exploration road map for Europe should be drawn by the end of 2010. The meeting reaffirmed Europe’s determination to remain a principal space-faring player in the face of rising Chinese and Indian space ambitions. But it also concluded that any meaningful future space exploration effort should be of a truly international nature in order to succeed. However, participants did not debate specific space exploration proposals. Of course, a key element of any meaningful new European space strategy should be a considerably increased budget for space activities. The European Commission currently has less than a €1 billion annual budget for space projects, which it hopes could triple for the seven-year budget period starting in 2014.[110]

Another field where increased EU-ESA cooperation appeared was in the development of the European Space Situational Awareness system (SSA). This programme is simultaneously funded by ESA (through its GST and SSA core element activities) and the European Commission (through its dedicated “Space” work programme of the FP7). Both institutions have initiated research and concept demonstration projects related to SSA. Therefore, achieving complementarity between the two programmes and avoiding duplication of development efforts has become a matter of the utmost importance. Consequently, coordination and interaction between the three programmes (SSA preparatory programme, GSTP and FP7) is crucial to minimising development risks and maximising benefit returns.[111]

2.3.2. European Union

The European Union maintained and augmented its engagement in space activities in the second half of 2009 and the first half of 2010, under the Swedish and Spanish EU Council Presidencies respectively. Key developments in this period included the entry into force of the Lisbon Treaty: the considerable progress made in the Galileo and GMES programmes, the increased cooperation between the European Commission (EC), the European Space Agency (ESA) and the European Defence Agency (EA), in promoting European non-dependence in critical space technologies and infrastructures, the meeting of the 6th "European Space Council" in May 2009, and the announcement of the third space-related call for proposals within the 7th Framework Programme for research and development in Europe.

As in many other areas of European Union activities, the key development from July 2009 to June 2010 was the entry into force of the Lisbon Treaty on 1 December 2009. The new Treaty on the Functioning of the European Union is the first document of its kind to set out an explicit EU competence in space activities. Under Article 189 of this Treaty, EU institutions are invited to draw up and implement a long-term European space policy, including the possible creation of a European space programme, in close cooperation with ESA.[112] One of the key features of this article was the fact that it referred not only to the exploration, but also to the exploitation, of space by the EU. This addition is thought to allow for the inclusion of a security dimension to EU space activities. It is also thought to push toward a closer cooperation between the EU, ESA and their respective Member States, as well as to create the necessary impetus for further developing the competitiveness of the European space industry on a global scale.[113]

In another development, on 29 May 2009 the European Space Council (the EU Competitiveness Council and the ESA Council meeting concomitantly) met for the sixth time in Brussels, in order to assess the progress made on implementing a common European space policy and to identify further objectives. The Space Council noted that the "structured dialogue" among all European institutional space actors was advancing well. It particularly took notice of the cooperation between the European Commission, the EDA and ESA on identifying critical space technologies in which Europe should become non-dependent on outside sources. The Council also noted the inclusion of the Multinational Space-based Imaging System (MUSIS) in the EDA programme list, as well as the adoption of the ESA Preparatory Programme for the development of the European Space Situational Awareness system (SSA).[114]

Another area of particular interest to the Council was the potential contribution of space related technological innovation and competitiveness to the overall

European Economic Recovery Plan (EERP) for 2010 and 2011. The Council recognised the importance of space activities to economic recovery and recommended their full funding from the €5 billion economic stimulus package of the EERP. The Council paid particular attention to the participation of satellite communications providers in the broadband connectivity promotion programmes of EERP, valued at €1.02 billion.[115] In the same context, it called for the full development of services based on the EGNOS, Galileo and GMES programmes. Especially as far as the latter was concerned, the Council identified the long-term funding of its space segment by the European Commission as a key objective. In this regard, it also encouraged closer cooperation between the European Commission, ESA and EUMETSAT.[116]

In a related development, the European Commission announced in October 2009 that it was considering maritime surveillance as its next major investment in space-based applications after Galileo and GMES. The European Commission was already financing pilot projects in this direction, including the space-based Automatic Identification System (AIS) that uses signals emitted by commercial vessels to determine their identity, destination, speed and cargo. However, in order for such systems to become useful for maritime surveillance in the field, they would have to improve their operational response times. All ships over 300 tons displacement are required to have such transponders by international maritime regulations. ESA has also developed two experimental AIS receivers that were launched to the ISS in September 2009.[117]

The increasing importance of space activities for the EU was made even more evident on 15 October 2009, when the European Commission President José Manuel Barroso made for the first time a speech dedicated entirely to the European Space Policy. During his presentation in a conference on this subject in Brussels, he reiterated the usefulness of space systems for EU policies and the need to achieve autonomy in relevant space technologies. An independent EU capacity in the field of Earth and near space observation should be a top priority, he said. Furthermore, he maintained that EU space activities should not be confined to producing direct financial results, but should seek to implement broader European policies as well. Finally, he called for further developing space programmes, including an independent European human spaceflight capability.[118]

Another European programme that saw considerable progress in 2009 and 2010 was the Galileo satellite navigation and positioning system. As early as June 2009, discussions among the European Space Agency (ESA), the European Commission and industrial partners were approaching their conclusion over the best way to contract the deployment of the system's satellites. The contract for the building of the 28 spacecraft required was due for signature later in the year and the two companies biding for it were Astrium Satellites and OHB System.

European authorities were disputing whether they should order all satellites from one manufacturer or split the contract between them. Another undecided issue was whether the entire constellation should either be ordered from the beginning, or split into two procurement stages to allow for last minute modifications. Finally, the question of which launcher to use remained open, with ESA preferring an exclusively Soyuz launching campaign and Astrium Space Transportation pushing for the use of Ariane 5 as well. Although a decision has not been reached yet, ESA and European commission seemed favourable to dividing the contract into two phases and splitting each phase between the two bidders.[119]

A decision on deploying the system's first fully operational constellation was repeatedly delayed, most recently in October, when ESA's Director General Jean-Jacques Dordain announced that last-minute satellite manufacturing difficulties were reported by its principal contractors, Astrium Satellites and ThalesAlenia-Space. The new timetable given foresaw a first launch in November 2010 and a second early in 2011, always onboard Soyuz rockets launched from the European Spaceport in French Guiana.[120]

In late December however, it was made known that the European Commission had finally decided to select OHB Technology of Germany to build at least the first 8 Galileo satellites for approximately €350 million. The European Commission chose the OHB-led consortium that included Britain's Surrey Satellite Technology Ltd. over its competitor EADS Astrium Satellites consortium that also included ThalesAleniaSpace, although it was widely considered lacking the industrial depth to build the entire 22 spacecraft constellation. However, the European Commission's decision to maintain competition in the programme weighed heavily on its decision. The situation was further complicated by its refusal to simultaneously award the remaining 14 satellites contract to the Astrium consortium, due to what was described as the company's non-compliance with the competition's bidding guidelines. Both Astrium and ThalesAleniaSpace had made considerable industrial investments at the early stages of the Galileo programme, when they were given sole charge of the project.[121]

The official announcement of the European Commission's decision came on 7 January 2010, when a team led by OHB Technology of Germany was selected to build the first batch of 14 Galileo navigation satellites. The consortium would also include small satellite specialist Surrey Satellite Technology Ltd. (SSTL). This decision was a setback for competitor EADS Astrium Satellites that was expected to get the order. The contract was valued at €566 million and launches were scheduled to begin in October 2012 and continue in three month intervals. Eighteen more satellites would be ordered in the near future, through a new open competition. In choosing OHB, the European Commission manifested its intention to double source the programme in order to minimise financial and technological risks.

Another key plank in the European space policy during the past 12 months has been the increased cooperation among different European institutions in the field of space security. The European Defence Agency's (EDA) increased participation in relevant space projects was particularly important. The inclusion of MUSIS in EDA development programmes already mentioned above was significant in this trend. Another positive development has been EDA's participation, together with the European Commission and ESA, in a joint task force to investigate European strategic non-dependence in space activities. EDA's contribution was particularly welcome in identifying key security related space technologies that should be developed in the near future within Europe in order to achieve this non-dependence.[122]

Furthermore, EDA increased its efforts to pool security related space services demand among EU Member States, in an effort to reduce their cost. In this regard, it signed an agreement with satellite communications provider Astrium Services to set up a European common contracting vehicle for commercial bandwidth procurement for military use. Astrium Services came under a €130,000 contract from EDA in November 2009 to pool European commercial satcom requests under a common contracting scheme known as the European Satcom Procurement Cell. This scheme would allow Astrium to sign longer term bandwidth lease contracts with commercial operators on behalf of participating European governments through London Satellite Exchange, its subsidiary that acts as an intermediary between commercial satcom suppliers and buyers. In this way EDA hoped to secure a 30 to 50% discount from currently used spot prices that European countries are usually paying. Until that time, five countries had confirmed their participation in this scheme, namely France, Italy, the Netherlands, Poland and the UK. The European Satcom Procurement Cell drew its operating principle from its French counterpart Astel-S, a commercial bandwidth pooling procurement contract set up between Astrium and the French Armed Forces in 2005.[123]

Finally, in July 2009 the European Commission of the European Union issued its 3rd call for proposals on space related R&D projects within the framework of the 2007–2013 FP7 research funding programme (for a total budget of €114 million). Proposed funding was divided into three categories of activities, namely space-based applications (€47 million), space technologies R&D (€58 million) and cross-cutting activities (€9 million). Space applications mostly referred to developing GMES products, with a special focus on providing multipurpose Earth observation services and down streaming them to European users, especially on a regional level. Space related R&D was approached through the scope of interoperability and harmonisation of products, as well as ensuring their long term sustainability. This budget line also included funding for developing critical technologies to support Europe's independent access to space. Finally, under the theme of cross-cutting issues the European Commission aspired to enhance space cooperation with third parties and especially Russia and African countries.[124]

2.3.3. Eumetsat

On 1 July 2009, the European Meteorological Satellite Organisation, EUMETSAT, decided in favour of a $90 million contribution to the Jason-3 ocean-altimetry satellite, a joint U.S.-France project. Securing funding was strenuous, as a number of Member States regarded Jason-3 as a bilateral U.S.-France programme. Additional support for the project has been secured from the European Union's GMES programme and ESA, which have both announced they would be purchasing Jason-3 data.[125]

On 9 December, the European Space Agency (ESA) bid-evaluation board failed for the second time to select a winner for the €1.4 billion contract to build six Meteosat Third Generation (MTG) satellites on behalf of Europe's meteorological satellite agency EUMETSAT. ESA will cover 75% of the project's budget and EUMETSAT the remaining 25%. The principal competitors for the contract were EADS Astrium and ThalesAleniaSpace. According to industry sources, the delay could be attributed to political pressure surrounding the contract as Germany which, together with France, is the programme's main financial contributor, was determined to acquire MTG prime contractor status for its industry.[126]

On 3 February 2010, an ESA evaluation board chose the consortium of ThalesAleniaSpace and OHB Technology to build the next generation of Europe's meteorological satellites for Eumetsat. The contract for the six spacecraft (four imaging and two equipped with sounding devices) known as Meteosat Third Generation (MTG) was expected to reach a value of €1.4 billion. The satellites were to include significant improvements compared to their predecessors, including a three-axis stabilization system instead of a simpler spin-stabilised design. Negotiations between the consortium and the European Space Agency (ESA), which had assumed the role of the contracting authority on behalf of Eumetsat, were expected to start immediately after the announcement of the decision. The full life-cycle cost of the programme was expected to exceed €3.3 billion over a period of 20 years, 75% of which would be covered by EUMETSAT. As the competition for the MTG satellites would likely be the biggest single satellite construction contract to be signed in Europe in 2010, the selection process proved to be a highly contested one. The ESA evaluation board had to convene three times before reaching a decision and the agency's Director General J.J. Dordain publicly admitted to unusually high political pressure surrounding the programme. The other contestant for the project was Astrium Satellites.[127]

2.3.4. National governments

2.3.4.1. France

In 2009 and 2010, France maintained its ambition to remain the leading European nation in space activities. This effort was reinforced by the successful reorganisation of its national space agency, CNES. Under a renewed six year contract with the French government through 2015, CNES is expected to consolidate France's position in the European space sector. As is the case elsewhere in Europe, the agency's strategic planning focuses on developing downstream services for users, improving the country's space industry competitiveness and leading multinational European programmes, but without duplicating ESA ones. The key strategic objective for France is to maintain its individual status as a principal space faring nation, while at the same time increasing its participation in international cooperative space projects.

More precisely, the French civilian space programme has three major tiers: space applications (with a particular focus on Earth observation), access to space, and space related research and development. In the field of space applications, French policy is evolving along two axes, with the development of high performance optical imagery satellites (such as Pleiades) on the one hand, and the outsourcing of lower resolution ones to commercial users through the management of the SPOT constellation by AstroTerra (now an Astrium-Spot Image joint venture) on the other. This policy is aimed at maintaining French technological capabilities in this field while reducing operating costs at the same time. In the same spirit of resource economies, dual use space systems and civil-private synergies are consciously and constantly being pursued. It should be recalled that CNES operates under a double mandate under the Ministry of Defence as well as the Ministry of Research.

Two examples of this dual approach can be seen in the following examples. In June, the French Military Intelligence Directorate announced it was leaning towards outsourcing future production of high-resolution satellite images, under a €750 million project to build world-wide digital terrain models by 2020.[128] Simultaneously, Astrium Satellites announced that it had secured a €66 million contract with the French military to start development on the next batch of two military observation satellites, set to be launched in 2014–2015. These were poised to replace the current Helios system. The new satellites are expected to be a higher resolution capability version of the civil-military Pleiades constellation. French officials assured, however, that in spite of this development France was still committed to the multi-national MUSIS program to build a common European satellite imaging system.[129]

At the same time, on 18 December 2009, Helios 2B, the French new generation optical imaging satellite, was launched onboard an Ariane 5 GS rocket from French Guiana. The 4.2 tons spacecraft was an identical copy of the Helios 2A launched in December 2004, except that it fielded a more accurate optical instrument, believed to provide up to 35 cm ground resolution. It was placed in the same 700 km altitude near-polar Low Earth Orbit as Helios 2A, at a 180 degrees distance from it. EADS Astrium Satellites built the spacecraft, while its principal optical imager was built by ThalesAleniaSpace. The two-satellite Helios system has an estimated cost of €2 billion. Apart from France, four other European countries participate in the programme with a 2.5% stake, namely Italy, Spain, Belgium and Greece. Germany also has access to its data through a bilateral agreement to exchange Helios data with radar imaging data from its own SAR Lupe spacecraft. The Helios satellites are controlled by the French space agency CNES trough its Toulouse facility, but their daily tasking is conducted by dedicated centres in the participating countries, each using its own encrypted data links. The system's expected life span is five years.[130]

Free access to space is also considered a critical national capability for strategic reasons. France maintains and covers a third of the French Guiana launch facility's operating costs. It is also heavily involved through CNES in the development of future more capable versions of the Ariane 5 rocket, in coordination with ESA (within the frame of the ARTA programme). Unrestricted access to space has been recognised as a priority in the country's national security strategy documents as well. Finally, space launchers have been an area of increased bilateral cooperation with Russia, especially in regard to the development of the country's next generation rocket.

The close cooperation with former soviet republics seems to be expanding, driven by broader government policies as well. During a state visit of the French President Nicolas Sarkozy to Kazakhstan in October, EADS Astrium closed a deal with the country's government to build and launch two observation satellites. The deal, which also included the construction of a satellite integration and test centre, the training of Kazakh satellite engineers and the integration of the satellites with the Spot Image services network, was valued at €230 million. Astrium will also oversee and cooperate with the Kazakh space program, which according to the country's officials should have a $253 million annual budget.[131]

Finally, close cooperation with neighbouring Germany on a bilateral level remained a key plank of French space policy. On 4 February 2010, for example, the French President Nicolas Sarkozy and the German Chancellor Angela Merkel held their regular joint inter-ministerial meeting in Paris. On that occasion, a number of important agreements between the two countries relating to space activities were announced. First, it was agreed that France and Germany would jointly build a methane concentration monitoring satellite, capable of measuring the results of greenhouse effects to global climate change. The 180 kg spacecraft called the CH4 Atmospheric Remote Monitoring Explorer (CHARME) would be based upon the CNES Myriade small satellite platform and it was expected to be launched on a Vega rocket in 2013 or 2014. France and Germany would equally share its €120 million cost.[132] Second, the two leaders concurred on the timely development of MUSIS, the Multinational Space-based Imaging System that will secure interoperability among European countries' Earth observation satellites ground segments. Third, they agreed that their two countries' space agencies would jointly study the development of a new generation of the Ariane launcher. This programme has been financed by the French government's economic stimulus package. The final decision for the building of the vehicle, known as Ariane 6, would be made by the ESA Ministerial Council in 2011.[133]

2.3.4.2. Germany

From July 2009 to June 2010, Germany continued its effort to position itself as the European space technology leader. In order to achieve this, the country has developed a two-fold strategy. On the one hand, it takes the lead in key European space technology development projects, both in the frame of the EU and ESA. On the other hand, it constantly increases the visibility and public impact of its technological capabilities, either through its participation in the International Space Station (ISS) or in initiating its own national space exploration programme.

The bulk of Germany's funding for civil space activities is still dedicated to ESA programmes (over 75% of its 2009 space budget). However, the country has been also increasingly allocating funds to exclusively national high visibility space exploration projects, such as a lunar orbiter mission planned for 2015 (LEO project). Although the budget of these projects is still marginal compared to Germany's participation in ESA programmes, they are however considered strategically important in Germany. The drive behind this trend is clearly to demonstrate the nation's technological capabilities through exclusively national projects, which in return would increase the German space industry's reputation and client base, creating profit returns for the entire sector.

The same approach has been adopted in regard to German participation in the ISS. Although European participation in the station is represented by ESA, the German government was keen on underlining its industry's key role in building ESA's Columbus ISS laboratory compartment. The ISS has indeed a crucial role in the overall German medium-term space strategy: it will facilitate the scientific research and development of the country's future space technologies, it will function as a demonstrator of the German space industry's capabilities, and it will increase the visibility and reputation of the country's national programmes.

Germany has made a strategic investment in the ISS. Although France is currently the European country with the biggest contribution to ISS activities, Germany has subscribed for more future scientific projects onboard the station and has become its largest operating costs contributor. It is therefore the country that has the greatest stake in keeping it in use for as long as possible. An example of this particular interest in ISS operations appeared on 27 January 2010, when the German space agency's (DLR) chairman openly disagreed with ESA's intention to limit the ISS operating budget in order to cut expenses. As ESA was contemplating reducing the number of ISS astronauts, or of its control centres (U.S., Russia, Europe, and Japan each have one), the official DLR position was against any change in future ISS operations.[134]

Another field of particular interest to German space policy is space applications. In the past 12 months, the country has increased its involvement both in communications (Satcom) and Earth observation (EO) satellites. As with space science and exploration, the German strategy consists of developing independent national space technologies and then incorporating them into a larger European system. This method enables German industry to expand its technical know-how and to increase its market share, while respecting the country's engagement in common European policies. This strategic development concept is consistent with German industrial planning and political aspirations at a European level. However, the balance between these two strategic objectives is not easy to maintain, and it could result in Germany duplicating on a national level space capabilities that are already under development on an EU or ESA level.

For example, Germany is a prime partner in MUSIS, the future multinational European optical Earth observation system. In October, however, German industrial and government sources disclosed that they were considering the development of a national high resolution optical Earth observation satellite as well. According to these sources, the system, known as the High Resolution Optical System (Hi-ROS), would consist of two or three spacecraft featuring a 70 cm ground resolution and a quick operational response time provided by onboard Ka-band communication terminals linked to geostationary data relay satellites. This development was presented as a logical next step in developing German space observation capabilities to complement its existing radar reconnaissance satellites. Hi-ROS was expected to profit from the R&D undertaken for the German-made optical observation payload of Korea's Kompsat-3 optical imaging satellite, which is expected to be launched in 2011.[135]

The same situation applies to a certain extent to telecommunications satellites as well. On 13 October for instance, ESA and DLR announced that they had reached an agreement on the management of the planned space based European Data Relay System (EDRS). This should replace the existing Artemis spacecraft. The proposed EDRS would consist of one dedicated GEO satellite and two payloads on commercial satellites. In 2008, ESA had requested €230 million to begin its development, of which Germany had agreed to contribute 50%. Negotiations included resolving the issue of intellectual property rights in the system's German-built laser communication terminals. The system is envisaged to cooperate with the EU's Global Monitoring for Environment and Security Sentinel satellites and could be launched as soon as 2013.[136] At the same time, however, Germany is also developing similar technologies on a national level, such as the Heinrich Hertz space broadband demonstration spacecraft and the very high data rate Laser Communication Terminal (LCT) payload. This would imply that complementarity in the Satcom sector between national and European levels is also compromised from time to time.

Finally, in the past 12 months Germany appeared to come closer to a decision to build its first in-orbit satellite servicing demonstrator. More precisely, satellite manufacturer OHB of Bremen announced on 24 February it had been selected as the prime contractor for a technology demonstration experiment of in-orbit servicing and de-orbiting of satellites. The programme, known as the German Orbital Servicing Mission (DEOS in German), is run by the German space agency DLR. Although it has been under study for more than a decade, it was recently promoted to more detailed design work. However, the decision to build the first demonstrator, which could cost up to €200 million, had not been made yet. Nevertheless, this development demonstrates the increasing interest of European countries in these potentially revolutionising technologies. In-orbit servicing of LEO satellites would permit the extension of their operational life span at a fraction of their replacement cost. De-orbiting technologies would allow for the clearing of saturated orbital paths from obsolete spacecrafts, thus helping to mitigate the problem of orbital space debris and to minimise the danger of collisions with operational satellites. The DEOS demonstrator envisaged would have the ability to track satellites, autonomously rendezvous with them in orbit and refuel or repair them by the means of a robotic arm. Furthermore, it would be able to capture the target satellite and guide it into a destructive re-entry trajectory. A significant number of critical technologies would have to be validated before an operational system becomes available. In addition to this, the legal problem of determining liability in the case of the servicing spacecraft accidentally damaging its target satellite would have to be resolved.[137]

2.3.4.3. Italy

The recently elected president of the Italian Space Agency (ASI), Enrico Saggese, announced in January that the global economic crisis would not affect the agency's budget for 2010 and 2011, which was expected to remain at approximately €700 million annually (excluding military-related expenditure).[138] However, this would also imply that no increase could be expected in the near future. Given these financial constrains, Italian space policy would probably focus on maintaining currently announced projects, without expanding to new ones. Happily for Italy, the current financial crisis coincides with a period of limited additional budgetary needs, as a number of key programmes are completing their development phase (e.g. the Vega launcher), while others are only starting theirs (e.g. the next generation Cosmo-SkyMed).

In general, Italian space projects evolve around space applications, with the bulk of the funding going to Cosmo-Skymed services. As is the case in other European countries, ASI has focused on increasing the Italian industry's share in efficiently down streaming services to customers. A major milestone for the commercialisation of Cosmo-Skymed products to market customers was achieved in the past 12 months with the creation of E-Geo, a joint venture between Telespazio and ASI set up for this purpose. At the same time, the Italian space agency has begun development of the next generation of Cosmo-SkyMed that is expected to fly in 2016 (for an estimated €600 million budget). As in the case of its other European counterparts, Italy also participates in the effort to effectively downstream Galileo and GMES products (for which the country covers 30% of the budget). For the Galileo data utilisation project, ASI estimates that €100 million to €150 million of additional funding would be required.

Earth observation is also the basis of Italy's bilateral cooperation with France in the framework of the Orfeo programme that should combine Cosmo-SkyMed SAR data with the French Pleiades system optical data. In the mean time, cooperation between the two countries has turned to Satcom projects as well. ASI's president confirmed the upcoming acquisition of two new satellites by ASI in close cooperation with its French counterpart, CNES. The new spacecraft would be the telecommunications satellites Athena-Fidus and Sicral-2 (the second intended for military use). Sicral-2, the newest of the Italian military Satcom spacecrafts, would host a separate French payload to complement its Syracuse 3 constellation.[139]

Finally, the Italian space industry is also heavily involved in the development of the future small-satellite launcher Vega, an ESA programme for which ELV SpA of Italy is the prime contractor. ASI is paying for 60% of the project's budget, a fact that reveals its strategic character for Italian space policy. According to ASI's president, the first launch of the rocket is expected in early 2011. After that, he said,

ASI would most likely invest further launcher development funding into the next generation Ariane rocket.[140]

Italian space exploration initiatives are more modest than that of France or Germany. However, the country has a key participation in the ISS and has adopted an approach similar to Germany's in favour of using ISS facilities to their maximum capacity in the service of space science and technology development.

2.3.4.4. The United Kingdom

The year 2009 was a significant one for UK space policy, as it witnessed the decision to replace the British National Space Centre (BNSC) with a dedicated space agency. This development will surely lead to a profound restructuring of UK space activities and their administration. UK space policy has long been oriented towards producing concrete technological and industrial benefits for the country. It is therefore apparent that this decision was driven by the desire to increase value for money in UK space activities and to help the space industry improve its position in the European space market. BNSC, which is not a space agency but a coordinating body, relies on funding from ten different government departments. Through this user-oriented approach, BNSC has not always been able to secure adequate funding for British industrial participation in ESA's space applications programs. Through the creation of a more traditional space agency, UK officials hoped to assure a prime contractor's role for British firms in future ESA projects. Characteristically, the announcement that the government was contemplating its creation was immediately applauded by Astrium Ltd., UK's largest space company.[141]

Creating a UK space agency was a gradual process. On 22 July, Lord Drayson, the British Science and Innovation Minister, announced a three months public consultation on the possibility of creating a national space agency in the UK. After several months, the British government finally announced on 10 December that it would create a national space agency to replace BNSC by the end of 2010. According to the rationale of the decision, a dedicated space agency would be a better vehicle for strategic decision making, handling multi-partner programmes, coordinating space-related research and securing long-term funding for it. It would also go further in securing British participation in ESA projects, where 90% of the country's £270 million space budget is currently invested. The decision met the immediate approval of UK aerospace industry officials, who expected it to boost their participation in ESA programs and increase investment returns for their companies from them.[142] It also seems to be related to a steady increase in UK participation in ESA budgets over the past three years (from £158 million in 2007 to £205 million in 2009).

On 10 February 2010, the committee of experts finally delivered its study on the future of United Kingdom's space policy. The document, entitled the Space Innovation and Growth Strategy (IGS), was commissioned by the British government to propose reforms in the functioning and scope of UK's space activities administration and policy. The study included 16 recommendations on how Britain could raise the political profile and industrial impact of its space programmes. A first step towards this direction was taken in late 2009, when the creation of a dedicated space agency to overview all of the country's space activities was announced. The current UK space budget stands at £265 million, over 75% of which represents the country's contribution to ESA. One of the report's key recommendations was to double the amount of the British contribution and to manage it in such a way as to maximise industrial returns for the country. In fact, the panel of experts suggested that UK contribution to ESA should reach the levels of its French and German counterparts, if the country was to achieve the full potential benefits of its participation in ESA. Furthermore, the experts concluded that Britain should increase and support the competitiveness of its national space industry. To that effect, it proposed the backing of UK's exports in this field by the country's Export Credit Guarantee Department.[143]

Another policy advice included in the document was to expand the future British space agency's area of responsibilities to include military space activities as well. In this sense, the report's authors looked up to the example of France and Italy which both have space agencies with dual civil-military role. For that reason, the committee recommended that the development of national military space application should be considered in the country's new Strategic Defence Review, expected to be published later during the year. Unlike other European nations, Britain does not have national Earth observation (EO) satellites yet. According to the report, this capability gap should be addressed by the creation of a dedicated EO services agency in the UK, which would develop and use its own satellites. Furthermore, the committee included in its recommendations the need to formulate a coherent and explicit national space policy document in order to assure the long term strategic guidance of space programmes.[144]

Finally, on 23 March the British government announced it had decided to establish the UK Space Agency on 1 April 2010. The Agency would operate along the general principles mentioned above. It would not however benefit of any additional budget, since the financial conditions in the UK would not allow it. Nevertheless, the new administrative structure was expected to produce immediate benefits for the British space industry's exports. In addition to this, it would certainly secure a more visible presence within the European Space Agency for the UK as well as better industrial returns from ESA programmes for British companies.[145]

2.4. The United States

The most significant development in the U.S. in 2009/2010 has been the announcement of a new Space Policy, on 28 June 2010. This was a document describing the general strategic and policy guidelines and priorities that all the different U.S. government agencies delegated to conduct space activities should follow. All U.S. Presidents since Eisenhower have issued such policy papers, recognising the unique place and importance of space activities to their country's international standing, economic development, scientific advancement and national security.

The announcement of the Obama space policy did not come as a surprise. Soon after his election, the new administration officials and the President himself identified space activities as an area of great significance to U.S. policy, attributing to it a high priority within its working plan. Interagency consultations on the drafting of the policy began already in the summer of 2009, one year prior to its release, based on the authorisation of the Presidential Study Directive No 3. Consultations on the policy's content were not limited within the U.S. government, but on the contrary included inputs from close friends and allies among space-faring nations. During this process, separate talks were held with EU authorities, which underlined the latter's increasing competence in the field of space policy.[146] At a later stage of the review process, other important space actors such as Russia, China and India were informed of its outline, making international cooperation one of the new policy's key elements, already during its making.

The new U.S. space policy itself is a 14 page document with a carefully balanced structure. The first 4 pages include a brief introduction and a 2 page declaration of the policy's key strategic orientations and objectives, labelled "principles" and "goals" respectively. Then, the rest of the paper is evenly divided into two parts. The first provides the broad policy guidelines that all government authorities conducting activities in space should observe ("intersector guidelines"). The second part lays down the more specific actions that they should undertake in order to achieve the policy's objectives, divided into three fields of activity: commercial, civil and national security ("sector guidelines"). In short, the new policy demonstrates a very clear and articulate methodological approach, moving from its broad strategic orientations to the narrower policy guidelines and then to the specific objectives that should be met in every sector.[147]

The key strategic orientations of the new U.S. policy include: the creation of a sustainable, stable and freely accessible near space environment for all nations; the reiteration of the U.S. leading role in space activities; the expansion of international cooperation in space; the improvement of the space industry's manufacturing and commercial competitiveness; the increase of U.S. space assets' resilience against interference; and the implementation of innovative scientific research and development, including exploration and space applications programmes, with a particular focus on Earth observation missions. "Intersector guidelines" in the policy address an important number of key issues, such as: maintaining and enhancing U.S. space capabilities; fostering international cooperation; preserving near space environment through the promotion of a more responsible use of space; implementing more effective export policies to the benefit of the country's industries; advancing research on space nuclear power; improving the management of radiofrequency spectrum and protecting national space assets from interference; and finally increasing the resilience of mission-essential capabilities.

If one takes a closer look at the various policy guidelines presented above, it appears that they all evolve around three principal thematic areas. The first is protecting and improving U.S. space scientific and industrial competitiveness. This prerogative includes reviewing barriers to the private space sector's development, such as strenuous and counterproductive export control procedures. This point is linked to the overall Obama administration policy that seeks to mitigate the effects of the present economic crisis by increasing U.S. exports, including a review of the State Department's International Traffic in Arms Regulations (ITAR).[148] The ITAR list includes most of space system's components and preparations for its revision have started at the same time as the consultations for the drafting of the new space policy. In the framework of the same effort to revitalise the country's space industry and to reduce its dependence from government expenditures, the new space policy also calls for increasing government funding into innovative research and development, modernising infrastructure in a targeted manner (for example giving priority to space launch capabilities) and relying as much as possible on commercial services for government space operations. In general, the new policy clearly sees the current publicly managed space business model as problematic (perhaps in view of the recent financial turmoil) and it clearly indicates a preference for private investments, or public-private partnerships in space that it regards as more cost-effective.

The second tier of the Obama administration's strategic vision for space is that of an increased international cooperation. Cooperation in space activities has always been appreciated by the country's space policies, since it was considered as a stabilising factor in international relations and a field where the U.S. could leverage its technological advancement into an increased diplomatic status and recognition of its global leadership role. International cooperation is envisaged for all areas of space activities: space science, research and exploration, space transportation and especially nuclear power related research. Furthermore, the new policy pays particular attention to two areas of cooperation: preserving near space environment and developing transparency and confidence-building measures (TCBMs) in space.[149]

Regarding the first issue, it calls for respecting the UN Space Debris Mitigation Guidelines, encouraging international cooperation in Space Situational Awareness (SSA) information, developing new in-orbit debris removal technologies and finally promoting Global Navigation Satellite Systems' (GNSS) interoperability, even including soliciting foreign GNSS to strengthen GPS resiliency. Regarding the second issue, it seeks to foster international consultations and encourage the responsible and peaceful use of space. In this respect, it does not exclude considering arms control concepts in space, provided that they are equitable, effectively verifiable and not detrimental to U.S. security interests. This point reverses the previous administration's policy of considering TCBMs as unnecessary restrictions to the U.S. freedom of action in space and brings its position back to where it stood under the Clinton and previous administrations.

The third and final tier of the U.S. space strategy is to assure and enhance current U.S. capabilities in space. This aspect of the policy mostly relates to the concept of Operationally Responsive Space (ORS), which is not explicitly stated in the document, but it is however described as the ability to operate in a "degraded, disrupted or denied space environment".[150] The ORS concept is also implied when the policy calls for assuring the mission-essential functions that are indispensable to meet the minimum U.S. government operational requirements, together with increasing space infrastructure protection measures. Finally, the new policy pays particular attention to improving the management of radiofrequency spectrum and limiting intentional or not interference, in close cooperation with international partners.

The general strategic principles part of the new policy is followed by a second half that presents the specific guidelines for its implementation along three activity areas: commercial, civil and national security space, which appear in that order in the text and are most likely prioritised as such. The new U.S. commercial space policy seeks to outsource to the private sector as much of government space activities as possible. In the pursuit of that objective, it does not simply envisage the use of currently available commercial capabilities, but it aspires to actively build upon and modify them in order to create new possibilities. For that purpose, it states its readiness to assume part of the investment risks through PPP funding mechanisms. Furthermore, it refrains from developing government space capabilities that could antagonise with their commercial counterparts. Finally, it places all existing government space infrastructure to the service of commercial users on a reimbursable but equitable basis with government agencies. Most importantly, the policy does not exclude using foreign commercial services' providers for government missions, or hosting public payloads on commercial spacecraft. Finally, it aspires to foster a global open trade environment for space services by encouraging U.S. companies to be more extroverted and minimising regulatory burdens that might hinder activities abroad.[151]

Civil space guidelines, on the other hand, are divided into three categories: a) space science, exploration and discovery, b) environmental and weather Earth observation (EO) and c) land remote sensing. As far as the first is concerned, the policy mostly sets long term objectives, perhaps in the light of the U.S. government's previous decision to cancel the Constellation project. They include keeping up with robotic exploration missions, developing next generation space launch systems capable of supporting human missions to Mars by the mid 2030's, continuing ISS operations at least until 2020 and further pursuing scientific missions to explore the Sun and accurately catalogue Near Earth Objects (NEOs). Finally, it calls once more for the creation of PPPs to develop private spaceflight capabilities and invest in advanced space technologies. With regard to EO missions, the Obama space policy divides them into environmental (including weather) monitoring and land observation. With respect to the first, it underlines the importance of satellite assets to sustained global climate change monitoring and stresses the need for international cooperation in this field as well.[152] Furthermore, it evenly divides the labour for polar-orbiting satellite based weather monitoring between NOAA and the Department of Defence. Concerning land observations, the document clarifies the competencies of the different services using space assets and calls for the increase of government EO data openness, availability and compatibility for commercial use.

Finally, the new U.S. space policy is concluded with the country's national security space guidelines, which follow almost entirely the lines of previous policies. They focus on maintaining crucial space capabilities relevant to defence and intelligence missions, including measures to increase the survivability of satellites in a cost-effective fashion, improve rapid replacement capabilities (according to the ORS concept) and assure strategic independence by supporting the domestic space equipment supplier base. Further priorities call upon increasing SSA integration and effectiveness through inter-agency and wider international cooperation, with special focus on keeping existing capabilities in pace with the constant growth of the satellite population and maintaining the capability to attribute disturbances to U.S. space assets. At the same time, the document attributes space related competencies and responsibilities to the Department of Defence and the Director of National Intelligence indentifying their mission areas without any significant departures from the views established by previous administrations.

This chapter will present a brief analysis of the Obama national space policy and comparison to these published by the Bush (2006) and Clinton (1996) administrations. A similarly comparative approach was also adopted by administration officials in promoting the new policy, which they described as returning in many aspects to the spirit of the Clinton space policy.[153] Given the limited scope of this paper, previous policies will not be presented in detail.

First of all, it seems pertinent to assume that the order in which sector guidelines are presented is significant of each administration's priorities. Indeed, in the Clinton policy civil space guidelines were given first, followed by national security and commercial activities, whereas in the Bush policy national security came first, followed by civil and commercial space.[154] In short, the new U.S. space policy is the first to place commercial activities first and national security last. Clearly, this is a sign of the increased importance that the current President attributes to the commercial sector, which appears at the top of the list for the first time in history. According to this analysis, the increased importance attributed to commercial space does not perhaps signify a change of paradigm in the U.S. space policy, but it clearly indicates the administration's changed priorities.

Indeed, encouraging private entrepreneurship in space is clearly the new policy's top priority. Although the Clinton administration also attempted to exploit the competitive advantage held by U.S. commercial companies in space activities, the Obama policy adopts a more energetic approach and seeks to actively support their further development. In order to do so, it accepts to finance part of their R&D costs through PPP funding mechanisms, something that was explicitly ruled out by the last two administrations.[155] Furthermore, it demonstrates a preference to purchasing commercial services to the fullest extent possible (depending on their affordability), instead of using their government owned counterparts. To that effect, it does not exclude the utilisation of foreign based services.

Through this policy, the U.S. government apparently seeks to create new commercial markets, as for example in the case of the private human spaceflight industry. In addition to this, it recognises the profound change that the global space policy environment has witnessed over the past years, marked by the constantly increasing proliferation of space capabilities and actors. In establishing adequate policy lines to meet the globalisation of the commercial space market, the Obama administration abandons the approach of its immediate predecessor that sought to protect the U.S. "advantage" in space through tight security measures and strict export controls. On the contrary, it returns to the principle of "open doors" and free trade in space of the Clinton era. Furthermore, it exceeds the latter in recognising that, under the present circumstances, the U.S. space industry needs a competitive boost from the government to face up to constantly increased competition.

It appears that the approach presented above also determines the administration's stance towards export control measures. Several government officials have linked the new policy to the revision of the export control regime on space-related items, which currently poses restrictions upon their free commercialisation. It remains to be seen if a significant number of such items will be removed from ITAR. Nevertheless, it is clear that such a decision would be dictated by the administration's favourable view of an open and extroverted commercial space industry, as the only way to ensure its competitiveness on the long term. This attitude constitutes a clear return to Clinton policies, prior to the inclusion of space technologies to the ITAR list. It also differs from the Bush administration's introverted view of space technologies as a crucial national security and industrial asset, only to be shared with selected allies and "on a case-by-case basis".[156]

The second novel characteristic of the U.S. space policy is the fact that it does not limit itself to describing broad strategic outlines, but on the contrary it goes into specific details on how guidelines should be realised and objectives reached. This detailed approach is an indirect recognition of the increased complexity of the international space activities environment, with its multitude of emerging actors. It also implies that the administration was inclined to clarify the strategic vision on which related policy decisions were based, such as cancelling Constellation and providing a new direction for NASA.[157] The detailed nature of the new space policy was underlined by U.S. government officials. On the other hand, it has also raised some criticism to the fact that it fails to mention the budget required for its programmatic declarations.[158]

Another key plank of the new policy is its focus on international cooperation and its consequent multilateral approach to space activities. This characteristic signifies a clear departure from the previous administration's more unilateral tone and it does seem to return to the views held by the Clinton policy, if not expanding them even further. Indeed, the thread of multilateralism runs through the entire policy document. For example, it manifests itself in the potential for GNSS cooperation, which was not present in the 2006 policy.[159] Furthermore, the administration approaches the space debris issue in a broader, more global and coherent way than its two predecessors. This is especially the case when it discusses international cooperation in SSA projects in a systematic and detailed fashion. By doing so, it moves the debate forward from simply dealing with the debris threat to creating a more sustainable space environment and engaging all space faring nations through the promotion of more responsible policies and behaviours in space. In this sense, it implies a truly global and long term vision, according to which multilateral, and not simply bilateral, cooperation in space could become a stabilising factor for international relations in general.

It is worth noting that the current administration's vision of international cooperation and security in space does not limit itself to describing U.S. policies towards it. On the contrary, it places its attitudes in the broader context of a new order in space activities, based on all nations' adherence to the principle of preserving a sustainable space environment and demonstrating a responsible behaviour in space in order to protect it. Consequently, contrary to the previous policy of protecting the nation's narrowly defined interests against foreign competition in space activities, the new one places the U.S. in the centre of a multipolar but stable international environment in space activities. Needless to say, the U.S. still reserves for itself a leading role in formulating the rules of international conduct in space. Nevertheless, it tacitly recognises the fact that an increasing number of emerging actors would have to accept them, if they were to be meaningful. In this sense, the U.S. space policy rediscovers the virtues of the Clinton era's indirect strategic approach of "soft power".

Last but not least, one should note Europe's improved bearing upon the formulation of this new international setting. Administration officials have pointed out the consultations with EU authorities that preceded the publication of the policy. More importantly, they singled out the EU proposed Code of Conduct as a good starting point for discussing and implementing such rules of behaviour in space, albeit on a strictly voluntary basis.[160] This development was good news for Europe, as it demonstrated its own capacity to influence its strategic environment regarding space activities and constituted in itself a significant recognition of its standing.

However, in addition to the above the Obama space policy couples its multilateral approach with an acute sense of pragmatism when it discuses arms control initiatives. In fact, the new policy accepts in principle to consider arms control in space, provided that it serves the country's national interests. In doing so, it reiterates the Clinton administration's approach and reverses the previous policy of discarding such initiatives as restrictive to U.S. freedom of action in space. However, the new policy goes even further in this direction by considering the possibility of TCBMs in space, a tool so far related to strategic arms negotiations. By mentioning the possibility of TCBMs for the first time, the U.S. government adopts a space security approach that is more sophisticated than before. Furthermore, it builds upon the experience of bilateral strategic talks and advances them to a multilateral level for space security purposes. Finally, it creates a linkage between space security and ballistic missile defence, acknowledging that the two issues are related in sharing their operational medium.[161]

In relation to space security and space defence missions as well, the new policy adopts a more pragmatic and sophisticated point of view than before. In doing so, it takes into account the increased number of emerging space actors and the proliferation of space capabilities and services. Admittedly, national security objectives in space remain unchanged and they represent a major constant throughout all three last U.S. space policies. Consequently, the Obama administration pays equal attention to protecting its own national space assets and capabilities as its predecessors. Nevertheless, when examining it in its entirety, the new policy clearly refines the Bush era's unilateral approach of security through space control and the right to deny access to space to adversaries. Instead of this, it emphasises the resilience of critical capabilities, which implies not only the ability to deter any attacks against space assets, but also to maintain core capabilities in the face of such an event. Consequently, it places ORS at the centre of its national space security concept, on an equal foot as deterrence and protection.

Finally, a less substantial but politically important change has occurred in the new policy's choice of words regarding counterspace operations. The Obama administration remains adamant in its right to actively protect its space assets in the face of threat. It considers this as inherent to its national sovereignty rights and consistent with the UN recognised principle of self defence. Nevertheless, the new space policy document states that such counterspace actions will be taken "if necessary", replacing the phrase "if directed" used by both previous administrations. It would be exaggerated to presume that this difference implies any kind of change to the rules of engagement applied in such a case. It does, however, create the impression that such operations (and especially of destructive nature) would be considered as a last resort, when deterrence or other options have failed.[162]

Contrary to other policy areas, civil space activities guidelines remain mostly unchanged in the new policy. Extensive passages, such as the ones referring to the development of nuclear space capabilities or the use of EO missions for environment monitoring are taken almost word for word from previous policies. Space science R&D objectives and guidelines also remain the same, with the significant exception of a new access to space policy focused on the use of commercial services. However, the Obama administration goes into much more details in describing how its policies will be conducted and which government agency will be responsible for them, for the reasons described above. A new element in the policy is the direction to create data bases of environmental observations monitoring climate change consequences and to make them available for public use. In doing so, it emphasises the usefulness of space services for achieving sustainable development on Earth.

Finally, another example of improvement upon previous policies is the case of radiofrequency protection and counter interference measures, which seem to preoccupy the policy more than before. In this field as well, the administration demonstrates its preference for international cooperation in mitigating interference and its willingness to protect U.S. commercial providers from it too. Apparently, it recognises the fact that the growing number of space actors makes a cooperative approach to this issue more appealing than before. It also goes into length in describing U.S. actions in this policy area, which will seemingly become more and more important in the near future.[163]

Just like the ones that preceded it, the Obama administration's space policy is a product of its time. It recognises that the existence of a multitude of new government and commercial actors will create a radically new international space activities environment. It anticipates the emergence of a more pluralistic and multipolar order in space and attempts to prepare U.S. commercial and government entities to face the increased competition and complexity it will entail. It does not aspire to protect itself from this new reality by safeguarding its technological advantage and unilaterally protecting its narrow interests in space like the Bush policy did before. But it does, however, claim for itself the leading role in setting internationally accepted standards and rules of good conduct in space.[164] By doing so, it aspires to further extent its strategic influence and "soft power" in the field of space activities. In this sense, it finds itself closer to the Clinton policy. But in reality it moves further than this in proposing a coherent international cooperation model, based on a multilateral approach rather than the separate bilateral discussions held in the past. In conclusion, it does not simply seek to protect U.S. interests in space, but it regards a new international order in space activities itself as the highest U.S. national strategic interest in space.

The change in U.S. space policy was preceded earlier between mid 2009 and mid 2010, with a complete revision of the U.S. space programme and a restructuring of the NASA budget. The principal characteristic of the new policy was the cancelation of the Constellation manned spaceflight programme and the diversion of considerable funds to public-private sector space technology development schemes. This change in NASA policy direction was mostly focused on access to space policies and it will be presented below

Another issue that preoccupied administration officials was the possible removal of satellites from the U.S. munitions list. On 25 June, Ellen Tauscher was confirmed by the U.S. Senate as the new Undersecretary of State for Arms Control and International Security. During her confirmation statement she suggested that one of the Obama administration's top priorities would be to revise the U.S. export control regime, known as the International Traffic in Arms Regulations (ITAR). She specifically acknowledged the need to consider removing commercial communication satellites from the U.S. Munitions List (USML). The inclusion of such satellites in the list was decided by the U.S. Congress in 1999, when commercial satellites' export control jurisdiction was moved from the Commerce Department to the State Department.[165] This situation has created serious obstacles for satellite operators, since clearing USML items for export can take up to 90 days. This has practically obliged satellite operators to launch all satellites built in, or using components made in the U.S., from U.S. territory. Mrs Tauscher said that including satellites in the munitions' list has impeded technological innovation from U.S. companies in this field and it has decreased their competitiveness on a global scale.[166]

Earlier in June, the U.S. House of Representatives passed the Foreign Relations Authorisation Act for 2010 and 2011 that enabled President Obama to remove commercial satellites from the USML. However, the bill was still awaiting approval by the U.S. Senate Foreign Relations Committee.[167] At the same time, the U.S. House Permanent Select Committee on Intelligence picked up the ITAR issue in its 2010 Intelligence Authorisation Act. In it, the Committee underlined that strict export control of commercial satellites and related technology has had the exact opposite effects from those anticipated. In fact, it had encouraged local research and development of banned items in foreign countries and it had particularly motivated European companies to establish an international non-U.S. collaborative research environment in order to produce them.[168] For that reason, the Committee proposed that the administration progressively removes space-related items from the USML.

The Obama administration returned to the subject again on 13 August, when the White House indicated on its official website that it would move forward with export-control reform. The same commitment was reiterated again in a public statement by Assistant U.S. Secretary of State Andrew Shapiro on 9 September. Mr Shapiro asserted that bureaucratic struggle to remove certain items from the USML would be long, but that it had the backing of key officials, including U.S. Secretary of State Mrs Clinton. In the meantime, he said, the U.S. State Department was working on simplifying and speeding up its export licence application procedure.[169]

The entire public discussion on export regulations was conducted amongst fears that the U.S. industrial and technology base might erode in the near future because of extensive market consolidation, an aging work force and the absence of new major defence programs. Senior Pentagon officials had already expressed concern over losing irreplaceably skilled defence-related workforce due to retirement. If this trend continues, it is estimated that in the near future the U.S. industrial base will not be able to support every procurement policy the Pentagon decides. And on the other hand, the U.S. Defence Department will have to be content with facing monopolies in several defence product markets.[170]

Throughout the year the debate on removing satellite components from the Munitions List continued in both U.S. legislative bodies. But it proved to be a highly polarising one, with national security "hawks" facing aerospace industry officials that saw their global market share suffer since ITAR controls were imposed in 1999. In this context, on 2 December the President of the U.S. Aerospace Industries Association (AIA) M. Blakey addressed a letter to President Obama urging him to consider loosening the export control restrictions.[171]

During his state of the Union speech on 27 January 2010, U.S. President Barack Obama confirmed that he considered the reform of the U.S. export control system as one of his top priorities. This reform was the objective of the administration's National Export Initiative, which foresaw the creation of an Export Promotion Cabinet to assist the President. Further recommendations on how to relax export restrictions were due to be circulated by 29 January within the government by an interagency working group set up for this purpose. The group was mandated to review the entire export control procedure from scratch. The principal objective of this initiative was to increase export volumes in the face of the global economic crisis, as well as to maintain the U.S. industrial competitive edge. The proposed review had reputedly received full backing from the Pentagon. This development was greeted on the following day by Marion Blakey, the president of the Aerospace Industries Association, who highlighted the initiative's positive impact on employment within the defence and aerospace sector.[172]

2.5. Russia

During our reporting period, Russia continued the modernisation of its space infrastructure according to the ten-year Federal Space Programme announced in 2005. This document provided the key objectives and guidelines for the Russian space programme through 2015 and is still in force.[173] One more space policy related paper was produced in 2008 by Russia's Security Council, updating the security related project priorities.[174] The implementation of this programme so far has demonstrated the country's commitment to developing full scale capabilities across all areas of space activities.

According to the goals set in 2005, Russian space activities should have three key objectives: stabilise the country's economy, develop space science for the benefit of society, and consolidate its defensive power. These objectives were in fact in hierarchical order. Space technologies were first to contribute to the country's economic recovery after the "lost decade" of the 1990's, increase the volume of external trade and exploit the Russian industry's competitive advantage in space in order to consolidate its market position. Secondly, further space technology development should be targeted on space applications in order to enhance domestic stability (for example by extending state television broadcasting throughout Russian territory) and facilitate government work. Furthermore, scientific research should be conducted in close concertation with the industry in public-private joint ventures and should aim at diffusing its benefits to the entire society. Finally, military space capabilities should be preserved and modernised as much as possible and with the widest feasible utilisation of dual use systems.

Overall, one could say that so far the Russian space programme has, to various degrees, attained most of its objectives. The main challenge that it had to face was marrying two inherently divergent objectives. On the one hand, there was the necessity to maintain a widely degraded and even obsolete space infrastructure inherited from the 1990's. On the other, major investments had to be made simultaneously for the development of new systems. Apparently Russia has succeeded in this undertaking, mostly thanks to increased funding made available from the country's flourishing energy and armament sector. Indeed, the Russian space budget has manifested the largest growth in the world during the past 10 years. In 2009 alone the country's civil space budget rose to Rub 88.64 billion, representing a 100% increase from 2008.

Finally, the Russian space industry's output grew by 18% in 2009, in spite of the financial crisis. This positive development has been the result of the carefully planned financial support that the Russian government extended to the country's space industry. More specifically, in the last two years the latter have received over 21 billion Rubles ($609 million) in public funds. When one considers that this stimulus package was carried out as planned in the midst of a major financial crisis, it becomes apparent that the country's space programme enjoys strong political support. In deed, the Russian Prime Minister Vladimir Putin himself has underscored the space industry's importance for the country's economy in several occasions. In order to achieve the objectives mentioned above, the Russian space programme has focused during the past 12 months on three main activity areas: completing the deployment of the Glonass satellite navigation constellation, initiating the creation of a new space port in the Far East, and restructuring the country's space industrial base. In conclusion, during this financial stimulus programme the country's space industry has had the chance to improve its production facilities.[175]

2.6. Japan

Between July 2009 and June 2010, Japanese space policy saw important developments and changes. A long transitional period ended with the creation of a coherent new space strategy for the country. This is forged by two key documents, the Basic Space Law (BSL) of 2008 and the Basic Plan for Space Policy (or Basic Space Plan – BSP) adopted on 2 June 2009. The first was Japan's first comprehensive national space law, aimed at regulating all space activities, public and corporate, and setting the strategic scope of the Japanese space programme. The second was mandated by the BSL as the country's fundamental space policy document, in order to define and materialise BSL into a coherent Japanese space activities roadmap. The BSP was issued by the Strategic Headquarters for Space Policy, the inter-ministerial governing body for Japan's space activities, also set up by the BSL.[176] At the same time, the Japanese government has continued the administrative reorganisation of space activities, with the explicit aim of simplifying governance by concentrating authority under the single roof of the Japanese Cabinet Office. The Japanese Space Exploration Agency (JAXA) will coordinate and finance all government space activities, following the example of NASA.

The BSP established a five year roadmap for 2009–2013, along six basic pillars. 1) The realisation of a "secure, pleasant and affluent society" utilising space. 2) The enhancement of Japan's national security. 3) The promotion of "space diplomacy". 4) The creation of a "vigorous future" by promoting space related research and development. 5) Fostering of strategic space industries for the 21st century. 6) Considering the environment.[177]

The Basic Plan clearly demonstrates the strategic importance given to space for the future prosperity and security of the entire country. Furthermore, it identifies the future areas of interest for the country's space policy. These are space applications, security, international cooperation, scientific development, industrial development and environment protection. It should be noted that security in its broader sense (military, diplomatic and economic) becomes the cornerstone of the new policy, as it is depicted in three of the six pillars. Other key policy objectives include achieving full autonomy in space technologies and increasing public-corporate synergies in space activities. Finally, space exploration (including independent manned flights) also receives special attention in the document.[178]

Japan's ambitious new space policy has also secured adequate funding, a fact that in the face of the current financial crisis underscores its importance in the eyes of the country's government. The proposed 2010 space budget foresees a 25% increase that will bring its total sum to ¥436 billion ($4.7 billion). This is the total inter-ministerial funding according to the Basic Plan for Space Policy. However, even this sum falls short of the ¥2.5 trillion budget through 2013, recommended by the BSP. This is due to the Finance Ministry's intervention, which usually curtails space funding in its cost revisions in December.[179]

The principal civilian space activities funded in 2010 include the Daichi Earth observation satellite, a satellite for the Quasi Zenith GPS augmentation system and the new Advanced Solid Rocket for launching small payloads. Funding for defence space capabilities includes space situational awareness development, a dedicated military communications satellite, research in infra-red missile warning sensor technology and microsatellites. In spite of an expected change in the Japanese government in September, space programmes seemed to have secured the necessary political consensus to continue.[180]

A report, requested by Japan's cabinet official for space policy, Seiji Maehara, and delivered on 20 April 2010, urges Japan's government to form a new space agency and close the existing JAXA. The aim is a better, quicker and more efficient response to Japan's national needs concerning space activities. Nowadays, JAXA's policy and planning is in the responsibility of the Ministry of Land, Infrastructure, Transport and Tourism, while the budget is part of the Ministry of Education, Culture, Sports, Science and Technology (MEXT) causing unnecessary complications according to this report. It proposes a new space agency controlled by a small executive committee under the direct authority of the Prime Minister and Maehara. As 2011's space budget has to be requested in August 2010, the report recommends the establishment before this date.[181]

2.7. China

China's space policy evolves around the country's five years' economic development plans, the current plan having been decided in 2006. China's space programme is therefore meant to support the country's overall development objectives, while maintaining a comprehensive set of objectives for space activities. The main challenge for the Chinese programme is to mix the desire to develop independent capabilities to the maximum with the need to participate as much as possible in international space cooperation. No key strategic document was published in the field of Chinese space policy in the past 12 months, as the current ones were published at the beginning of the five-year plan and are valid until the end of 2010. New documents may be expected at the beginning of the new planning cycle.

China has a long-standing, full fledged space programme which spreads over a variety of activities. However, its main focus lays on manned space flight, space applications and the further development of its Long March family rockets. It is difficult to distinguish Chinese civilian and military space activities, as several systems have dual use and some key civilian projects (such as the Shenzhou spacecraft) are in fact under military control. Another cause of confusion is that Chinese authorities have a strict confidentiality policy regarding their space systems and they withhold information on the costs and annual budgets of most of their programmes. Although no accurate data is publicly available, the Chinese space budget is thought by most experts to increase constantly over the past few years.

China has also been increasing its involvement in international space cooperation, albeit through bilateral rather than multilateral agreements. On 17 November 2009 for example, the U.S. and China announced that the heads of their respective space agencies would exchange visits in 2010, in an effort to step up cooperation in space exploration, including human space flight. This joint statement was issued during the official visit of U.S. President Barack Obama in Beijing on 15–18 November. An agreement between the two agencies to talk at least once a year was reached during NASA Administrator M. Griffin's first historical visit to China, but the Chinese ASAT test in January 2007 postponed its implementation. In spite of this, NASA and the China National Space Administration (CNSA) have formed joint working groups for space science and space-based climate research since 2008.[182]

On 14 April 2010, Wang Wenbao, head of the China Manned Space Engineering Office, gave an overview of China's plans to build a 30-ton space station until 2022. The first steps would contain the launch of Tiangong 1 target and Shenzhou 8 doing docking and rendezvous experiments in 2011. This is meant to be the beginning of further docking, rendezvous and refuelling experiments including manned spacecrafts like the Tiangong 3 spacelab which was to be launched between 2014 and 2016. While these launches were to be conducted by Long March 2F rockets, the three modules in order to construct the space station between 2016 and 2022 were to be launched by China's new heavy-lift rockets yet in development. China's space station should be able to host three astronauts and it is planned to last 10 years.[183]

2.8. India

Indian policy traditionally aims at achieving social and economic development through space activities. The Indian space programme currently operates under the guidelines of the current 11th five year plan (2007–2012) which focuses on creating space applications capable of providing tangible products that improve life conditions in the country. Self reliance and space services oriented projects are the corner stones of India's space policy. These include two operational space systems, one for satellite communications and television broadcasting services and one for Earth observation.

As the five year plan develops, Indian space budgets have been increasing accordingly. In July 2009, the Indian Space Research Organisation (ISRO) announced that its 2010 budget would stand at Rp 49.79 billion ($1.04 billion), an increase of 40% from the previous year. This is the largest increase ever recorded in ISRO's history. The largest share of funding was set for launch vehicle development, followed by satellite technology and space applications. Rocket development plans include the Mk 3 version of the Geosynchronous Satellite Launch Vehicle (GSLV), with a lift capability of four tons. Its first flight is expected in 2010 or 2011, soon after a new indigenous cryogenic upper stage engine has been tested. The Indian human space flight program also acquired additional funding for the development of a two person capsule capable of attaining a 400 km orbit. In fact, ISRO reaffirmed its intention to proceed to its first manned flight by 2015.

Satellite technology research also secured funding, driven mainly by the Regional Navigational Satellite System. This is a proposed seven satellite navigation and positioning constellation to cover Indian national territory and neighbouring states. Other space applications projects include the environment monitoring missions Oceansat-3 and Resourcesat-3, as well as the new cartography satellite Cartosat-3.[184]

2.9. Emerging space powers

From 7 to 9 December 2009, the Third African Leadership Conference on Space Science and Technology for Sustainable Development took place in Algeria and as a result, Algeria, Kenya, Nigeria, and South Africa signed an agreement to build up the African Resources Management Satellite Constellation. The purpose of this constellation is to help with environmental monitoring, public health, and water as well as land usage.[185]

The Nigerian space programme is one of the most advanced of Africa. On 17 July 2009, the director of the Nigerian Space Agency signed a Memorandum of Understanding defining a roadmap of cooperation with the German based Infoterra in order to prepare a radar mission that Nigeria is considering.[186]

Nigeria is also planning to launch its second satellite, the NigeriaSat-2, in the fourth quarter of 2010 on a Dnepr rocket. The last milestone of this mission was reached in October 2009, when the spacecraft passed its flight readiness reviews.[187]

When it comes to South Africa, although a bill was signed in January 2010 by South African President to establish the South African National Space Agency (SANSA) in 2009, it is not done yet. However, in autumn 2009 the nominations for a board were approved. Once the space agency is established, it seems likely that one of its core programmes will be dedicated to Earth observation.[188]

Good news can be reported from the Sumbandila project – the first South African government-owned satellite that was launched on 17 September 2009 with delay. It was developed by SunSpace, the company that launched also the satellite-project of the University of Stellenbosch called SunSat in 1999. Sumbandila passes four times a day over South Africa and will collect images for governmental use for instance in water management, agriculture, and urban planning.[189]

On 22 February the Thaicom satellite fleet operator of Thailand announced a small 1.7% drop in its 2009 revenues. This result was mainly attributed to a 28% fall in broadband satellite terminals sales. Thaicom provides satellite broadband services through its dedicated Thaicom 4 (also named Ipstar) spacecraft. The firm expected this downward trend to reverse in 2010, as it planned to expand its operations to India and Taiwan. On the contrary, Thaicom's conventional satellite communications services' revenue increased by 1.6%, in spite the retirement of its Thaicom 1A spacecraft at the end of 2009. The company currently uses one more satellite, the Thaicom 2, which is also approaching the end of its expected operational life span. In 2009 Thaicom reported $134.7 million in total revenues.[190]

The conflict in Thailand had impacts on Thaicom as the government forced Thaicom to shut down broadcasting of the People Channel Television (PCT), although PCT was not a direct Thaicom customer. After demands to its customer were not accomplished, Thaicom started to jam PCT on 7 April, but the channel moved to another Thaicom customer based outside Thailand. After threats of Thaicom to shut down the whole C- and Ku-band capacity of its customer, the latter stopped broadcasting on 23 April. Thaicom found itself between government's instructions, "company's employees and assets" and decided in favour of the government.[191]

In January 2010, the question of satellite broadcasting interference came under the spotlight in the Middle East. In particular, France's National Frequencies Agency (ANF) asked the Geneva based International Telecommunication Union (ITU) to intervene with Iranian authorities in order to persuade them to cease jamming of BBC World Service's Farsi speaking programme over Iranian territory. French authorities were involved as the programme was transmitted through the Eutelsat Hot Bird 6 satellite. According to French sources, jamming had started during the Iranian election campaign in the spring of 2009 and had not stopped since. After repeated failed attempts to address this issue directly with Iranian authorities, ANF appealed to the ITU. However, since the latter is a purely regulatory organisation with no means to enforce its decisions on its Member States, it was unclear whether its involvement would bring about any results. In the meantime, the BBC had decided to move its broadcasting to another Hot Bird 6 beam and ultimately to another Eutelsat satellite over the region, in an attempt to overcome the interference.[192]

In the past 12 months, Brazil has continued to expand its scope of bilateral cooperation agreements on space activities in an effort to develop necessary technologies for its ambitious space programme. On 8 October for example, the country's space agency AEB signed a technology exchange agreement with Belgium's Liege space centre. This four year agreement includes a wide area of technologically advanced fields of activity, from nanosatellites to space instruments validation.[193]

Finally, joint Brazilian-Chinese cooperation in building the next generation of Earth observation CBERS satellites for the country suffered a set back in February 2010, when AEB announced the postponement of the CBRS-3 spacecraft launching to 2011. The delay was attributed to technical problems identified during the system's design review.[194]

3. Worldwide space budgets and revenues

Although it was not clear if the trend of increasing institutional space spending would continue in times of financial crisis, it can be observed that the previous year's trend persisted during the last 12 months. The institutional spending on space activities is estimated to have reached $68 billion. A more detailed view on institutional budgets can be found in the following paragraph 2.2.

In terms of commercial revenues of space activities, the Space Report 2010 indicates the total revenue of commercial satellite services to have been about $91 billion comprising telecommunications, Earth observation and positioning services. The revenue of space-related commercial infrastructure including manufacturing of spacecrafts and in-space platforms, launch services as well as ground equipment is estimated to have reached around $84 billion. In conclusion, the commercial space revenues of 2009 can be sum up to $175 billion.[195] A closer look on the commercial revenues is provided in paragraph 2.3.

3.1. Overview of institutional space budgets

The total institutional spending on space in 2009 can be estimated to be approximately $67.8 billion, a figure which shows a nominal increase of 9% compared to 2008.[196] This space spending is comprised of $36 billion in civil expenditures (or 53.1% of the total) and $31.8 billion in defence expenditures (or 46.9%). Consequently, the share remained virtually the same compared to last year's figures. Out of the estimated $31.8 billion of defence related space expenditures worldwide, $28.7 billion were spent by the United States, representing a share of 90% and indicating a minor percentage decrease compared to the year before. These funds came from the Department of Defence (DoD), the National Reconnaissance Office (NRO), the National Geospatial-Intelligence Agency (NGA) and other government entities. It should be borne in mind that not all relevant funding is made public, resulting in a degree of uncertainty regarding the exact figures of expenditures on defence space activities.

Adding up civil and defence space expenditures, the United States had the biggest institutional space budget in 2009, spending $48.8 billion ($20 billion civil expenditures and $28.7 billion defence expenditures). The total U.S. public space budget constituted 72% of the global institutional spending in 2009. The next largest space budget does not belong to a state but to the European Space Agency, which coordinates civilian space programmes on behalf of its Member States. ESA's budget in 2009 reached approximately $4.8 billion. The next largest national space budgets are furnished by Japan ($3.0 billion), Russia ($2.8 billion), France ($2.8 billion), China ($2.2 billion), Germany ($1.4 billion) and Italy ($988 million), all a considerable distance from the United States. A noteworthy fact was the enormous increase in Russian space spending that nearly doubled, compared to the $1.5 billion spent in 2008. Looking at the rest of the European countries, their accumulated total public spending on space activities in 2009 reached $7.2 billion, representing a 10.6% share of global institutional spending in 2009.

Combined, the United States and Europe accounted for almost 83% of the global institutional spending on space in 2009.

Consulting the absolute numbers alone only tells one side of the story, as comparisons between countries with different economic conditions like prices or

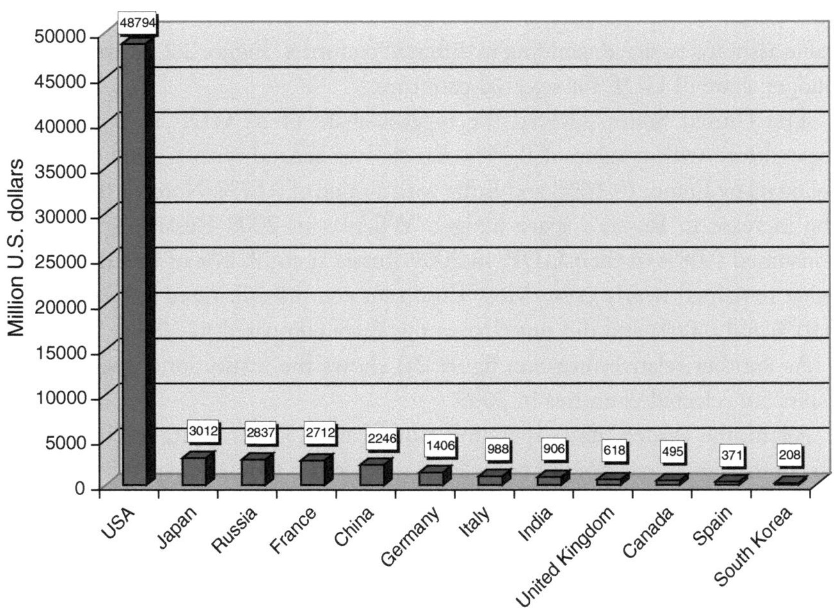

Fig. 3: *Public space budgets of major space powers in 2009 (Based on Euroconsult data).*

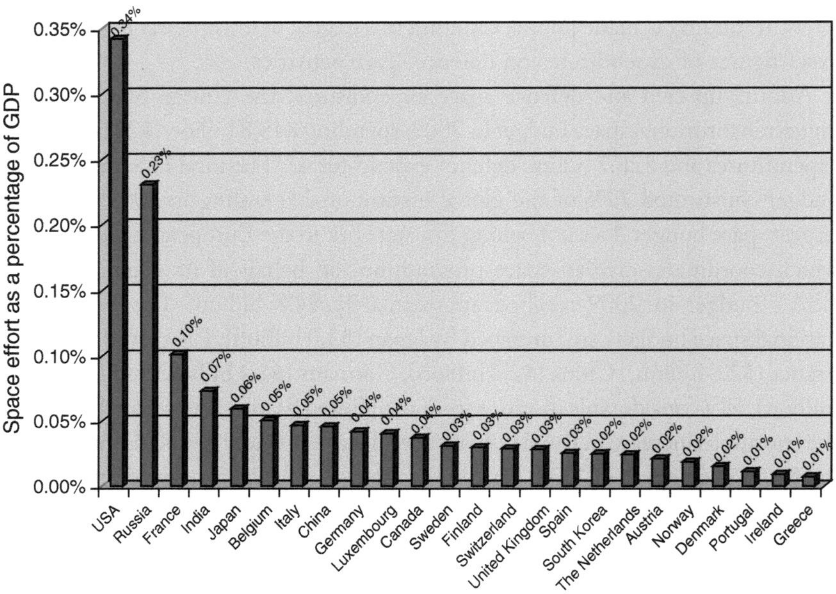

Fig. 4: *Public space budgets (selection) as a share of nom. GDP in 2009 (source: Euroconsult data, IMF*[198]*).*

wages levels can be misleading. It therefore makes sense to relate the amounts spent to the GDP or to population size. This gives a better indication of the comparative value assigned to space spending in different countries. Figure 2.2 shows the space budget share of GDP for selected countries.[197]

The United States devoted the biggest share of its GDP to public space expenditure with a value of 0.34%. Russia was second with a share of 0.23%, followed by France (0.10%) and India with a value of 0.07%. Noteworthy again is the increase in Russia's space budget. Whereas in 2008 Russia's space effort consumed 0.08% of their GDP, in 2009 Russia spent 0.23% of its GDP whose value remained nearly even. Most European countries featured values between 0.01% and 0.06% and did not change the share compared to 2008.

As another relative measure, figure 2.3 shows the institutional spending per capita for selected countries in 2009.

Again, the United States spending more than $155 per capita in 2009, led by far. France and Luxembourg completed the podium with each spending around $43 per capita. Japan and a number of European countries spent in the vicinity of $20 per capita. It is also possible to rate the GDP share of public space funds against the public space funds per capita. This is done in figures 2.4 and 2.5, with the latter excluding the United States and Russia to display the other countries more clearly.

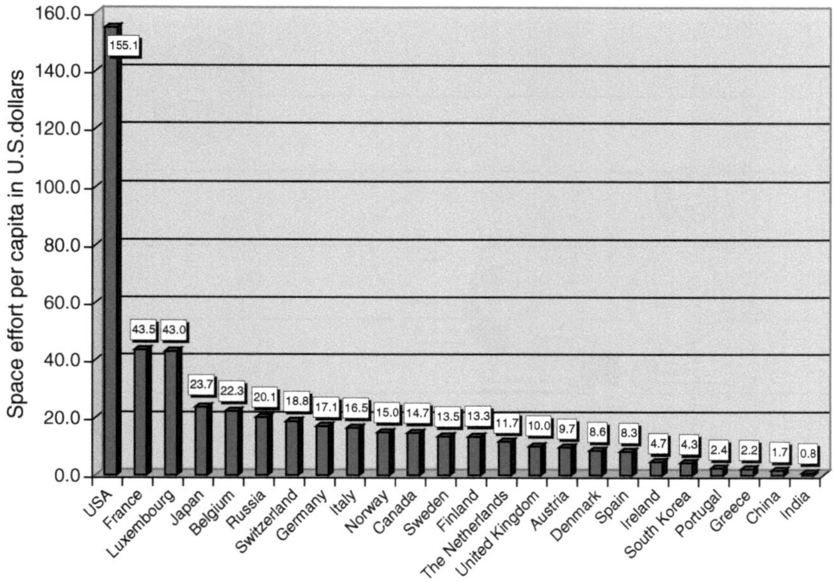

Fig. 5: *Public space budgets per capita (selection) in 2009 (source: Euroconsult data, UN World Population Prospects*[199]*).*

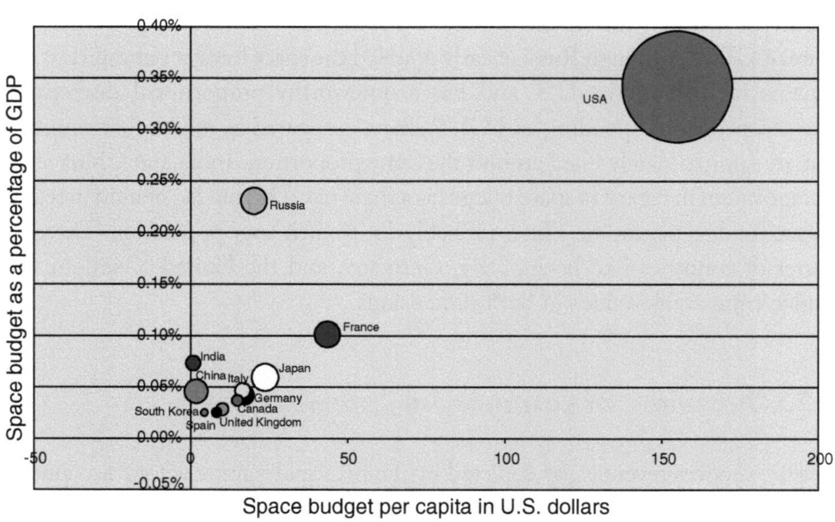

Fig. 6: *Public space budgets as share of GDP mapped against space budgets per capita in 2009 with the bubble size indicating the absolute space budget (source: Euroconsult data, UN World Population Prospects*[200]*).*

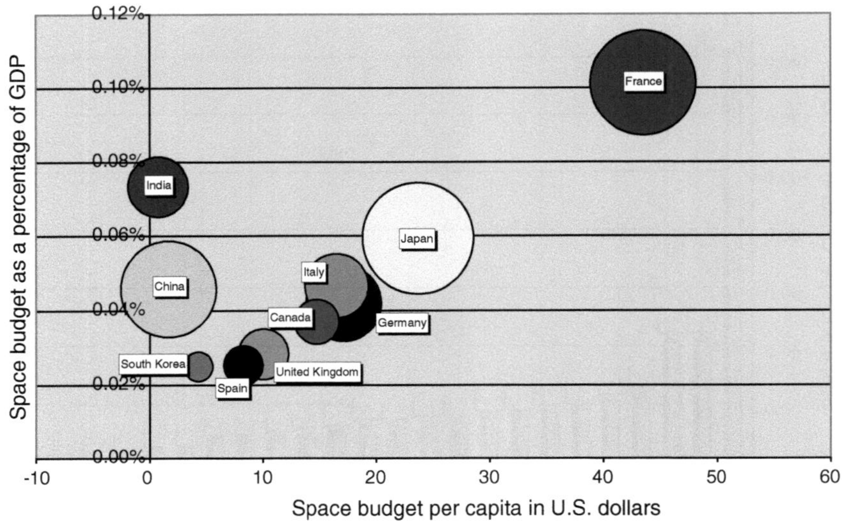

Fig. 7: *Public space budgets as a share of GDP mapped against space budgets per capita in 2009 with the bubble size indicating the absolute space budget, excluding the U.S. and Russia (source: Euroconsult data, UN World Population Prospects[201]).*

The United States is unique by evidently excelling in both dimensions, i.e. in the public space fund share of GDP and in space budget per capita. France and Russia follow. While France is also way ahead of the other countries in both dimensions, Russia has made big efforts and now has a very high and remarkable position in the share of GDP. Although Russia nearly doubled the space budget per capita, it still remains far behind the U.S. and has a noteworthy proportional discrepancy between per capita spending and GDP share, compared to most other countries that are approximately lined around the same proportion. India and China show average values in regard to space budget as a share of GDP, but lag behind in regard to space budget per capita. This is probably due to their large population. There is a cluster of countries like Japan, Italy, Germany, and the United Kingdom that display comparable values in both dimensions.

3.2. Overview of commercial space markets

Satellite services revenues, as depicted by the principal market actors' accounting statements, manifested robustness and resilience to the financial crisis' effects throughout the last 12 months.

On 5 November for example, Eutelsat, the world's third largest commercial satellite operator based on revenue, announced an 11.6% increase in revenue in the

3rd Quarter of 2009. This result was mostly attributed to a 46% increase in multi-usage revenues to €22.9 million, driven by increased sales in the Middle East and Central Asia regions. The announcement confirmed that major commercial satellite operators were largely untouched by the financial crisis.[202]

On 16 November EADS Astrium also announced €3.23 billion revenue from the first three quarters of 2009, an increase of 17.4% over the previous year. Pre-tax profits however were down to 4.8% of revenue (€155 million) from 5.1% during the same period in 2008. According to the same announcement, Astrium Space Transportation accounted for 41% of total revenue, Astrium Services for 38% and Astrium Satellites for 21%. Astrium Services were severely curtailed by the drop in the British Pound, as the company's principal revenue source is its contract with the UK military to provide the Skynet 5 satellite communication services.[203]

Earlier, in July, EADS Astrium had reported a 29% revenue increase (at €2.19 billion) and a 22% backlog increase (at €15.6 billion) in the first half of 2009. These results were attributed, among other factors, to the acquisition of Spot Image the previous year. Astrium's pre-tax profit margin however was down to 4.5% from 5.2% the previous year. This decrease was attributed by Astrium officials to restrictions due to government procurement rules relating to space systems.[204]

3.2.1. Direct Broadcast Services

Direct Broadcast Services (DBS) refer to direct-to-home satellite television and radio broadcasts. This business category manifested considerable improvement in the period under examination, fuelled mostly by growing demand in Europe and emerging markets, such as India. On the contrary, demand in the U.S. market showed signs of stagnation, probably because of its rapid expansion during recent years, in conjunction with the financial crisis' effects.

Eutelsat for example, the world's third-largest satellite fleet operator, announced a 7.2% revenue increase (at €940.5 million) in the 12 month period from June 2008. According to the company's statement, this increase could sustain 7% annual growth throughout 2012. According to the same sources, satellites operated by Eutelsat were filled to capacity (over 97%) and efforts were undertaken to reposition some of them in order to reduce the fill rate to 80%. Strong television demand in Europe and U.S. government bandwidth demand in the Middle East contributed mostly to these financial results.[205]

The second largest satellite fleet operator, SES, also reported record gross profit margins in July and announced that it was still on track with its target to attain a 5% annual growth rate throughout 2010. However, the company's recurring revenues in the first half of 2009 were up by only 2% from last year, to €843 million. These results were attributed to SES's lower than expected growth in North America, the opposite of its satisfactory performance in Western Europe. In total, 81% of the company's revenues were attributed to satellite lease operations.[206]

On the other hand, satellite services and space hardware provider ThalesAlenia-Space reported a flat revenue growth rate and a drop in pre-tax profits for the same period. New orders however were up 30% from the previous year, at €904.6 million. Company officials attributed these mediocre results partially to the destruction by earthquake of one of its satellite component facilities in L'Aquila, Italy.[207]

In another development, the Indian space agency ISRO announced in July an 18% increase in the number of direct-to-home (DFH) pay television subscribers over the first quarter of 2009. Although India has been long considered one of the greatest potential DTH markets, its protectionist regulatory environment and excessive import taxes have limited foreign access to that market. Antrix, ISRO's commercial arm, has only recently introduced non-Indian satellite systems to the market, on the condition that it acts as their reseller.[208]

3.2.2. Fixed Satellite Services

Fixed Satellite Services (FSS) refer to the use of spacecrafts that utilise land terminals in fixed positions to broadcast. They include broadband internet, communications and network televisions and radio broadcasts. FSS business demonstrated considerable resistance to the crisis' consequences. Demand was fuelled by long-term agreements for communication services that were largely unaffected by the crisis, as well as a steady increase in satellite broadband internet demand. However, risk aversion created among service providers by the general financial conditions led some of them to reconsider their medium-term plans for broadband internet development, mainly because of the considerable up-front investments related to it.

In a 22 September report to the London Stock Exchange for example, satellite broadband service provider Avanti Communications of Britain announced a return to pre-tax profits during the 12 months ending in June 2009, after reported losses the year before. The company reported £8 million annual revenue (up from £5.9 million) and pre-tax gains of £1.8 million, from £1.4 million losses last year. Avanti also reported it was receiving penalty payments from Astrium Satellites for the delay in the delivery of the Hylas broadband satellite. Hylas would be the first dedicated broadband satellite in Europe, scheduled to be followed by Eutelsat's Ka-Sat later in 2010. In the meantime, several European governments have announced financial incentives to stimulate commercial satellite broadband projects.[209]

The same trend was evident in the U.S. as well, where in May 2009 the Federal Communications Commission issued a report on the future expansion of broadband services in rural areas in that country. The report made an appeal to President Obama to increase the project's budget beyond the $7.2 billion already allocated to it under the American Recovery and Reinvestment Act of 2009. The report also underlined the importance of satellite communications in reaching the project's objectives in the near future, reminding that their use would be indispensable in order to achieve full coverage of rural areas. Finally, the Commission urged greater coordination among U.S. federal agencies involved, as well as for a review of all existing broadband programmes.[210]

In general, the global financial crisis has tightened credit supply throughout the satellite industry. As a result, national export-credit agencies were stepping in more and more often to help finance commercial space projects, usually by guaranteeing their bank loans. In September, for example, France's export-credit agency Coface approved a loan guaranty to Gazprom Space Systems of Moscow for acquiring two Yamal-400 communication satellites. The spacecrafts would be built by ThalesAleniaSpace and launched onboard an Ariane 5 rocket. Coface's U.S. counterpart, the Export-Import Bank, is also engaged in similar schemes. Although both agencies are set up as private companies (Coface is owned by Natixis) they often finance projects based not only on financial reasons, but also on the need to support critical industrial space infrastructure in their respective countries.[211]

In a related development, SES and Eutelsat announced on 31 July that they were rethinking their involvement in providing S-band satellite services in Europe. The two companies had formed the joint venture Solaris Mobile and won one of the two European Union Commission licences to operate S-band satellite services in Europe. After a failure in deploying an S-band antenna onboard Eutelsat's W2A satellite in April however, the two companies would have to launch a new satellite in order to abide by their licence obligations. Representatives from both companies expressed their concern over making such an investment and announced that they would wait to receive their insurance claim before they made their decision.[212] One possible solution to this financial impasse, they said, would be a possible merger of the two licensed S-band providers in Europe, Solaris Mobile and Inmarsat of London. Such a solution however had been explicitly ruled out by the European Commission at the time of the licensing process.[213]

In October, the market research firm Tauri Group issued its annual report on the state of the personal spaceflight industry in the U.S., commissioned by the Commercial Spaceflight Federation. The report found that although the sector grew by a modest 6% (to $261 million) in 2008, total investment rose by more than 20% (to $1.46 billion). Of that, individuals provided more than 50%, private equity funds 30% and government only 15%. However, government clients still accounted for over a third (or $126 million) of the industry's revenues. The report's findings confirmed the growing investor interest in commercial spaceflight, as well as the increasing involvement of bigger companies that enter the industry as vendors or services providers.[214]

At the same time, NGA is moving forward to increase the amount of imagery provided by commercial optical reconnaissance satellites. In April 2009, Director of National Intelligence Dennis Blair announced that NGA would be purchasing images from at least two satellites equipped with 1.1 metre apertures. DigitalGlobe and GeoEye are among the principal candidates for acquiring these contracts.[215] In September 2009, NGA also announced that it would launch a new contracting vehicle for acquiring image data, called EnhancedView. This project will be similar to the current NextView programme, with the difference that NGA will be requesting 0.25 metre resolution products, instead of 0.5 metre that was the usual standard until now. This would require operators to fly their satellites in lower altitudes, at the expense of their operating life span. It would also complicate operations for private companies as U.S. legislation prohibits the sale of better than 0.5 metre resolution images outside U.S. government agencies.[216]

In spite of this development however, U.S. satellite broadband providers expressed scepticism over how much funding they would get from the broadband stimulus package. In a 30 September submission to the Commission, the U.S. Satellite Industry Association (SIA) urged officials not to focus on local labour-intensive terrestrial installations. In all, satellite operators have applied for $2.2 billion in loans and grants under the programme, out of a total $28 billion funding requests. Among them, Echostar has applied for $483 million under a joint venture with ViaSat Inc. intended to launch a dedicated broadband satellite by 2012, as well as for $530 million for a similar venture with WildBlue Communications of Denver. Other contenders include Skyterra, Spacenet Inc. and AtContact Communications LLC.[217]

Faced with financing difficulties and unsure government backing, companies involved in broadband satellite services have shown a tendency to consolidate their market positions. Apart from the joint ventures mentioned above, ViaSat Inc. announced in September the purchase of satellite broadband services provider WildBlue. The transaction cost $443 million in cash and $125 million in stocks. With this purchase, the future market for broadband services in the U.S. will most likely be contested by two companies, ViaSat and Hughes Network Systems. Both companies have scheduled launches of dedicated broadband service satellites, ViaSat of ViaSat-1 in 2011 and Hughes Network Systems of Jupiter in 2012.

Both Satellites are developed by Space Systems/Loral and they should field an over 100 GB Ka-band capacity.[218]

In August, the U.S. government announced its intention to simplify contracting arrangements for purchasing commercial communications satellite capacity. The plan involves consolidating the government and defence purchasing contracts into one common scheme by 2011. By doing so, U.S. government will be able to buy bandwidth directly from commercial satellite operators, something that is not currently possible. The new contracting vehicle is known as "Future Comsatcom Services Acquisition" (FCSA) and it will greatly simplify contracting procedures. In 2008, the U.S. government spent $397 million on satellite transmission contracts, of which $350 million were for defence purposes.[219]

3.2.3. Remote Sensing

Remote sensing refers to commercial companies that provide optical and radar images to the open market, mostly to government entities that have been increasingly outsourcing such capabilities over the past few years. Although image procurement is usually conducted through short-term agreements that acquire data at spot market prices, their demand was not considerably affected by the crisis, mostly because of growing corporate and public demand for these products. This trend has also led to an increasing simplification of related public procurement policies and a consequent ease of existing export controls applied to satellite image data.

The U.S. government announced on 7 October, for example, that it would relax its commercial radar satellite data restrictions. This should allow commercial operators to offer high quality images of up to 1 meter ground resolution to the open market, instead of the existing 3 metre limit. Northrop Grumman Aerospace Systems of Los Angeles would be the first company permitted by the U.S. Department of Commerce to operate a Synthetic Aperture Radar (SAR) image satellite under the new regulations. Company officials claimed that the proposed satellite, called Trinidad, would be based on components from the Israeli TecSAR satellite, tested by the USAF earlier in the year. According to the same sources however, the satellite's development would require firm government purchasing commitments to start. With this development, European SAR image commercial providers will be facing U.S. competitors within the next two years.[220]

3.2.4. Mobile Satellite Services

Mobile satellite services (MSS) relate to applications delivered to mobile terrestrial terminals such as ships, aeroplanes, automobiles, cell phones etc.

According to a market forecast published by Euroconsult of Paris in October, mobile satellite services revenue growth is expected to average 8% over the next decade, reaching some $2.5 billion by 2018. Satellite broadband services should account for the bulk of this increase, as they were projected to rise by 25% annually over the same period. Smaller growth rates were expected in machine-to-machine applications (+16%), maritime mobile satellite services (+7%) and aeronautical services (+13%),[221] among others.

In July, mobile satellite services provider Globalstar Inc. completed a life-saving financial rescue package, which included a $586 million credit from a consortium of French banks guaranteed to 95% by Coface, the French export-credit agency. This package would allow Globalstar hardware providers Arianespace and ThalesAleniaSpace to resume deliveries of the 24 second-generation Globalstar satellites due for launch onboard Soyuz rockets in 2010.[222] After this development, Globalstar announced to investors on 8 July that it expected to return to a 30% annual revenue growth rate by the end of 2010, when its new 32 satellite constellation (including eight existing ones) would become operational, putting an end to the service degradation of the past three years caused by satellite problems.[223]

The market for mobile satellite telephone calls is expected to experience intensification in the next months. Thuraya, United Arab Emirates-based satellite telephone provider, is offering cheap handhelds, the new Thuraya XT model, and also Inmarsat is to step into the handheld market with the aim to gain a 10% share.[224]

The satellite manufacturing revenues in 2009 experienced a significant increase compared to 2008. The total revenues of satellite manufacturers that built satellites both for governmental and commercial launches are estimated to have reached $16.15 billion in 2009 which indicates a rise by 48% from the $10.94 billion gained in 2008. It can be observed in Figure 2.6 that this augmentation marks an abrupt end to the trend of slightly decreasing revenues from 2006 on. The increase of $5.21 billion is primarily due to the construction of high-value defence satellites, whereas the share of the manufacturing revenue of the commercial satellites slightly decreased from $5.2 billion in 2008 to $5.14 in 2009.[225]

It is difficult to assess the exact annual revenues for launch services or the allocation between partners or countries. This is due on the one hand to the often complex package of financing mechanisms and industrial structures in some countries, and on the other hand to the reduced visibility of revenues from national institutional launches. These often draw on military budgets and, in addition, commercial launch service prices are usually not disclosed. The Federal Aviation Administration (FAA) estimates commercial launch revenues for 2009 at $2.49 billion. This represents an increase of $520 million from 2008 commercial launch revenues. Europe again had the lion's share, with more than $1 billion, representing 42% of the total annual revenues, followed by Russia ($742 million and 31% of the total revenues), the United States ($298 million and 12% of the revenues) and Sea Launch and Land Launch ($280 million and 12% of the revenues – Figure 2.7). As a whole, commercial launch revenues grew steadily between 2004 and 2009, witnessing an increase of almost 150% from roughly $1 billion in 2004 to almost $2.5 billion in 2009.[226]

Ground equipment revenues include infrastructure elements, such as mobile terminals, gateways and control stations, and consumer equipment, such as very small aperture terminals (VSAT), ultra small aperture terminals (USAT), DTH broadcast dishes, satellite phones and digital audio radio satellite (DARS) equipment.

Portable Navigation Devices (PND) form one of the sub-segments of end-user electronics incorporating GPS chip sets. Although the PND market grew by more than 30% in 2008, it decreased in the last quarter of 2008. Indeed, growth was affected by the crisis as the PND business is very dependent on the automotive sector. TomTom and Garmin are the two leaders in the PND market. Although both companies experienced reduced revenues in 2009, their expectations for 2010 are optimistic and they assume growing markets. Furthermore, TomTom notes the upcoming threat to their business by free turn by turn navigation on smartphones, as for instance offered by Navigon in cooperation with telecommunication providers.

TomTom reported $1.48 billion revenue in 2009, which represented a 12% decrease compared to 2008 and indicates a trend of diminution since 2007.[227] Garmin had total revenues of $2.95 billion in 2009, a 10% increase compared to 2008 ($3.49 billion). It sold 16.6 million units in 2008, which indicates a small decrease. Also, three of its four activities areas (i.e. automotive/mobile, aviation and marine segments) experienced a retracement; only the consumer-related area of outdoor and fitness increased. Accordingly, their revenues considered by region also dropped by 16–18%, however, their small asset in Asia increased by 3%. Garmin itself names the economic crisis and the "depressed global economy" as reasons for its performance and expresses optimism concerning improvements in the global economy and therewith revenues in 2010.[228]

As the space industry continues to demonstrate increased hardware reliability, low accident rates and booming growth in recent years, insurance costs have been decreasing. In fact, the repeatedly good performance of insured commercial space assets has attracted new insurance capacity into the market, pushing premiums to historically low levels. This trend has continued uninterrupted in the past 12 months. For example, insurance companies that cover space launches have demonstrated an increased appetite for risk by raising the maximum underwriting value for a single space launch. In March 2010 in fact, the launch of two communications satellites onboard an Ariane 5 rocket broke all records, with an accumulated liability coverage value of $700 million. Given that commercial space launches are expected to grow in the following years and the technologies involved have proven their worth in practise, one can expect this trend to continue. Nevertheless, this unusually long period of higher insurance limits and lower rates is beginning to raise concerns among insurers on the long-term viability of their business. Indeed certain commercial space underwriters, such as Paris Re, have announced that insurance rates were approaching a level where they may no longer support the assumed risk. In that case, some insurance brokers may consider their withdrawal from the market all together. The global space insurance market currently has a total coverage value of approximately $17 billion, distributed across 175 insured satellites.[229]

Also interesting is the development of space insurance activities in the coming years in and with Islamic organisations and countries. In 2009, the satellite fleet operator Yahsat of Abu Dhabi agreed with underwriters to sign the first Shariah compliant satellite insurance package. This could be a first step for further insurance agreements and also more space activities of the Islamic World due to higher security.[230]

3.3. Evolution of the space industry

3.3.1. Industrial evolutions in Europe

On 10 June, the French commercial image satellite operator Spot Image assured its customers that it would be willing to field two new medium resolution satellites, Spot 6 and Spot 7, starting in 2012. The two new satellites should complement the highly successful Spot 5 that currently accounts for the majority of the company's revenues and is expected to operate until 2014. However, Spot Image declined to give more specific details on the project, as the financing decision had not yet been made by its principal shareholder, Astrium Services, which was still negotiating on this issue with the French government.[231]

A few days later, on 18 June, Astrium Services announced that financing issues had been resolved and that the company would cover the approximately $500 million cost of the two satellites, without asking for any financial support from the French government. However, Spot Image officials asked for some kind of commitment from the French government to buy images from the future satellites. Companies in the United States (such as GeoEye and DigitalGlobe) already operate in this fashion. In general, it appears that the commercial image industry is in the middle of a transition from a fully government financed business model to a mixed private-public financing, based on guarantees of future government contracts.[232]

On 1 July, satellite services provider Telespazio and the Italian Space Agency ASI agreed on the creation of a joint commercial platform for selling data from the Cosmo-SkyMed radar imaging satellite constellation. A new company called e-Geos has been set up for this purpose in Rome, with Telespazio holding 80% of its shares and ASI 20%. Under the agreement, the former transferred its Earth observation division to the new company and the latter its rights to commercialise Cosmo-SkyMed images. According to company officials, e-Geos was close to striking a deal to supply images to clients in the Middle East.[233]

In a related development, e-Geos announced on 19 November that it had secured a contract from the European Space Agency (ESA) to provide radar and optical images to the EU Global Monitoring for Environment and Security project (GMES). The €3.5 million deal included providing radar data from Italy's Cosmo-Skymed constellation and optical data from U.S. commercial operators GeoEye and DigitalGlobe.[234]

On 20 July, ThalesAleniaSpace secured a €17.9 million contract from ESA to build the Experimental Re-Entry Test Bed (EXPERT). This will be a 440 kg bullet-shaped re-entry capsule that will be launched for a sub-orbital flight onboard a Russian submarine launched Volna rocket, as early as October 2010. The capsule is expected to offer valuable data for the development of ESA's Advanced Re-entry Vehicle (ARV). ARV will be an enhanced, atmosphere re-entry capable design of ESA's Automated Transfer Vehicle (ATV) that carries supplies to the International Space Station (ISS). ARV first flight is scheduled for 2016.[235]

In October, NATO announced it would be conducting an open competition to acquire additional satellite communication services for its troops in Afghanistan. Since 2004, NATO's Consultation, Command and Control Agency (NC3A) has had a $659 million agreement with a consortium using bandwidth on Skynet, Syracuse-3 and Sicral satellites to provide SHF and Ultra-high frequency communications. Extra bandwidth is indispensable, however, as Afghanistan communication infrastructures are very limited. Due to security reasons, the competition would be limited to NATO Member States only. In addition to current providers, Germany, Greece, Spain, Turkey and the United States all have satellites that could be used to fill the gap.[236]

On 30 November, satellite fleet operator SES Astra announced its intention to order four direct broadcast television spacecrafts from EADS Astrium Satellites, at a total cost of approximately €500 million. This contract would constitute the biggest single satellite order ever made by SES and it would account for more than

15% of 2009's commercial satellite sales worldwide. The spacecrafts would be delivered at 6 month intervals starting in 2012. Apart from featuring more than 60 Ku-band transponders for direct broadcast services each, the spacecrafts would also have 2 to 4 Ka-band payloads. This would be a step forward in the company's effort to develop next-generation broadband communication services in Europe. Furthermore, the series' fourth satellite was expected to carry a navigation payload for the European Commission's Egnos programme, designed to enhance GPS signal reliability.[237]

In a parallel development, SES announced that it was also negotiating with Astrium Services the sale of its German-based ND SatCom subsidiary. The company held a 25% participation in MilSat Services, the satellite communication provider contracted by the German military to operate its two SatComBw communication satellites. EADS Astrium Services, an Astrium Satellites sister company, held the remaining 75%.[238]

On 21 December 2009 Avanti Communications of London announced it had secured a $309 million in loans from U.S. and French export credit agencies (the Export-Import Bank and Coface respectively) to build its next generation Hylas 2 Ka-band broadband services satellite. The spacecraft, which was scheduled for launch in 2012, would weight 3.1 tons and provide an 8.28 gigahertz capacity to up to 1 million subscribers, representing a significant capabilities increase from its predecessor Hylas 1. According to the package's terms, Avanti would sell 21.5 million shares in order to raise £86 million on its own to finance the project.[239]

In a related development, commercial satellite fleet operator SES also announced on 21 December 2009 that it had secured a €522.89 million loan, backed by the French export-credit agency Coface, to pay for the construction of 4 new satellites by EADS Astrium. The spacecrafts, named Astra 2E, 2F, 2 G and 5B were scheduled for launch between 2012 and 2014. SES officials claimed that the company, which had a €4 billion accumulated dept, could find credit in the open market and did not necessarily need Coface's backing. However, they added that its support was welcome as it allowed them to borrow on lower interest rates and to spread the repayment period onto a longer time period throughout 2022.[240]

On 26 January 2010, OHB Technologies chief executive Berry Smutny confirmed that although EADS Astrium did not obtain the prime contract for building the first phase of the Galileo NPT constellation, it would still secure up to 50% of the contract's work as a sub contractor of OHB. In fact, the official clarified that EADS subsidiary SSTL of Britain would build most of the satellites' electronics payload, which would account for up to 40% of the spacecraft's total value. He also said that OHB would be using for the Galileo project a derivative of the satellite bus developed for the German SAR-Lupe constellation.[241]

On 9 February the French space agency CNES signed a €280 million contract with ThalesAleniaSpace for the construction and launch of the Athena-Fidus satellite. Athena-Fidus will be a 3 ton Geosynchronous Earth Orbit (GEO) telecommunications satellite based on Thales' Spacebus 4000 B2 platform and using EHF and Ka-band frequencies. It is a joint project evenly funded by the French and Italian space agencies. Its mission will be to offer high speed (up to three GB) communications to military and civil-protection agencies. Its technical specifications would allow it to transmit real time video data from unmanned aerial vehicles (UAV). The satellite will be operated by France and it is expected to be launched onboard an Arianespace rocket by 2014 at the latest.[242]

In February 2010, Surrey Satellite Technology U.S. (SST-US), a subsidiary of the British Surrey Satellite Technology Ltd (SSTL), proposed the replacement of the older commercial imaging spacecrafts Ikonos and Quick-Bird by a constellation of small satellites. Ikonos was launched in 1999 and it is operated by GeoEye of Dulles, whereas Quick-Bird flew in 2001 for DigitalGlobe of Colorado. Both companies have been the principal commercial image providers for the U.S. Department of Defence. SSTL officials have been promoting the use of a constellation of 5 SSTL 150 Kg imaging spacecraft instead, for a total price of less than $150 million. SSTL had already built a similar constellation for the German commercial imaging provider RapidEye AG. The ground resolution offered by these small satellites would be approximately 1 m in black and white and 4 m in colour images, which is comparable to Ikonos' and slightly inferior to Quick-Bird's performance. A constellation of small satellites also has the advantage of shorter revisiting times, but on the other hand it offers a much narrower swath path due to its fixed camera. From a military point of view, a constellation of small satellites also shows greater survivability and redundancy to interference than a single spacecraft. Given that the U.S. Pentagon is already developing such formation flying satellite systems for military use, one could argue that private satellite services providers that work with the U.S. military would soon follow suit, offering commercial spacecraft comparable to the dedicated military ones. In other words, private companies seem to incorporate to their new satellite systems technical characteristics that imitate the standards set by government spacecraft, in an attempt to secure contracts.[243]

Eutelsat, the world's third largest satellite fleet operator, reported on 18 February that its 2009 financial results were good beyond all expectation. More specifically, the company's officials announced a 9.6% revenue increase for the 2nd half of 2009, which brought total revenues in the aforementioned period to €508 million. This figure represented a gross revenue margin of 81% (EBIT-DA), which was well above the firm's objective of a 77% annual average for the period from 2009 to 2012. Thanks to this performance, Eutelsat foresaw to exceed for the first time in its history the €1 billion revenue mark during the 2010 fiscal year. Furthermore, the firm's officials now expected its average growth to surpass the 7% benchmark on an annual basis through 2012. Eutelsat's backlog also exceeded expectations, marking a 19% increase from the previous year and standing at €4.2 billion at the end of 2009. The firm's positive financial performance was attributed to several factors, including an increase to its available satellite transporters from 489 to 532, a raise in satellite lease prices in Europe and, most importantly, Eutelsat's bullish presence in the emerging Russian, Middle East and African markets.[244]

On 23 February Dutch Space of Leiden, Netherlands, announced it was selected to provide solar arrays for the first four GMES Earth observation satellites, of the Sentinel 1 and 2 series. The company was chosen by ThalesAlenia-Space under a € 13.4 million contract to build panels for Sentinels 1A and 1B (equipped with a C-band Radar payload). Simultaneously it was also contracted for the sum of €10.3 million by Astrium Satellites for Sentinels 2A and 2B (equipped with a super-spectral imaging instrument). The company consequently subcontracted Astrium to build the solar cells and Airborne Composites of Hague to provide the carbon-fibre panels.[245]

On 24 February ThalesAleniaSpace announced it was chosen by the French space agency CNES to build the Jason-3 ocean altimetry satellite. The Jason series satellites are a joint effort by the U.S. National Oceanic and Atmospheric Administration (NOAA), Europe's Eumetsat meteorological agency and CNES. The Jason-3 budget was estimated at €252 million, including launching and three years of operation costs. ThalesAleniaSpace would be responsible for providing the spacecraft's Proteus platform, its principal instrument (the Poseidon 3B altimeter), as well as for the system's integration, testing and launch preparations. NOAA would provide secondary payloads and it would be in charge of its launch, which was scheduled for July 2013. The satellite was planned to work in tandem with Jason-2, already in orbit since July 2008.[246]

Due to problems of a Russian mobile gantry, the first launch of the European Soyuz version is scheduled to the end of 2010, not allowing a second flight in 2010. Also this means additional costs for ESA that will be asked for an additional $50 million funding. The delay also led Arianespace to reactivate its Starsem affiliate to launch four Soyuz rockets from Baikonur Cosmodrome in Kazakhstan beginning in September. The first launch of Arianespace's new light-lift rocket is likely to be not before 2011. After successful demonstration flights of the two new rockets, Arianespace expects to launch annually six to seven Ariane 5 rockets, three to four Soyuz and one or two Vega spacecrafts from the European spaceport in French Guiana.[247]

3.3.2. Industrial evolutions in the United States

One of the most noteworthy events in 2009 was the bankruptcy of the commercial launch services provider Sea Launch Co. LLC. Sea Launch was struggling to meet its 22 June deadline of paying back $245 million in maturing bank loans. Efforts to refinance this debt continued throughout June. These included loan guarantees by two of its main shareholders, Boeing Co. and ASA Group of Norway. In the meantime Sea Launch was continuing to lose contracts, as several of its clients (such as SkyTerra) transferred launches to its competitors, Arianespace and International Launch Services, amongst rumours that Sea Launch would not be able to meet its scheduled launch dates.[248]

Unable to refinance its $2 billion debt (half of which was to Boeing, its main stockholder and hardware supplier), Sea Launch had to file for Chapter 11 bankruptcy protection on 22 June. The suspension of its activities for an entire year in 2007 (following an on-pad launch failure) together with a precipitous rise in raw material prices finally took their toll on the company. As U.S. legislation excluded Sea Launch from launching U.S. government payloads, maintaining a stable revenue base in the long-term proved impossible. Sea Launch's bankruptcy weighed heavily on its main shareholder, Boeing, which was facing up to $478 million in pre-tax charges related to the uncollected debt.[249] Sea Launch's bankruptcy also increased fears of a possible escalation in launch prices from the other two major launching services contractors, ILS and Arianespace.[250]

In the face of such fears both Intelsat and SES, the two most important commercial fleet operators, announced in September that they were willing to support Sea Launch's exit from bankruptcy by guaranteeing future launch contracts with it. Intelsat, which had several launch contracts pending with Sea Launch prior to its failure, reiterated its commitment to using Sea Launch services. Boeing and Space Systems/Loral also voiced support for the company. At the same time, Intelsat and SES also asked for the U.S. government to address the space launch services gap that Sea Launch's bankruptcy has created. In their opinion, this should include reconsidering the U.S. ban on satellite exports to China and encouraging the return of the Atlas and Delta rockets to commercial activities.[251]

Soon after Sea Launch's bankruptcy disputes arose among its principal shareholders on sharing payments due to its creditors. Boeing, which prior to the bankruptcy had reimbursed $448 million to Sea Launch creditors, demanded that other shareholders participate in it proportionally. The Norwegian participating company Aker agreed to do this, but Russian and Ukrainian owners declined. On 19 October Boeing filed a request for arbitration on this issue with the Stockholm Chamber of Commerce. In addition to this, Boeing had loaned another $523 million directly to Sea Launch and demanded that the rest of the owners participate in it as well. The Sea Launch bankruptcy was the main cause for a 13% decline (to a total of $672 million) in profits that Boeing manifested in the first three quarters of 2009.[252]

On 11 June Lockheed Martin, Northrop Grumman and Raytheon were each awarded a $30 million contract from the U.S. Air Force to start development work on the ground-based radar segment of the future USAF space surveillance system, known as Space Fence. Preliminary plans include the construction of three S-band radars, one of which will be located outside the U.S. The system is expected to have higher resolution than the existing one. The contracts awarded included system design review, present trades and analysis data, operating simulations and lifecycle cost estimates. The prime contractor for the project is to be selected in 2012.[253]

On 10 June, Lockheed Martin was awarded a $1.49 billion contract to build the third of a total of six USAF Space Based Infrared System satellites (SBIRS). They are set to replace the existing U.S. ballistic missile launch warning constellation. The system includes the geosynchronous SBIRS satellite, as well as a highly elliptical orbiting payload. The launching of the constellation's second satellite is scheduled for 2011. At the same time, Lockheed Martin also announced the commercialisation of a low cost GPS enabled wireless tracking device that can be monitored across radio frequency identification and satellite communication networks. This product is addressed to civilian as well as military users.[254]

On 19 May 2009, the USAF launched TacSat-3, its first hyper-spectral reconnaissance satellite. TacSat-3's main payload was the Advanced Responsive Tactically Effective Military Imaging Spectrometer (ARTEMIS). The ARTEMIS spectrometer can distinguish the spectral signatures of different materials, enabling it to identify camouflaged objects or to detect freshly dug earth. TacSat-3 is part of the U.S. Defence Department's Operationally Responsive Space (ORS) programme. It has an on-board computer that enables it to process data before transmitting it directly to the battlefield and its tactical response time should be less than ten minutes. The ARTEMIS payload was built by Raytheon within only 15 months and on a $15 million budget. Off-the-self commercial technology was used extensively in its construction. The entire programme's budget was $90 million, including launching costs. Should TacSat-3 prove the reliability and effectiveness of the technologies involved, USAF is planning to launch a constellation of similar satellites, able to cover multiple operational theatres simultaneously.[255]

On 15 June, Northrop Grumman announced that it had delivered the second of a pair of the long-delayed Space Tracking and Surveillance System (STSS) satellites to the U.S. Missile Defence Agency. These were built in 1999, under a cancelled system demonstration project. The programme was resumed in 2002.

When launched, the satellites will enable the USAF to track ballistic missiles in every stage of their flight, something that current space-based assets cannot do. Their payload includes a multi-band infrared tracking sensor that other missile tracking satellites lack. STSS is a technology demonstrator that will determine the usefulness of a constellation of such satellites for U.S. ballistic missile defence. The post-2002 cost of the programme was approximately $1.35 billion.[256]

On 21 June Northrop Grumman and Israel Air Industries concluded a three week technology demonstration test of the Israeli Synthetic Aperture Radar (SAR) reconnaissance satellite TecSAR for the U.S. Defence Department. The test included tasking, downlinking, processing and delivering TecSAR images from a mobile control station inside a van, within less than 15 minutes of their request. The demonstration was conducted in Key West Florida under U.S. Southern Command's Project Thunderstorm, an initiative seeking to utilise next-generation imaging capabilities to counter asymmetric threats.

Intelsat and SES, the world's largest commercial satellite fleet operators, announced on 23 July that they would jointly try to "persuade" Washington to allow them to launch commercial satellites from China and India. The two companies were expected to be joined by Space Systems/Loral, the largest U.S. builder of commercial communications spacecrafts. The two companies noted that ITAR procedures and the Sea Launch bankruptcy had practically reduced launching services providers to only two (Arianespace and ILS). They therefore insisted on lifting the ban on Indian launches of U.S. satellites, which the U.S. department has maintained in spite of the latest U.S.-India bilateral cooperation agreement on defence and technology trade. However, Congressional sources noted that this lobbying initiative could result in effects on the Hill exactly opposite than expected.[257]

In September, the two companies were joined by satellite operators Telesat and Echostar and intensified their lobbying by hiring former U.S. Senate Armed Service Committee Chairman John Warner to address to executive branch officials, although ethics rules forbid ex-Congress members from lobbying. Warner would try to convince top officials that the absence of a U.S. launch provider is harmful to its national security interests, and this should be addressed by lifting the ban on Chinese launchers and facilitating commercial flights on board the Atlas and Delta rockets.[258]

On 11 August, Raytheon Space and Airborne Systems unveiled its new infrared light-wave detector that is four times larger in dimension than the current one. Its 4 k×4 k focal plane array comprises 16 million pixels in 4,096 columns and rows, laid on a 64 square centimetre surface. The new detector promises greater sensitivity and higher frame rates, while at the same time simplifying systems design and lowering construction costs.[259]

On 1 July Terrestar-1, the world's largest commercial spacecraft ever built, was lifted to orbit onboard an Ariane 5 ECA rocket. The 6,910 Kg satellite will provide mobile voice and data communications services in North America, using the 2 GHz and S bands.[260] The satellite's most striking feature is its unfurlable 18-metre-diameter s-band antenna. Concerns about whether the antenna would deploy properly had delayed its launch since early June.[261] The satellite will be operated by TerreStar Networks Inc. of Reston, VA.[262]

On 24 September, Space Exploration Technologies (Space X) Corp. announced it would launch a prototype of its reusable Dragon cargo capsule onboard the maiden flight of the Falcon 9 rocket. Dragon is a reusable capsule under development since 2006 for conducting cargo flights to the International Space Station (ISS). In December 2008 Space X won a $1.6 billion contract from NASA's Commercial Orbital Transportation Services to conduct 12 such flights to ISS by 2016. Dragon's launch has been delayed since 2007 because of problems with the development of the Falcon launcher.[263]

In an important development, Raytheon announced on 26 October that it had been awarded a $3 million contract to study the integration of the new U.S. Missile Defence Agency sensors to USAF's Space Surveillance Network. USAF is already using MDA's fixed radar stations for space surveillance purposes, but it would like to add its new mobile radars (like AN/TPY-2 X-band radar) to the mix. According to Raytheon it was the company itself that came unsolicited to the Air Force with a proposal to develop an open command and control architecture able to merge all available sensors into one dual-purpose system. Full-scale development of the programme could begin in 2012. At the same time, Raytheon has been in contact with MDA and the intelligence community in order to allow the dedicated Space Surveillance sensors to be used for early missile launch warning and intelligence gathering.[264]

In a separate development, representatives of the Israeli Space Agency suggested during a Space Security workshop on 3 November in Tel-Aviv, that Israel's planned Arrow-3 high-altitude ballistic missile defence system could be easily adapted to an Anti-Satellite (ASAT) role as well. According to workshop participants, Arrow-3's agile last stage exoatmospheric hit-to-kill vehicle could be modified to intercept LEO satellites and the system's Green Pine radar could be used for tracking them. During the workshop, Iran's future deployment of earth observation satellites was identified as a possible motive for Israel to acquire an ASAT capability.[265]

In the meantime, the U.S. Operationally Responsive Space (ORS) Office disclosed in November that it was making a capability assessment of the German-built LAPAN-TUBSats. The ORS office could order up to 8 such satellites to complement U.S. tactical imaging capabilities, at an estimated cost of $60 million.

The 50 Kg low-cost spacecrafts are built by the Technical University of Berlin (TUB) and they are equipped with a near real-time remote-controlled digital video camera, with an 8 to 10 metres ground resolution. TUB had already built such satellites for Indonesia and Morocco, among others. The system's main operational advantages were its good price-capability ratio and its remote-controlled camera that allows for shorter re-tasking times.[266]

On 23 November, mobile satellite operator Inmarsat announced the acquisition of the U.S. communications services provider Segovia Inc. for $110 million. In 2008 Segovia had reported a net profit of $18 million on $67 million total revenues. With this purchase, Inmarsat was expected to boost its position in the U.S. government contract market. The company was already a customer of Inmarsat's L-Band services. It also operates a network of Very Small Aperture Satellite (VSAT) satellite Earth stations for the U.S. Defence Department, which is relevant to Inmarsat's involvement in DARPA's "Persistent Broadband Ground Connectivity for Spacecraft in Low Earth Orbit Effort" programme. According to Segovia's Chief Executive M. Wheeler, 80% of the company's business was with the Pentagon. Prior to its acquisition, Segovia had also begun to provide services from Inmarsat's competitor, Iridium Communications. With this acquisition, Inmarsat could expand its BGAN broadband service to new distributors.[267]

In a related event, on 23 November Cisco Systems Inc. saw its first space-based internet router launched aboard a commercial communications satellite. The router was built for the U.S. Defence Department's Internet Routing in Space (IRIS) technology demonstrator, launched aboard the Intelsat 14 satellite. IRIS was the first dedicated U.S. military payload to reach orbit on a commercial satellite. In 2008, SES Americom had also signed a contract with USAF to host an experimental missile warning sensor aboard a communications satellite scheduled for launch in 2011. Although the experiment is funded by the Pentagon, Cisco has property rights over the router that it hopes to commercialise with satellite communications service providers. Intelsat also had expressed its interest in adding internet routers to its future spacecrafts. The demonstration acquired renewed interest since the cancellation of USAF Transformational Communications Satellite (T-Sat), which was also set to feature space-based routers.[268]

On 7 December Virgin Galactic unveiled its passenger carrying SpaceShipTwo suborbital space-plane, during a gala presentation at the Mojave Air and Space Port in California. The vehicle is designed to carry 6 passengers and a 2 member crew. The company's officials expected commercial operation to begin in 2011 from the Space Port America field, currently under construction in N. Mexico. Until then the space-plane was scheduled to contact extensive flight testing.[269]

In January 2010 the Denver-based United Launch Alliance (ULA), a Lockheed Martin-Boeing joint venture operating the Atlas 5 and Delta 4 rockets, announced it had posted a strong campaign in 2009. The year ended with a total of 8 successful launches, whereas 10 more were scheduled for 2010 for the two vehicles combined. ULA is the principal space launch services provider for the U.S. Department of Defence and civil agencies and it was created in 2006. Both the rockets it uses were developed under the USAF's Evolved Expendable Launch Vehicle (EELV) programme. In spite of the high launching rate maintained in 2009, ULA has a backlog full of government launches that would not leave room for commercial payloads before 2012 at the earliest. This queuing up of payloads was caused by launch stagnation in 2008, when both rockets were grounded for technical reasons. This situation had prompted ULA to take measures such as reducing launch intervals by 20–25%, developing a dual payload adapter for its rockets and delaying the assignment of payloads to 6 months prior to each launch.[270]

In the mean time, a series of legislative and administration delays within the U.S. Department of Defence has caused a considerable revenue decrease in 2009 for U.S. commercial satellite image providers. On the one hand, Pentagon's existing contracting vehicle called NextView was only renewed in December 2009, three months later than expected. As a result, the U.S. National Geospatial Agency (NGA) that is the contracting authority on behalf of the U.S. government could not procure any images from commercial companies in the aforementioned period. Furthermore, this situation was not expected to change in the first half of 2010, as a new contracting vehicle called EnhancedView was not yet put into place, in spite of the fact that the existing one was set to expire in June 2010. Although NGA assured commercial providers that it would renew it on a monthly basis until the new one came into force, it was clear that they could not expect any increase in revenues from U.S. government contracts in 2010.[271]

In the face of these events DigitalGlobe, one of the country's major commercial satellite image providers, announced it would have to count on foreign customers to support its growth in 2010. DigitalGlobe projected a 22% total revenue increase in 2010, even assuming that U.S. government contracts would remain in 2009 levels. The additional income necessary to boost the company's growth rate was expected to come from sales to commercial and foreign government customers. Commercial sales income in particular was thought to increase by 15% in 2010, in spite of its 3.6% decline in 2009 (amounting to approximately $50.9 million). Given that U.S. government contracts account for almost 75% of the firm's revenues, the bulk of its growth would have to come from foreign government customers. To increase that source of income, DigitalGlobe had set up a mechanism known as the Direct Access Programme (DAP). Under this scheme, foreign countries approved by the U.S. government could downstream images from DigitalGlobe's satellites directly to ground stations in their territory for an annual fee of roughly $35 million. DigitalGlobe planned that 25% of its spacecrafts' operational time would be used by foreign governments under this short of lent-lease agreement. The company announced that it had already secured five such customers and it was negotiating with another two, the identity of which it did not disclose.[272]

Should this development materialised, it would bring about a significant change in the company's revenue sources. Overtime DigitalGlobe expected these customers to pay on average $50 million for its services. Given that the firm's total revenues in 2009 amounted to $281.9 million, it becomes obvious that in the future DigitalGlobe would rely solely on these clients for its growth. Even at a $35 million annual fee, the seven foreign government clients would provide the company with $245 million in revenues per year, which would be over 60% more than the $150 million that it is currently receiving from its U.S. government. In this sense, one could argue that the DAP programme signifies a change in the U.S. commercial satellite image providers business model, obliging them to rely more on their foreign customers than on their domestic ones. It is also worth noticing that all DAP clients would have access to DigitalGlobe's latest and more accurate WorlView-2 satellite, while the U.S. government usually purchases products from the older Quickbird and WorldView-1 spacecraft.[273]

3.3.3. Industrial evolutions in Russia

In September, Roskosmos was obliged to postpone its Phobos-Grunt mission to the largest of Mars's two moons. The mission, which is Russia's first interplanetary mission since its failed Mars orbiter and lander mission in 1996, was originally scheduled for October. However it was postponed until the next launch window in 2011, due to delays in the final testing of the spacecraft. The Phobos-Grunt mission consists of an unmanned lander and a sample-return craft.[274]

In September, the French export-credit agency Coface announced it would guarantee the necessary loans for the construction of two Yamal-400 communication satellites from ThalesAleniaSpace on behalf of the Russian operator Gazprom Space Systems. Both spacecrafts would be launched aboard an Ariane 5 rocket. However, the programme was criticised by Russian authorities that are currently reviewing the country's satellite communication sector, because of its lack of Russian technology content.[275]

On 17 October, Roskosmos published a summary of its planned Yasny spaceport in the Russian Far East. The document stated that development of the site would require building an entire 30,000 people city almost from scratch. The Russian federal government has estimated that total construction costs might reach as high as $13.9 billion. At the same time, the site still faced long launch delays because of rocket-debris cleanup disputes with Uzbekistan. In October, the launch of two Swedish Prisma satellites and the French Picard were delayed for the same reason. Apparently, the new launching site could face the same availability issues as the Baikonur Cosmodrome has faced because of similar disputes with Kazakhstan.[276]

On 2 March 2010 Russia successfully launched three Glonass-M navigation and positioning satellites. All three were lifted to orbit onboard a Proton M rocket launched from the Baikonur Cosmodrome. The launch was postponed from the previous October due to the need to inspect a critical component suspected of malfunction. Russia is currently running a programme to replenish and modernise its Glonass satellite constellation. Another three space craft were launched in late 2009 and three more were expected to follow in August 2010. After that, the initially planned 24 satellite constellation would be fully operational.[277]

3.3.4. Industrial evolutions in Japan

HTV-1 was successfully launched on September 10 from the Tanegashima Space Centre and arrived at the ISS on 17 September. Apart from carrying various provisions for the ISS crew, HTV-1's payload included two highly sophisticated earth observation instruments on behalf of the U.S. Navy. These were the Hyperspectral Imager for the Coastal Ocean (HICO) and the Remote Atmospheric and Ionospheric Detection System (RAIDS). Both instruments are experimental. They were developed by the U.S. Naval Research Lab and were mounted on the external payload platform of the ISS's Japanese Experiment Module Kibo. Both instruments are intended for military and civilian purposes alike. HICO will provide high resolution real-time imaging of coastal areas and RAIDS will monitor the earth's Ionosphere and Atmosphere in order to provide space weather data. These are the first high-performance observation instruments mounted on the ISS.[278] On 2 November, HTV successfully completed its 59-day mission and was destroyed as planned by re-entering the atmosphere above the Pacific Ocean.[279]

In September, the U.S. headquartered company BB Sat acquired a licence to provide satellite broadband services in Japan. The company has concluded agreements to use Ka-band capacity onboard a Japanese satellite already in orbit and to cooperate with Japan's Internet Service Providers (ISP) in handling sales and customer services.

On 28 November, Japan successfully launched the first of its new generation earth observation satellites aboard an H-2A rocket. The space craft, called the Information Gathering Satellite (IGS) Optical-3, would provide optical imagery of 60 cm ground resolution, a considerable improvement on Japan's current 1 metre capability. The $562 million satellite was the first of a total of four reconnaissance satellites, two optical and two radar, scheduled for launch through 2012. Further spacecrafts with even higher resolution capability are planned for 2014.[280]

3.3.5. Industrial evolutions in China

The Chinese launch-service provider China Great Wall Industry Corp. (CGWIC) is expecting to be responsible for more than ten launches per year during the next two years. An aim of the company is to attract Western Business. Helpful could also be the low insurance rates of the well-proven Long March Series that are not far from Ariane 5 or Proton rockets. CGWIC is pretending that customer satellite teams have the full control over the facility – including access-permissions. Due to the U.S. International Traffic in Arms Regulations (ITAR), it remains nearly impossible to bring U.S. satellite components to Chinese launch bases.[281]

3.4. Industrial overview

In order to get a more detailed overview of the main developments of the space industry in 2009, a segmental appraisal will be undertaken in the following section. Three main activity areas will be presented: the launch sector, the satellite manufacturing sector and the satellite operators. These three segments make up the two main components of the overall space industrial business, namely the launch sector and the satellite industry. The two strands of the business are closely interrelated, as none of these industry branches can prosper without the other. Indeed, satellite manufacturers and satellite operators need a guaranteed and stable access to space, whereas launch providers rely on orders from the satellite industry to sustain their activities.

It is important to clarify some central concepts which will be at the centre of the analysis in the following sections, in particular the notions of commercial launch and commercial payload. Indeed, since the commercial space industry is growing in significance and progressively replacing the traditional forms of government-operated space activities, it has become more difficult to define and interpret what commercial launches and commercial payloads encompass. In the following section, a launch classification differentiating commercial and non-commercial launches and payloads will be used. A commercial payload is described as having one or both of the following characteristics:[282]

3.4.1. Launch sector

Despite its crucial importance for the satellite industry, the launch sector is an enabler rather than a significant economic activity. The revenues it generates are far less important than the ones originating from the satellite manufacturing and satellite services business.

The year 2009 was an even more active year for the launch sector than 2008, with a total of 78 launches conducted by launch providers from Russia, the United States, Europe, China, India, Japan, North Korea, South Korea, Iran and the multinational consortium Sea Launch. Three non-commercial launches failed: a Taurus XL launch in February 2009, a North Korean Taepodong 2 launch in April 2009 and a South Korean KSLV-1 launch.[283]

The main events for the rocket industry in 2009 were the successful launch of RazakSAT on the Falcon-1 vehicle by the privately-funded Space Exploration Technologies Corporation (SpaceX) from the company's Kwajalein pad, and the collision of the Iridium satellite and Cosmos 2251 spacecraft in February 2009. Furthermore, the Sea Launch Company's bankruptcy and the following protection actions had a major influence on the launch industry sector. Additionally, 2009 was marked by Iran's first orbital launch in February, successfully deploying its payload into LEO on a Safir three-stage rocket. As mentioned above, South Korea also performed its first orbital launch. However, the deployment of the satellite named STSAT 2 A in August 2009 failed. In 2009 North Korea also conducted its first launch since 1998, albeit an unsuccessful one.[284]

When looking into specific countries, Russia was again the world leader in the launch sector, representing approximately 37% of the total number of launches. It was followed by the United States (app. 31% of the total), Europe (app. 9%), China (app. 8%), Sea Launch (app. 5%), Japan (app. 4%), India (app. 3%) and South Korea, North Korea and Iran which launched one vehicle each, or approximately 1% of the total launch figure (Figure 5.5).[285]

Russia launched 29 vehicles in 2009, using eight different systems (as much as in 2008) whereas the United States conducted 24 launches with eight different launch systems as well (three more than in 2008). China used five different systems for six launches, Japan two systems for three launches, Europe one system for seven launches, Sea Launch one system for 4 launches, India one system for two launches and North Korea, South Korea and Iran each used one system for their respective launches. A total of 29 different launch systems were used in 2008, eight more than in 2008 (Table 5.1).[286]

Regarding the share of payload launched in 2009, Russia, the United States, Europe and India launched more than 80% of the total payload units launched in space in 2008. When considered in detail, Russia launched 37.8% of the total, while performing 37.2% of the launches. The United States accounted for 30.8% of launches and 25.2% of payloads. Europe launched 12.6% of the payloads with a share of only 9% of the total launches. China accounted for 7.7% of the launches, but carried only 8.1% of total payloads in orbit (Figure 5.6). The difference between the share of launches and the share of payloads carried by Europe is attributed to the fact that continent's main contractor Arianespace concentrated on the launch of heavier payloads. Indeed, the Ariane-5 vehicle can carry two GTO satellites at a time, thus sending heavier payloads in orbit with fewer launches.[287]

In total, 111 payloads were launched in orbit in 2009, five more than in 2008. Russia was the world leader again, as it launched 42 payloads. It was followed by the United States, which launched 28 payloads and by Europe which launched 14 payloads. India launched nine and China seven. The remaining Sea Launch consortium, Japan, North Korea, South Korea and Iran accounted for 11 payloads. In 2009, Russia took the lead in terms of commercial payloads as well (12 payloads), followed by Europe and Sea Launch (eight and three payloads), the United States (two payloads) and China (one payload – Figure 5.7).[288]

There is a fairly equal distribution of payloads among the different mass classes. In the period from April 2009 to March 2010, micro, small, intermediate and large payloads were roughly equally distributed, as each class made up around 20% of the total number of payloads launched. Large spacecrafts represented around 13% of the total number of payloads, and heavy ones accounted for only around 10% (Figure 5.8).[289]

Micro payloads are mainly science satellites, technological demonstrators or small communications satellites, like the Orbcomm series. Small payloads are very often Earth Observation satellites, such as SAR-Lupe, Jason or the RapidEye series. Medium payloads feature the most diverse set of satellites, including small satcoms in geostationary orbit, Earth Observation satellites, and most of the Russian military satellites from the Kosmos series. Intermediate payloads encompass medium satcoms and big scientific satellites. Large payloads refer to big satcoms, as well as to the Soyuz and Progress spacecrafts flying to the ISS. Lastly, heavy payloads are all linked to the ISS: the modules Kibo and Columbus, as well as the cargo spacecrafts ATV and Leonardo.[290]

Of the total launches conducted in 2009, 69% were non-commercial, representing 54 launches, and 31% were commercial, representing 24 launches. Only five actors performed commercial launches, whereas five actors performed non-commercial launches. As a whole, there was a decrease of commercial launches from 28 in 2008 to 24 in 2009, after an increase in 2008 by comparison to the previous year.

As in 2008, GEO launches were again the top commercial activity, and the whole space transportation market was largely driven by the demand for GEO satcoms. This trend is likely to continue in the near term. Commercial launches were particularly important for Europe and Sea Launch. U.S. launch services in contrast, continued to rely heavily on the governmental market, with only four out of 24 U.S. American launches being of a commercial nature. This was the case in Russia as well, where the relatively important domestic institutional demand continued to support the launch sector as ten out of 19 payloads related to government activities. India and Japan focused on non-commercial launches, as well as the newcomers North Korea, South Korea and Iran. Russian launchers conducted ten commercial launches, followed by the European Ariane-5 with five commercial launches. Sea Launch and the U.S. conducted four commercial flights each and China one.[291]

Regarding the launch service providers, Arianespace again dominated the market, as its Ariane-5 vehicle (ECA and GS versions) flew seven times in 2009.[292] Arianespace has won more than 50% of the commercial launch contracts worldwide in the last two years.[293] In 2009 it placed 14 payloads into orbit in seven launches, totalling 35 successes in seven years and confirming its technological maturity.[294] A core feature of the company is the ability to carry two satellites at a time, a characteristic which maximises the benefits of using Ariane-5, but which also makes the company more vulnerable to satellite schedule slips.[295] In 2009, the company's sales are estimated at €1 billion. At the beginning of this year, Arianespace announced that 6 to 7 launches are scheduled for 2010.[296]

As for the U.S. American launch providers, Boeing Launch Services (BLS) conducted two commercial launches. The first one was used to orbit the weather satellite GEOS into GEO in June and the second to place the WorldView 2 satellite into LEO. Lockheed Martin Commercial Launch Services (LMCLS) launched one Intelsat satellite into GEO.[297] The two companies traditionally do not compete in the commercial launch market, as their launch vehicles would not be cost-competitive for such an endeavour and as they can count on steady revenues from governmental demand.[298] SpaceX carried out its second successful commercial flight of the Falcon 1, transporting the Malaysian RazakSat satellite into LEO. Orbital Sciences Corporation (OSC) performed two non-commercial launches in 2009, one of which failed in February 2009.[299] United Launch Alliance (ULA) and United Space Alliance (USA) carried exclusively non-commercial launches.[300]

The Sea Launch Company launched only one single satellite from its sea-based platform in 2009. This mission was carried out for the Italian army, deploying a communication satellite in GEO. Additionally, the company conducted three Land Launch missions from Baikonur. All in all, Sea Launch launched only four rockets in 2009 compared to six launches in 2008. This was due to the company's bankruptcy in June and the resulting effort to restructure its finances.[301] During the first quarter of 2010, the company planned to submit a reorganization concept as a step toward emerging from bankruptcy status.[302]

As far as the Russian launch service providers are concerned, International Launch Services (ILS), International Space Company Kosmotras (ISC Kosmotras) and Eurockot Launch Services carried the ten commercial launches in 2009. ILS launched seven commercial Proton rockets in 2007, carrying mostly communication satellites. Additionally, ILS conducted three launches for its prime contractor Khrunichev, for example taking three GLONASS navigation satellites into their orbits in December 2009.[303] ISC Kosmotras launched one Dnepr-1 rocket and Eurockot one Rockot.[304]

Total commercial launch revenues in 2009 amounted to roughly $2.5 billion, about $500 million more than in 2008. Not surprisingly according to its market share in commercial space flights, Europe takes the lead accounting for approximately $1 billion in revenues, followed by Russia (app. $750 million) and the U.S. (app. $300 million).[305]

An estimated 29 satellite launch contracts were signed in 2009 for geostationary spacecrafts. The two main actors in the sector were the same as in 2008, namely Arianespace and ILS, followed by China Aerospace Corporation and Mitsubishi Heavy Industries as minor actors (Figure 5.9).[306]

Arianespace had a very solid year again in terms of contracts signed, winning more than half of the open competition commercial contracts. These "Services and Solutions" contracts include for instance the launches for satellite owners Hispasat, Arabsat, Intelsat, Inmarsat and Avanti Communication.[307] Furthermore, Arianespace signed a contract with the ESA in June 2008 for two Soyuz launch vehicles in order to orbit the first four operational Galileo satellites from Europe's Spaceport in French Guiana.[308] The company plans to continue its steady launch rate in the near term.[309] Arianespace also signed a contract with Astrium for the production of 35 Ariane-5ECA rockets in February 2009, at an estimated €4 billion value. With this contract, Arianespace has a total of 49 Ariane-5 in its backlog.[310]

ILS also signed 13 launch contracts for GEO satcoms in 2009, 6 more than in 2009 and as much as Arianespace. The contract partners include satellite owners AsiaSat, Intelsat, SES World Skies, Intelsat and Eutelsat, among others.

Sea Launch signed six launch contracts in 2008. Due to its mid-2009 bankruptcy, the company lost its status and the ability to sign contracts for future launches. The loss of Sea Launch as a main provider resulted in a market duopoly of Arianespace and ILS, a situation about which commercial satellite fleet operators expressed great concern. Sea Launch might emerge from bankruptcy by the end of this year.[311]

Among the remaining actors in the launch sectors, the China Great Wall Aerospace Corporation won a contract for the Nigerian Nigcomsat-2 satellite and the Apstar 7 of APT Satellite Holdings. Mitsubishi Heavy Industries on the other hand, signed a contract with the Korea Aerospace Research Institute for the launch of the Kompsat-3 satellite.[312]

The main feature of the launch industry in 2009 was the more or less unexpected bankruptcy of Sea Launch and its side effects. The two remaining principal launch services providers Arianespace and ILS had to fill in the gap for former Sea Launch customers, which led to the creation of a duopoly market. The reaction among satellite operators led by Intelsat and SES was to increase pressure for loosening U.S. export controls on spacecraft launched in China. Arianespace and ILS however claim that they did not see much growth in 2009, in spite of the elimination of Sea Launch as a competitor.[313] Nevertheless, launch prices worldwide have been increasing in the past two years. This process continued in 2009, especially due to the bankruptcy of Sea Launch.[314] Therefore, satellite operators have become increasingly worried by the launch market's diminishingly competitive nature.

3.4.2. Satellite manufacturing sector

Satellite services represent the most mature and lucrative market in the space sector. Indeed, space based communications is the core business for satellite service providers and satellite manufacturers alike. Therefore, looking at the market share of satellites launched and ordered in a given year is not only a good indication of the vitality of domestic space industries, but it also helps assessing the global trends in the space industry.

In 2009 111 payloads were launched. Only 23% of the payloads were commercial, significantly less than in 2008, when they represented 40%. 38% of the launched payloads were manufactured by Russia, 25% by the U.S. and 6% by China. Europe accounted for 13% of the payloads launched (Figures 5.6 and 5.7).[315] Ninety-eight satellites were launched in 2009, nine more than in 2008. Most of them were manufactured by U.S. companies with 39 of the satellites launched (40% of the total figure), followed by Europe (24 satellites representing 24.5% of the total), Russia (15 satellites or 15.3%) and China (13 satellites or 14.5%).[316] When looking at the performances by the bigger satellite manufacturing companies, ThalesAlenia had a very active year as 11 of its satellites were launched in orbit in 2009. Other European manufacturers such as Surrey and EADS Astrium accounted for two and four satellites respectively. The two top U.S. satellite manufacturers were Space Systems/Loral (SSL) and Lockheed

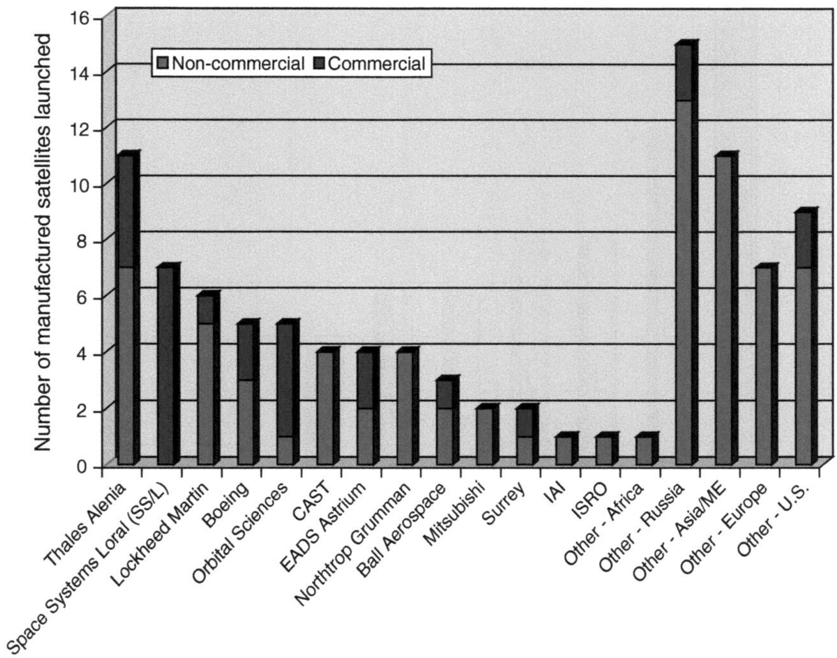

Fig. 8: *Satellites launched in 2009 by manufacturer and commercial status (source: Futron).*

Martin with seven and six satellites respectively, followed by Boeing and Orbital, with five spacecraft each (Figures 5.10 and 5.11).[317]

Of the 98 satellites launched, 26 were commercial. The main part of the commercial satellites launched were European or U.S. built: seven of the commercial satellites were European, representing only 27% of the total number of commercial satellites launched, whereas 17 commercial satellites were manufactured in the U.S., accounting for 65% of the total. Twenty-eight satellites were launched to GEO and 70 into other orbits. When looking at GEO satellites, Europe lost the lead it held in 2009: 25% of the GEO satellites launched in 2009 were European (three made by EADS Astrium and four by ThalesAlenia). In contrast, 61% of the GEO satellites launched were U.S. built, whereas Russian ones accounted for 11% of the total figure.[318]

2009 was an extremely successful year in terms of satellite contracts awarded. 41 commercial GEO satellites were ordered, nearly twice as many as the 23 orders in 2008. Manufacturers from the U.S. won 19 contracts, whereas European manufacturers signed 12 and Russian manufacturers 5. There were also satellite orders with two co-prime contractors: The two Arabsat 5C and 6B satellites will be built by ThalesAlenia and EADS Astrium, and the OverHorizon satellite of OverHorizon AB by Orbital and ThalesAlenia (Figure 5.12).[319]

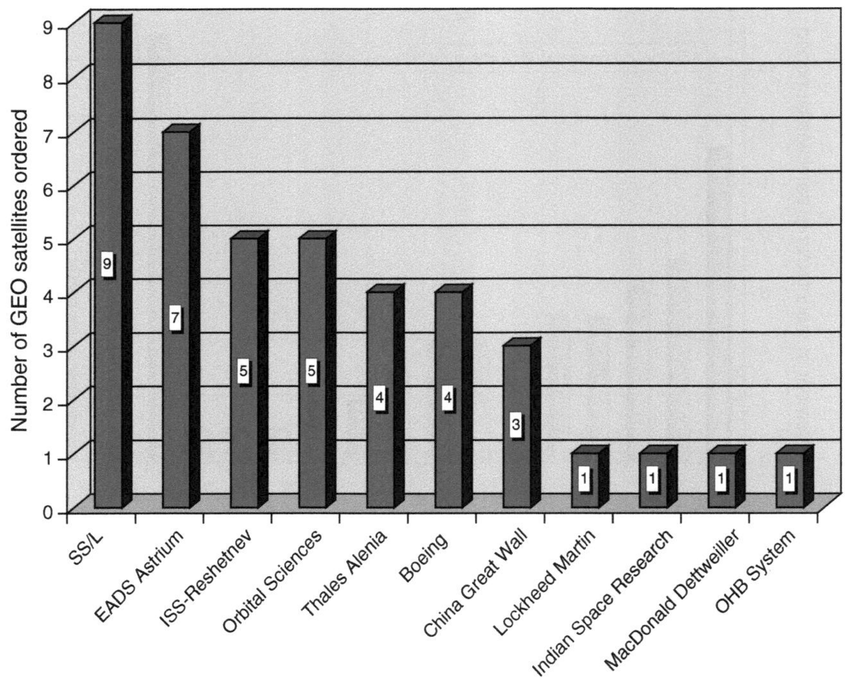

Fig. 9: *GEO satellite orders in 2009 by manufacturer.*

As for market trends compared to the previous year, in 2009 the sector witnessed the entry of two new actors in the commercial export business. The German-based OHB System AG won a contract for its first Small-GEO mission. The building contract for the Spanish Hispasat AG-1 communication satellite has a value of €48 million.[320] Additionally, OHB attracted the industry's attention by winning the contract to build 14 Galileo-satellites for €566 million.[321] The second newcomer was the Canadian MacDonald Dettweiller and Associates Company, chosen to build a communication satellite system for the National Space Agency of Ukraine, which also includes a GEO communication satellite. This satellite will support direct broadcast television and high-speed Internet access in Ukraine.[322]

The Russian manufacturer ISS-Reshetnev won, among others, a contract with the Russian Radio Research and Development Institute for developing the Express AM5 and AM6 satellites in May 2009. For this project, ISS-Reshetnev cooperates with ThalesAlenia. Additionally, the company negotiated a contract on a TELKOM-3 telecommunication satellite for PT Telekomunikasi Indonesia.[323] The rising number of contracts and the international involvement of Russian companies are indicators of the increasing integration of Russia's satcom industry into the global market.

In 2009 China's presence in the satellite market was also confirmed by contracts attributed to Chinese companies from Nigerian satellite services providers, the Asia-Pacific Mobile Communications Satellite Company Limited and the Laos National Authority for Science and Technology.[324] However, there is still some uncertainty regarding the long term sustainability of the Chinese market presence, due to a Long March 3B rocket failure in August. The rocket's malfunction did not allow it to place the Indonesian commercial communications satellite Palapa-D into GTO.[325]

Despite the emergence of the new actors, the traditional satellite manufacturers from Europe and the U.S. had a very good year. As a whole, the business stayed stable despite the crisis, and the main buyers of satellites remained the major satellite fleet operators: SES and Intelsat alone accounted for more than a third of the orders. If the growth of the sector continues at this pace, one could expect a stable rate of replacement satellite orders, at around 20 spacecrafts per year. The growing commercial capabilities in Japan, South Korea, India and elsewhere could have further positive effects. However, the main commercial satellite providers such as SES, Intelsat and Eutelsat have already promised to subside investing now that they have completed their hardware replacement expansion cycle.[326]

Looking at the European satellite manufacturers in more detail, EADS won seven contracts in 2009, whereas ThalesAlenia had four orders, excluding the co-orders with EADS Astrium and Orbital Sciences. Newcomer OHB System AG won one contract with Hispasat. As for the U.S. satellite manufacturers, Boeing was contracted by Intelsat to build four new 702B satellites, which could give the company a chance to rapidly re-enter the commercial market within the course of this year. Orbital Sciences won five contracts, including orders from Intelsat and OverHorizon. Loral Space and Communications has invested $350 million in the past few years into its SSL division. This enabled the company to sign nine contracts in 2009. Loral is the only satellite manufacturer which depends entirely on commercial contracts. Consequently, it is also more vulnerable to market fluctuations than other manufacturers.[327]

With the exception of Loral, no satellite manufacturer relies on commercial orders alone. Therefore, institutional orders constitute an important segment for the satellite manufacturing sector, particularly in the United States. For example, Orbitel announced that its advanced space programme division increased revenue by 15.6 percent in 2009 and will continue to grow in 2010, mainly thanks to manufacturing classified and unclassified satellites for the U.S. government.[328] In Europe, the biggest non-commercial order was that of the 14 Galileo satellites to the German OHB company, leaving behind the main competitor, EADS Astrium. In 2009, only a little more than 25% of all launched satellites was commercial. Consequently, the government related satellite manufacturing business still represents the greatest revenue source for the sector's companies.

In terms of future perspectives, 20 to 25 satellites are expected to be launched annually in the next few years. Rising technologies e.g. HDTV are keeping up the demand for new satellites.[329] However, the current "high cycle" of the manufacturing market should last until 2013 and corresponds to replacement of older satellites.

3.4.3. Satellite operators sector

In 2009 Satellite services remained the single largest satellite industry segment[330] led by satellite TV growth for a global increase of 10.7% of revenues. This year saw a global increase of revenues and some falls are also noticeable for certain of them. Other consequences should be expected after the financial crisis which had at this time not yet hit the sector. The podium stays globally the same with a steady and quite similar growth of the revenues for Intelsat and SES (6%) while Eutelsat experienced a significant increase of 10% especially boosted by the growing market in South America. Telesat confirmed its fourth place with an impressive augmentation of 28.9% of its revenue. The rest of the ranking knew a certain upheaval with the fall of the Russian satellite communication from the sixth to the

Tab. 1: *Top 10 FSS operators in 2009.*[331]

Rank	Company	Country	2009 revenues in million $	Satellite in Orbit	Satellites on order
1	Intelsat	Luxembourg	2500	52	8
2	SES	Luxembourg	2440	44	11
3	Eutelsat	France	1410	26	6
4	Telesat	Canada	750	12	3
5	Jsat Corp.	Japan	362.7	13	2
6	SingTel Optus	Australia	236.6	5	0
7	Hispasat	Spain	216.4	4	3
8	Russian Satellite Communications	Russia	200	11	4
9	Star One	Brazil	193	7	1
10	Arabsat	Saudi Arabia	189	4	3

eight one, with a significant drop of his revenues of 14%. A phenomenon that would be explainable with the fall of the rubble in 2009 and the expansive renewal of the fleet. A significant effect affected Gazprom system.

In the same way Jsat Corp saw its revenues drop of 11% but keeps its place followed SingTel Opus and Hispasat both in progression. At the bottom of the ranking, Star One and especially Arabsat with 20% had a significant increase in their revenues on this period.[332] Just behind the top list, the Norwegian Telenor Satellite Broadcasting could next year join the club with an increase of its revenues from $125 million to 177 in 2009. The Satcom market was so on the roll between 2009 and 2010 and saw a fierce competition within it with candidate to access to the ten first places while the five leaders consolidated their positions.

4. The security dimension

4.1. The global space military context

The following chapter briefly presents key developments in the field of military related space activities. These include the military space government programmes and related spending, the industrial achievements in military space technologies and the evolution of space security doctrines of all the major space-faring nations. Nevertheless, one should take into account that studying military space activities based on open sources is always a difficult task, given the confidentiality clauses that usually apply to them. Consequently, all the verifiable facts and figures given below cannot provide a fully detailed picture of all related developments. They do however put them in perspective by providing an overall estimate of the general trends in this field.

Space-related military spending remained generally stable from mid 2009 to mid 2010. While the worldwide military spending on space rose slightly by 7% from $29.4 billion in 2008 to $31.8 billion in 2009, the share nearly remained the same. Again, the United States led by far with a share of 90% of worldwide space-related defence expenditures, but the sum of the other countries now occupy a bigger share of 10% compared to 5% in 2008. However, the European space spending dedicated to military use decreased in 2009 by 32% to approximately $752 million.[333]

It has to be stated that money spent on dual-use programmes or research is not included in this overview. It should also be kept in mind that spending is not always clearly allocable, because some budget positions can be assigned to various categories. The standard confidentiality and secrecy along with potential opaqueness can further complicate matters. Different purchasing powers and work force costs add another degree of ambiguity, calling for relativisation of numerical budget values

4.2. Europe

4.2.1. National initiatives

Although European cooperation in dual-use space technologies is increasing, military space activities in 2009 and 2010 remain predominantly within the national government's field of competences. Consequently, European cooperation in such matters remained into the sphere of bilateral or at best multilateral arrangements among the major European space-faring nations.

The domain of Space Situational Awareness in particular has been one of the areas that attracted particular attention in the last 12 months. The German Armed Forces for example, proceeded to the creation of a separate military command to coordinate space surveillance activities, known as the German Space Situational Awareness Centre (GSSAC). The command's facility was inaugurated in 2009 and it was projected to reach the planned staffing ceiling already within 2010. The centre's creation was deemed necessary not only to facilitate German military participation in the future European space-monitoring capability, but also to develop the country's national competencies in this field.[334]

Furthermore, Germany and France are the only EU members that have a limited space surveillance capability through the use of two nationally operated Radars. These are the German TIRA, which is capable of tracking objects in orbit and determining their nature, and the French Graves, which is more suitable for wide area surveillance. Given the complementarity of the two systems there has been a strong incentive for bilateral cooperation on their use. Therefore, the two countries have initiated in 2009 an exchange of surveillance data programme, enabling them to coordinate the use of both systems and to multiply their operational usefulness. Nevertheless, even the combined systems' accuracy is not always sufficient for the accurate identification and tracking of objects in orbit. Therefore, both countries still have to rely on USAF's space monitoring network for more accurate measurements in order to avoid losing their military space assets to collisions with space debris.[335]

However, since military space still remains predominantly under strict national jurisdiction, European cooperation does not proceed at the same pace in all areas of activity. Contrary to the dual-use SSA where cooperation has been increasing, negotiations on the MUSIS system stagnated throughout the reporting period.

MUSIS is a six-nation European effort to build a common ground segment for its members' Earth observation satellites. Some participating countries would contribute their optical reconnaissance satellites to the common system, others their Radar satellites and others would only share its operating financial burden. This complicated arrangement however has made it difficult to quantify each participating country's contribution to and expected returns from MUSIS. Consequently, negotiations among the participants have been fruitless for the past 12 months and the programme's future was threatened as certain participant countries could be inclined to proceed independently with their national projects.[336]

In spite of the difficulties that the multilateral MUSIS programme has encountered, the bilateral cooperation between Germany and France in the field of Earth observation (EO) satellite data exchange moved forward in the second half of 2009 and the first half of 2010. On 4 May, satellite manufacturer OHB of Bremen announced it had secured a €14 million contract in order to provide ground stations to both countries permitting access to data from each other's EO spacecraft. Under this Franco-German bilateral agreement known as the Europeanisation of Satellite-Based Reconnaissance programme (E-SGA), France would be granted access to data from the German radar EO SAR-Lupe satellites and Germany from the French Helios 2 optical EO satellites. Ground stations in both countries were already under construction and they were expected to become operational by July 2010.[337]

The increasing importance attributed to military space activities in Europe was also reflected in their increasing budgets, which showed remarkable resilience to the past 12 months' adverse financial conditions. On 25 November 2009 for example, French Defence Ministry officials announced they were on track to increase military space spending by 8% annually through 2014, in accordance with the defence policy announced by President N. Sarkozy in 2008, before the financial crisis's consequences were fully felt. This would mean that French military space budget would rise to €600 million by that year, from €380 million in 2008. An additional €200 million would be channelled to the French space agency CNES for dual-use technologies development. At the same time, the country's joint defence staffs was set to create a Joint Space Command by July 2010 in order to coordinate military space assets use better. At the same time, France was actively seeking participation from other EU countries in its future optical and electronic surveillance satellites.[338]

At the same time in the UK, Defence Ministry officials announced they were considering a revision to the Skynet 5 contract with Paradigm, an EADS Astrium Services subsidiary, to allow for the addition of a forth satellite to the Skynet 5 constellation. According to the same sources, tactical imagery transmission demand from UK forces in Afghanistan has brought the existing three satellite constellation to its capacity limits. Furthermore, demand was expected to double in the next 5 years, or quadruple by comparison to the initial bandwidth demand models made back in 1998, on the basis of which the contract with Paradigm was concluded. For the time being, UK officials said they were dealing with the situation by scrutinising transmission demands and using the Skynet 5C spacecraft, which was originally planned as a back-up spare satellite. Military satellite communications providers in France and other NATO countries admitted facing the same problem as well.[339]

Finally, the decision to deploy a fourth Skynet 5 satellite was officially announced on 9 March 2010. The original Skynet 5 contract with Paradigm was valued at £3.5 billion through 2020 for the lease of three communications satellites and related services. This contract was extended through 2022 for an additional cost of £400 million, allowing Paradigm to launch a fourth satellite by 2013 in order to meet the increased demand. Since the Skynet 5 contract was originally set up as a public-private partnership (PPP), with Paradigm maintaining the ownership of the spacecrafts and leasing their services to the military, the extension costs would only burden the UK government towards the end of its duration. Thus, British armed forces would be able to use the fourth satellite and only pay for it in 2020. This outcome clearly demonstrated the advantage of PPP's adaptability to changing operational needs, especially at a time of budgetary difficulties for the UK.[340]

The satellite manufacturer ThalesAlenia Space and the ground-services provider Telespazio are to build the Sicral 2 military telecommunications satellite, going to be launched in 2013. The satellite will be used by both the Italian and the French defence ministries, containing separate UHF- and SHF-frequency payloads for both countries. However, this joint military satellite project marks a clear break to previous European politics where countries have their own military communications satellite systems. Noteworthy is the fact that both contracting companies are in a joint French-Italian ownership. The satellite is to be included in France's Syracuse 3 system and will enforce Italy's capacities to contribute UHF bandwidth to NATO.[341]

4.2.2. European Union level

Between July 2009 and June 2010, the European Union has been increasing its involvement in dual-use military applications that could potentially have a military usefulness. EU security related space activities are mostly managed through the European Commission, the European Defence Agency and the European Union Satellite Centre. The European Space Agency also acts as a programme coordinator and procurement authority for most of these projects. The EU's principal security related programmes include the Galileo navigation and positioning satellite constellation, the Global Monitoring for Environment and Security (GMES) Earth observation programme and the Space Situational Awareness (SSA) project. As all of the aforementioned programmes are of potentially dual civil-military usefulness, they were already presented in details in chapter 3 of this report. However, a brief analysis of developments related more closely to their space security aspects would be pertinent in this chapter.

4.2.3. European Space Agency

One of the principal trends during the period in question has been the increasing involvement of the European Space Agency (ESA) to the development of dual-use space capabilities for Europe. This tendency has led to an increased level of cooperation with the European Defence Agency (EDA), which is the only other European institution tackling this issue.

This trend was especially evident in the development of the future European Space Situational Awareness (SSA) system. Both agencies were scheduled to start official discussions in 2010 on the subject of how the European dual-use SSA should be created and managed. Although the SSA started as an ESA programme, EDA became quickly involved, mostly upon the insistence of European governments. In fact, the past 12 months have witnessed increasing pressure in favour of the active participation of EU military personnel in SSA's development and operational use, with certain voices even calling for the complete militarisation of the system's operating concept. This trend has been also induced by the rising number of European military space assets in use, as well the growing realisation of their vulnerability to space debris collisions described above.[342]

Another example of ESA-EDA coordination appeared in February 2010, when they both issued parallel contracts of a €400,000 value and 6 month duration each for the realisation of two preliminary concept studies investigating the use of Unmanned Aerial Vehicles (UAVs) for domestic and maritime security purposes in Europe. Although the two agencies demanded separate studies, they nevertheless issued a common request for proposals. This fact undoubtedly illustrates the gradual expansion of ESA's mandate into the field of security-related activities. Although it is forbidden for UAVs to fly over European civilian airspace in order to avoid accidents, it is anticipated that this regulation will change as UAVs demonstrate an increasing degree of operational reliability. The future UAV system envisaged would use satellite communications to downstream surveillance data and it is expected to operate complementary to Earth observation satellites, offering shorter revisit periods than space assets.[343]

Finally, ESA involvement into space security applications became greater in May 2010, when the agency began the concept development of a system of systems focused on utilising existing and future European space capabilities for security purposes. The programme, known as the Global Integrated Architecture for Innovative Utilisation of Space for Security (or Gianus), intends to integrate all existing European Earth observation, communications and navigation satellites into a single interface. By creating a unified operational system from all European space assets, Gianus would facilitate its use for security purposes and eventually make it more adaptable to specific military operational needs. At a time when military budgets across Europe are strained by the ongoing financial crisis, dual-use space systems developed could be a viable option for the majority of its member-states.[344]

4.3. The United States

As the Operationally Responsive Space concept is maturing in the United States, more and more major U.S. space contractors, such as Boeing or Lockheed Martin, are becoming interested in it. Boeing's Phantom Work's Advanced Network and Space Systems are currently working on small satellites that they qualify as "disruptive technology". Analysts in the U.S. expect 17 more small satellites to be built within the next decade, for a projected total value of $1.4 billion (a 40% increase from current levels). The Pentagon is moving forward with purchasing smaller, simpler and cheaper satellites to cover ORS needs for a number of reasons, including their affordability, survivability and launch-on-demand capability.[345]

In a related development, the U.S. National Reconnaissance Office (NRO) has started a new program to use tiny satellites, known as "cubesats", as in-space test platforms for future satellite technologies. "Cubesats" satellite buses, which typically measure 10×10×10 cm, are relatively cheap, they can be purchased in bulk and be ready on demand to serve as test platforms. They could help NRO to validate new instruments, missions and capabilities, such as hyperspectral sensors, attitude control systems, or radio-frequency modules. NRO has reportedly asked for acquiring approximately 20 to 50 "cubesats" at a $250,000 unit cost, over a two-year period beginning in 2010. One more advantage of using "cubesats" would be that they could be launched in a timely fashion aboard the first available launcher. NRO was even considering placing them on top of NASA's boosters.[346]

In a parallel development, U.S.A.F. Space and Missile Systems Centre (SMC) issued on 13 November a Request for Information (RfI) on the possible development of its first cubesat demonstrator mission. SMC was particularly interesting in validating the operational utility of a cubesat constellation. The demonstrator would be used to collect space weather data, it should have an operational life span of 1 year at a 400 km altitude and it should de-orbit within 5.[347]

Furthermore, the U.S. Missile Defence Agency (MDA) was also reported to seek complementing its future Space Tracking and Surveillance System (STSS) with a constellation of small satellites, known as the Precision Tracking and Surveillance System (PTSS). MDA has been persistently asking the Congress to fund a 1,125 Kilogram demonstration satellite for several years now.[348]

On the other side, the U.S. Senate Armed Services Committee's 2010 defence authorisation bill more than doubled the budget for ORS that the Pentagon had requested. In fact, Senators appropriated an additional $170 million to the Air Force's $112.9 million request for the ORS Office in 2010. Extra funding included a program to build prototype, low-cost, half-metre ground resolution imaging satellites within 36 months. The ultimate goal would be to field a large constellation of such satellites and to acquire them on a fixed-price basis for no more than $100 million a piece, including launching costs.[349]

In another development, the U.S. House of Representatives approved on 30 July an $80 million funding for Northrop Grumman to continue development of its Kinetic Energy Interceptor (KEI), in spite of the Defence Department's repeated attempts to cancel the project.[350]

On 18 August, U.S.A.F. announced the results of the Schreiver War Games 5 space defence exercise, held at Nellis Air Force base in April 2009. During the exercise it became clear that a sophisticated space-faring nation could deny key U.S. military space capabilities in case of conflict. The exercise scenario also established the need for improved space surveillance capabilities, especially as far as tracking small satellites is concerned. In fact, U.S.A.F. admitted that during the exercise they were often unable to determine the nature and source of events involving small satellites.[351]

On 11 August the U.S. National Geospatial-Intelligence Agency (NGA) announced its plans to issue up to four $85 million and 5 year duration contracts to commercial providers of radar satellite images. The request included Synthetic Aperture Radar (SAR) images both in the X and the C band. Since the U.S. lack commercial radar satellites, demand will have to be met by European (Cosmos Skymed and TerraSAR-X) and Israeli (TecSAR) satellites. The request represents a considerable budget increase from the $10 million that NGA has been paying annually to the Canadian Radarsat-2 operator for similar services over the past ten years. Its broad requirements would also indicate that contracts will be most likely split among several operators.[352]

EnhancedView is part of a wider satellite imagery strategy for national security purposes announced by Director of national Intelligence Dennis Blair, on 7 April 2009 and approved by President B. Obama. The concept, known as "two-plus-two", has two piers. One is the purchase of two highly sophisticated imaging satellites by the National Reconnaissance Office (NRO) for strategic intelligence purposes, scheduled to be build by Lockheed Martin Space Systems. The other is the contractual purchase of lower resolution images from two less capable commercial satellites by the NGA, for geospatial and tactical military use. This approach was adopted in the U.S. House of Representatives' 2010 Defence Appropriations Bill.

On the contrary, the Senate's Defence and Intelligence Authorisation Bills asked for the deployment of a larger constellation of cheaper and less technically advanced satellites.[353] On 16 September the Senate's Defence Authorisation Bill upheld this option and effectively turned down the version of the Bill preferred by the Obama Administration. NRO's Director Bruce Carlson his opposition to the Senate's proposed plan, which he deemed technologically riskier. Low technological risk was the primary driver behind Lockheed Martin's selection, after Boeing's failure to develop a more advanced system under NRO's Future Imaging Architecture (FIA) in 2005. According to press reports, the two companies' struggle for the contract could be behind the Senate's decision, as the two Senators that opposed NRO's plans the most have important Boeing facilities in their constituencies.[354]

On 17 September the U.S. Defence Advanced Research Projects Agency (DARPA) proceeded to an open call for new concepts for removing space debris from Low Earth Orbits (LEO).[355] The announcement was made as DARPA and NASA were preparing to host o joint conference on this subject later that year. According to NASA's Office of Safety and Mission Assurance, coming up with innovative and inexpensive solutions could prove difficult. They were particularly concerned by the threat to space environment posed by increasing numbers of tiny satellites, such as cubesats. Since cubesats are currently almost undetectable, their proliferation creates uncontrolled space debris and increases the danger of cascade collisions. However, finding ways to remove debris from LEO could also run into diplomatic disagreements, as any such system would inherently also be capable of working as an orbital Anti Satellite weapon (ASAT).[356]

On 9 October U.S.A.F. issued a request for information for the development of its next generation of space-weather monitoring satellites. U.S.A.F. is currently using sensors onboard its Defence Meteorological Satellite Program spacecrafts. Nevertheless, similar payloads have been eliminated from its successor, the civil-military National Polar-Orbiting Operational Satellite System, because of budget cuts. This request for information would be a first step in exploring alternatives for a space-weather monitoring system that could involve payloads onboard commercial satellites, or small dedicated satellites. The program has secured an initial $15 million budget and launch is expected in 2015.[357]

On 22 October U.S. Defence Secretary R. Gates announced during an official visit to the Republic of Korea (RoK) that the U.S. would extend its missile defence shield over that country. Under this doctrine of "extended deterrence", he said, the U.S. would use the full range of its military capabilities (conventional, nuclear and missile defence) to defend RoK against any incursion from the North. The announcement was made as RoK is moving forward with its own autonomous ballistic missile defence system, based on Aegis destroyers and Patriot Advanced Capability-2 interceptors.[358]

In an interesting technology development, the U.S. Defence Advanced Research Projects Agency (DARPA) issued a Request for Information (RfI) on 22 October for a system that could provide internet connectivity to Low Earth Orbit (LEO) satellites. The Request, issued by DARPA's Tactical Technology Office, called for a 100 Kilobit per second broadband connexion to all LEO satellites flying at a 500 km altitude and in any orbit. The RfI also asked for 95% operational availability and the capacity to be used not only for downloading data, but also for telemetry and controlling the satellite. Information from government laboratories, universities and private companies were due in 5 November.[359]

Following on this RfI, DARPA issued on 3 February 2010 a Notice of Intent to award a sole source contract to develop and build the necessary hardware to Inmarsat plc of United Kingdom. The contract, under a program called the "Persistent Broadband Ground Connectivity for Spacecraft in Low Earth Orbit Effort", solicited the Broadband Global Area Network (BGAN) service provided by Inmarsat's I-4 communications satellite. The project's objective would be to provide near-24/7, very low latency, on-demand ground-to-spacecraft connectivity for LEO satellites. From an operational perspective this would greatly enhance Operationally Responsive Space capabilities, allowing performing of missions such as time-sensitive satellite control directly from the theatre of operations, rapid data transfer and direct-to-theatre data delivery on small portable devices.[360]

For this purpose, a space-based version of Inmarsat's BGAN airborne terminal would be developed, tested, certified and integrated into the planned F6 fractionated spacecraft demonstration cluster, scheduled for launch in 2013. According to the same announcement, Inmarsat has been selected as the sole source contractor for this program because of its considerable expertise in the field of end-to-end satellite broadband services.[361] Inmarsat itself had sought to boost its business with U.S. government agencies, when it acquired the U.S. based communication services provider Segovia Inc. on 23 November 2009.[362]

On 10 October Lockheed Martin Aerospace launched the scaled prototype of a rocket plane from Space Port America in New Mexico, U.S. The self-propelled

90 Kg winged vehicle was launched on a vertical ramp provided by UP Aerospace of Denver and it reached the altitude of 915 metres. The test was the third since 2007 and the second successful. The vehicle is considered a demonstration of an advanced reusable launcher, under the U.S. Department of Defence Operationally Responsive Space concept. Further details on its development were restricted.[363]

On 19 December 2009 U.S. President B. Obama signed the 2010 Defence Appropriations Act that included funding for military space programmes. This Bill, approved by Congress, called for the formulation of a long-term space investment strategy through 2025, which should be delivered to its defence committees by 1 May 2010. Accompanying documents of the legislation particularly demanded that the U.S. maintains a robust space launch capability, by ensuring the Evolved Expendable Launch Vehicle's (EELV) utilisation through 2030. This effort would also include the development of a common upper stage for both of EELV's variants, notable the Delta 4 and Atlas 5 rockets. Other major space programmes that secured funding were the civil-military National Polar-orbiting Operational Environmental Satellite System (NPOESS) and USAF's Third Generation Infrared Surveillance system. On the contrary, the Bill did not foresee budgeting of the seventh Wideband Global Satcom communications satellite.[364]

On 29 December 2009 the U.S. National Geospatial-Intelligence Agency (NGA) awarded three contracts to commercial radar satellite imaging companies. These included EADS North America, Lockheed martin Space Systems and MDA Geospatial Services of Canada. The first two would provide imaging from European commercial spacecraft, notably the German TerraSAR-X and the Italian Cosmo-Skymed (through its commercial brunch e-Geos) spacecrafts respectively. The third would use imaging from Canada's Radarsat-1 and -2 satellites. The total value of the contracts was estimated to reach $85 million over a period of 5 years. Commercial images would act complementary to data acquired through more sophisticated U.S. military radar observation satellites.[365]

In an interview on 14 December 2009, the U.S. Defence Information Systems Agency (DISA) Director Mr Bruce Bennett announced that DISA had plans to lease the Netted Iridium Service on behalf of the U.S. Navy. This service would allow ground military operators in Afghanistan not only to communicate from point to point as before, but also to transmit voice and data broadcasts to several recipients simultaneously and at much faster speeds. Netted Iridium is a retooled variant of the standard Iridium service using special radios built by NexGen Communications LLC of Dulles, an ITT subsidiary. DISA spent approximately $70 million in 2009 acquiring bandwidth for military purposes on the 66 LEO satellite constellation owned by Maryland based Iridium Communications. This figure was expected to rise to $80 million in 2010. The Netted Iridium Service became the U.S. Navy's only option for obtaining narrow-band communications capacity after continued delays in the development of its new generation communications satellites, known as the Mobile User Objective System (MUOS).[366]

In the same interview, DISA's Director announced that his agency had spent over $400 million to acquire commercial satellite communications services in 2009, $50 million more than the year before. In fact he admitted that commercial operators contribute the bulk of U.S. Armed Forces satellite connectivity worldwide. In the mean time satellite operators increasingly deploy Ka- and X-band capacity to complement their standard Ku-band transponders and DISA is expected to triple its demand of X-band capacity in 2010. Finally, DISA Director announced the implementation of a new contracting vehicle for communications satellite capacity called the Future Comsatcom Services Acquisition. Under this new contracting scheme, DISA would be able to acquire bandwidth directly from commercial operators, without having to pass through intermediate private companies, as it is the case today.[367]

At the same time, back in the United States commercial satcom operators were asking of the Defence Department to adopt a more comprehensive and long term approach in commercial bandwidth acquisition. At present the Pentagon is funding these acquisitions, which account for nearly 80% of its global satellite connectivity demand, through supplemental war funds that are approved by Congress on an annual basis along with each year's defence budget. This would mean that no long term acquisition planning would be possible under the current purchasing scheme. To complicate things further, supplemental war funding, which represents roughly 20% of U.S. defence spending, is expected to be curtailed over the next few years.

At the same time, USAF's plans to develop its next generation military communications satellite were set back by the cancellation of the Transformational Satellite system in 2009. In response to this situation, commercial operators have asked in January 2010 for the creation of a dedicated regular line in the U.S. defence budget to cover satcom services procurement on a longer term and not on the current year to year basis. Such a decision, they said, would enable them to make the necessary investments to respond to the Pentagon's growing commercial satcom use, as well to offer services that are more adapted to specific military needs. However, according to industry officials the problem remained that commercial capacity demand was highly unpredictable, as it depended mostly upon unforeseen geopolitical events and contingent operations.[368]

On 13 January 2010, the California based telecommunications company Cisco announced it had completed a successful in-orbit test of its space internet router component onboard Intelsat's IS-14 satellite. The router was part of the U.S. Defence Department's technology demonstration project on Internet Routing in

Space (IRIS). Although the Pentagon was funding the experiment, the router itself was owned by Cisco, which planned on commercialising the devise immediately after the conclusion of its testing period in April 2010.[369]

On 26 January 2010, the U.S. Defence Department announced that a consortium led by satellite operator Intelsat had secured a five year $542.7 million contract to provide end-to-end satellite communication services to the U.S, Navy. The project, known as Commercial Broadband Satellite Program (CBSP) called for the procurement of 449 megahertz of Ku-band, 329 megahertz of C-band and 82 megahertz of X-band capacity, which should be accessible to 95% of the Earth's populated regions. For that reason, the consortium included no less than 17 companies (including major providers such as SES), in order to achieve constant global coverage. Intelsat's proposal included as many as 20 satellites and 8 teleports. This contract is expected to replace the U.S. navy's existing one for the procurement of L-band mobile services from Inmarsat of London.[370]

The U.S. Missile Defence Agency (MDA) successfully launched two new generation Space Tracking and Surveillance System (STSS) demonstration satellites on 25 September 2009. The spacecrafts included considerable improvements form previous systems, including the capability to track cold bodies in space for the first time. However, their initial testing period run into serious delays due to numerous technical problems. In fact, officials of Northrop Grumman Aerospace Systems of Los Angeles, the satellites' manufacturer, confirmed only on 21 January 2010 that the system's testing and sensor calibration had resumed. By then, the entire programme came under scrutiny by the Congress for budget overruns and poor programme management, which resulted in the curtailing of several scheduled tests of the system. As a result, MDA's medium term Integrated Master Test Plan that was completed in mid-2010 did not include even a single test related to STSS. It should be noted that although MDA estimated the programme's total cost since 2002 at $1.35 billion, the U.S. Government Accountability Office (GAO) raised it in its own project assessment to $ 3.1 billion. According to MDA sources, development of an operational version of the system would not begin before the completion of current demonstrator testing.[371]

On 1 February 2010 the U.S. Air Force announced its budget appropriation request for 2011. Although the overall USAF budget request was up 3% from 2010 (at $170.8 billion), space procurement and development expenses were curtailed by more than 8% (to approximately $8 billion). This development ended the upward trend in USAF space activities spending over the last few years. In fact, the proposed budget focuses on capitalising on the operational use of existing space systems and foresees only limited funding for new development projects. Even transformation programmes that were given priority until now, such as these run by USAF's Operational Responsive Space (ORS) Office for example, suffered budget cuts. The disproportionate decrease in defence space appropriations was attributed by government officials to the general trend of limiting R&D spending in the 2011 defence budget. Since R&D related expenses are proportionately higher in space activities than in other USAF programmes, the same officials claimed, the resulting budget cuts were also greater. Among the most important defence budget cuts, one can single out the termination of the $26 billion Transformational Satellite communications system, the scaling down of the Third Generation Infrared Surveillance (TGIRS) programme and the 25% decrease in funding for ORS projects ($94 million in 2011 from %124.3 million in 2010).[372]

As it seems, operational programmes already in use or in their final development stages represent the bulk of the requested appropriations. These include $426.5 million for space situational awareness projects ($188.1 million more than in 2010) and the purchase of additional spacecraft for a number of USAF programmes, like the fourth SBRS satellite, the seventh Wideband Global Satcom and the fifth Mobile User Objective System narrowband communications one. The USAF space surveillance system stands out among the programmes with the greatest increase in spending. Funds are mostly diverted to upgrading the system's ground-based radar network known as Space Fence, as well as developing follow-on to the Space Based Space Surveillance satellite due for launch in June 2010.[373]

At the same time the Obama administration requested on behalf of the U.S. Missile Defence Agency (MDA) $8.4 billion for Ballistic Missile Defence (BMD) in 2011, representing a $500 million increase from the previous year. The planned budget was heavily influenced by the new BMD posture announced on 1 February 2010 through the Ballistic Missile Defence Review Report published by the Pentagon. This new BMD orientation adheres to the deployment of SM-3 interceptor equipped warships in European waters, in order to compensate for the cancellation in September 2009 of the more capable ballistic missile interceptors that would be installed in Poland. At a later stage, the new planning also calls for the development of a land based version of SM-3 (with an initial $281 million budget for 2011). Furthermore, the new doctrine calls for longer development and operational evaluation periods for the new ABM systems, as well as for increased international cooperation in this field. This new policy document also gives considerable attention to developing new ballistic missile warning sensors, both airborne and in space. The space component under development is known as the Precision Tracking Space System, for which MDA has requested $67 million. The programme should benefit from initial work done over the Space Tracking and Surveillance System demonstration satellites, three of which are in orbit since 2009. In an effort to limit development costs and streamline the required system tests, MDA has thoroughly reviewed its testing programme in late 2009 and it has requested increased funding for such activities in 2011 ($1.11 billion compared to

$823.3 million in 2010). Finally, funding for the development of the experimental high-power chemical Airborne Laser is terminated. The budget only foresees $99 million for the transformation of the system into a ground based test bed for directed energy weapons.[374]

In the U.S., the Department of Defence disclosed that it was experiencing mounting problems related to poor hardware manufacturing and quality control. Space and missile defence programmes in particular suffered from repeated delays and failures caused by deficient components. For example, a failed Missile Defence Agency (MDA) test of the Ground-based Midcourse Defence system on 31 January 2010 may have been caused by faulty components. In February, Pentagon officials confirmed that USAF increasingly discovers hardware and software flaws in satellite and launcher components during their final assembly and testing. According to the same sources, deficient parts included crucial pieces of hardware, such as gyroscope and reaction wheels on satellites. These manufacturing flaws were attributed by government officials to the retirement of older generations of space industry skilled workers and the low rate of their replacement. At the end of this process, the U.S. space industry could be facing a permanent loss of space systems manufacturing know-how. On the other hand, this situation could partially explain the resistance that the new US space policy has met within certain department of Defence circles.[375]

On 11 February 2010 the U.S. Missile Defence Agency (MDA) announced a successful test of the Airborne Laser system. A Boeing 747 aircraft carrying a high power chemical laser tracked down and destroyed its intended target, a short-range liquid-fuelled ballistic missile. However, in a second test later the same day the weapon malfunctioned and failed to destroy a sounding rocket that simulated a solid-fuelled ballistic missile. Nevertheless, MDA qualified the test as a success, since ABL had reportedly achieved to intercept a similar solid-fuelled target in a separate test on 3 February. In spite of its success as a system demonstrator, ABL has been deemed as an operationally non-viable and logistically expensive weapons platform. Therefore, MDA had apparently decided to terminate all funding for the programme in 2011 and to invest in other directed-energy technology research projects. As for ABL itself, it would be transferred to the Pentagon's Office of Defence Research to serve as a test bed for laser technologies.[376]

On 17 February, the satellite platform for USAF's first non experimental operationally responsive satellite was delivered by ATK Space Systems of Maryland. The spacecraft, known as ORS-1, was built in 16 months and it was transported to Goodrich ISR Systems of Connecticut, the programme's prime contractor, where the integration of its Earth observation payload would take place. Its launch was scheduled for the second half of 2010. ORS-1 is the first operational satellite developed under the Operationally Responsive Space (ORS) concept, a USAF project aimed at developing small satellites cheaper, faster and ready to launch on demand. The programme's objective is to achieve a quick launch capability for Earth observation satellites, as a response to urgent intelligence demands from units engaged in military operations. Although the satellite's exact cost had not been disclosed, government official's confirmed that the project had met its budget and timetable objectives. ATK announced that the platform's cost was approximately $34 million, whereas the Pentagon's Operationally Responsive Space Office, which is the programme's contracting authority, had previously announced that the total budget could reach as high as $162 million. The satellite bus of ORS-1 was based on the one manufactured by ATK for TacSat-3, Pentagon's latest experimental operationally responsive satellite launched in May 2009. In spite of the programme's proclaimed success, the question of its exact cost remains pertinent, as the affordability of the satellites used is undoubtedly one of the most crucial parameters of the ORS concept.[377]

Although the budget for science and technology development programmes at the U.S. National Reconnaissance Office (NRO) was cut off half during the last five years, it intends to reverse the trend and come to the historical budget amount, according to NRO Director Bruce Carlson. Due to the classified character of the NRO's budget he didn't disclose any concrete figures on his speech of 14 April 2010. Nevertheless, the NRO will conduct its most aggressive launch campaign of the last 25 years until late 2011.[378]

Finally on 22 April 2010 USAF launched the X-37B Orbital Test Vehicle, a winged unmanned spaceplane demonstrator. The vehicle was expected to perform evaluation manoeuvres for as long as nine months, before re-entering the atmosphere and landing as a conventional airplane. The 8.9 metres long spacecraft built by Boeing's Phantom Works of California was lifted in its maiden flight by an Atlas 5 rocket. It has a cargo bay similar to the space shuttle but of smaller dimensions, capable nevertheless of accommodating two small satellites. USAF hopes that when the spacecraft becomes operational it will add a quick and affordable satellite launch capability to its inventory. The programme was originally started by NASA in 1999, before it was passed on to DARPA in 2004 and finally to USAF's Rapid Capabilities Office, which currently has overall supervision of the project. A second test flight was scheduled for 2011.[379]

4.4. Russia

The Russian military space programme remains highly classified and all available open source information regarding it is scarce. Consequently any attempt to analyse it should be considered by definition only indicative. Nevertheless, the available information does permit us to draw some conclusions on its overall nature and strategic orientation.

The basic lines of the Russian military space strategic plan follow in broad terms the orientation of its civil space programme. These are the following: to re-establish Russia as a major military space power on a global scale, limiting its capabilities gap with the U.S.; to revitalise the country's space industrial base in order to make it competitive on an international level and even export-oriented; and to use space related technologies' research and development as a pivotal point for the overall growth of the Russian scientific research and economy, especially in the field of electronics.[380] In spite of its long military space tradition, the Russian military space programme has been increasingly utilising dual-use systems, marking a possible change in its operating concept. For example, one should not forget that the Glonass satellite navigation constellation was initially meant to be a military system, whereas today the Russian authorities put more emphasis to its commercial use as well. As another example, Russian officials announced on 15 April 2010 that they were developing a meteorological satellite scheduled for launch in 2011, disclosing at the same time that it would also be of military use, as it would be capable of detecting submarines as well.[381]

Nevertheless, the Glonass satellite navigation constellation remains a militarily crucial space asset and its expected full operational deployment by the end of this year should be considered a major evolution step for the Russian military space capabilities. Apart from this project, the Russian armed forces proceeded to the launch of at least two publicised military satellite launches in the past 12 months, the mission of which was kept classified. In total, Russian military authorities publicly admit to operate a fleet of over 60 spacecraft dedicated to military missions, the bulk of which is used for Earth observation purposes.[382]

In addition to this, another system of inherent dual-use nature that should greatly upgrade Russian military space capabilities is the development of the Angara heavy launcher, capable of lifting up to 24.5 tons into orbit. The rocket is scheduled to replace both the Rockot and Proton vehicles, thus improving the Russian fleet's homogeneity. It also encompasses the new dual-use oriented approach in space systems' development, as it would not only greatly improve the country's capacity to lift military payloads, but it is also expected to have a broader commercial launch services use as well. The rocket was planned to use the Plesetsk space centre instead of the Baikonur Cosmodrome in an attempt to reduce Russian space transport dependence on third countries and its first launch was expected in 2011. However, the vehicle's debut could be postponed for a year due to a cut in its 2009 budget expenses that were allocated to the construction of additional launching facilities at Plesetsk. Finally, in a related development the Khrunichev State Research and Production Space Centre that manufactures the launcher announced it had requested additional funds of 10 billion Rubbles ($290 million) over the next three years in order to complete the project.[383]

Finally, in 2009 and 2010 the Russian armed forces have continued to develop the country's Earth-based counter space capabilities under their long-term air defence modernisation programme. This project, which treats air and space as an operating continuum for the purposes of air defence, aims at creating an integrated weapons system consisting of ground-to-air missile systems with anti-aircraft and anti-missile capabilities, as well as modernised MIG-31 supersonic interceptors.[384] In addition to these, the system is expected to include some kind of Earth-based anti-satellite (ASAT) capability as well, although this has not been officially confirmed. Nevertheless, there have been official statements to the effect that the Russian armed forces are in fact developing a new type of ASAT based on what was described as a "fundamentally new weapon [technology]".[385] At the same time however, the Russian government also declared that it was continuing to oppose the development of co-orbital ASATs or any other form of space weaponisation, insisting that any such future systems would be exclusively Earth-based.[386]

4.5. Japan

The Japanese national security strategy is evolving rapidly since the creation of a Defence Ministry in 2007 and space systems have a pivotal role in this transformation process. Japan is bound by its Constitution to pursuit only the "peaceful" use of space. Nevertheless, the country is now in the middle of a policy shift that seeks to define "peaceful" not as "non-military", as it was the case until now, but as "non-aggressive". This will enable the regularisation of the use of dual-purpose satellites by Japan's armed forces and it will reinforce the cooperation between the country's space agency JAXA and the Ministry of Defence. The further development of space systems for national security purposes is one of the pillars of the new Japanese security strategy published in August 2009. The defence against ballistic missile attacks and the improvement of the country's intelligence gathering apparatus through the use of Earth observation satellites are among the conditions identified as crucial for the realisation of the new strategy.[387]

For that purpose, Japanese space activities have been increasingly focusing on the development of their space applications segment. Japan is actively pursuing the development of dual-use space assets across all fields of applications. These include a new generation of Information Gathering Satellites (IGS) with greatly improved ground resolution, a satellite positioning system of regional coverage known as the Quasi-Zenith Satellite System (QZSS), an early warning system to cooperate with its ballistic missile defences and a space situational awareness system similar to the one currently used by the U.S.

The IGS programme is the most mature of these projects, both technologically and operationally. Japan currently operates four of these spacecraft, two equipped with optical and two with radar Earth observation apertures. Although their constellation was only completed in 2007, a new generation of improved IGS spacecrafts is at the final stages of development and they were scheduled for launch between 2009 and 2014. The first of these satellites was successfully launched on 28 November 2009. The IGS programme's total budget from 1998 to 2014 is estimated to reach approximately ¥1 trillion, with an expected average spending of ¥60 billion annually through 2014.[388]

The flight testing of the Quasi-Zenith Satellite System (QZSS) is also expected to begin in the summer of 2010 with the launch of the programme's first demonstrator spacecraft, QZS-1 (nicknamed "Michibiki"). Upon the successful conclusion of the testing schedule, QZS-1 should be joined by two more satellites to form the system's complete operational constellation. It should be noted that QZSS is not indented to be an autonomous satellite navigation system; its mission would rather be to enhance the accuracy and redundancy of the GPS signal over Japan. Nevertheless, from a technological point of view QZSS's development will greatly enhance Japanese know-how on building such systems, possibly allowing it to pursuit the development of its own independent system in the future.[389]

In general, Japan's ambitions in developing a full range space security apparatus are adequately backed by sufficient budget resources. Out of the ¥348.8 billion reserved for space activities in the 2009 budget, ¥213billion (or almost 40% of the total) was related to dual-use space applications. The IGS programme received the most funding (¥66 billion), followed closely by BMD related systems (¥58 billion), QZSS (¥14 billion) and GX rocket development (¥11 billion).[390] The latter's development was however cancelled on 16 December 2009 by the Japanese government, due to its continued budget overrides and unsure commercial prospects.[391] It is highly likely that Japan's early warning satellite programme, which appears to be the nation's next priority in military space capabilities development, would be funded from the BMD related budget line.

4.6. China

As it is the case with Russia, the Chinese military space programme is also classified. Very little and always unofficial information exits the country regarding these projects. What it should be noted is that the Chinese military space programme also evolves around a dual-use concept. This means that all space assets are conceived from their very beginning as military use compatible. These include the country's communications satellites and the Beidou satellite navigation system, which will emit a government-only signal alongside its commercial one. However, the great difference with other space-faring powers is that the Chinese dual-use space assets are mostly run directly or indirectly by the Chinese Army, due to that country's unique political and administrative structure. The Shenzhou manned spaceflight and the Long march launcher programmes are among the most significant examples.

The Shenzhou manned spacecraft programme, also known as Project 921, did not see any significant changes in the past 12 months. After the spacecraft's successful 3rd flight in the September of 2008, the programme's focus has shifted to improving Extra Vehicular Activities know-how and testing in-orbit docking technologies for use in the future Chinese space station. This seems to be the next step in the Chinese human space flight programme. CAST has already three space docking stations under construction, with the first (Tiangong-1, meaning a palace in Heaven in Chinese) scheduled for launch in 2011.[392] The next two space stations are scheduled for launch before 2015. Each one will have an expected life span of only two years. They should be primarily used to demonstrate docking technologies, in cooperation with the three Shenzhou space craft also scheduled for the same period.[393]

4.7. India

India does not have a dedicated military space programme. However, future dual use satellites will have an inherent military utility and ISRO does not place any restrictions on their use. Nevertheless, an important new dimension for the country's nascent military space capabilities emerged earlier this year. During a press conference on 3 January 2010, the Director-General of India's Defence Research and Development Organisation (DRDO) V.K. Saraswat disclosed that his country had begun development of an Anti-Satellite weapon system (ASAT). Mr. Saraswat said that the system had just entered its initial development phase and that it would not be made operational unless his country decided that it "needed" it. He predicted that the ASAT would consist of two components, a laser used for tracking targets and a direct ascent interceptor missile of a 120–140 km range, equipped with an exoatmospheric direct impact kill vehicle. Initial testing of the missile was expected to begin in September 2010, while the full system should be developed by 2014.[394]

4.8. Other selected space actors

The spatial sector has seen over the past years the emergence of a new series of actors in a world ever more globalized. The growing importance of these protagonists can be easily assessed through their often ambitious space programmes. Canada in addition to be an associate member of ESA has thus seen its involvement to space activities gradually augment. Firstly, this movement is noticeable by the regular rise of its budget to reach approximately €395 millions in 2010.[395] Major programmes are under development such as the Radarsat Constellation Mission (RCM). RCM has become a key strategic mission for the Canadian government, costing between or in military initiative, the Remote-sensing Situational Awareness (URSA) component will focus on mission planning, tactical reconnaissance, target acquisition and battle damage assessment. The Canada's next-generation radar Earth observation system, suggesting that the budget delays that have slowed the project's development are a thing of the past. A fresh cash infusion has been made in 2010 of 397 million Canadian dollars. The three-satellite Radarsat constellation missions, should ensure that Canada maintain its role as a world leader in aerospace technology. It is also a key asset to defend their Arctic sovereignty. The first of the three satellites could be launched in mid-2014, with the two others launched in 2015 knowing that the total cost of the system is around 600 million Canadian dollars. The European Space Agency (ESA), of which Canada is an associate member, is in negotiations with Canadian officials on providing user interoperability of the three Radarsat Constellation spacecraft with ESA's Sentinel-1 C-band radar satellite, set for launch in 2012.[396] Concerning Space exploration, in the framework of future international space exploration effort. Canada's MDA Corp.awarded a contract to design and build two lunar rover prototypes for yet-undetermined missions under a contract with the Canadian Space Agency.[397]

Iran tries to join the concert of space faring powers by testing its home made rocket. A capsule containing live animals and featuring a camera mounted on the vehicle that provided a live video stream of the rocket's ascent has been successfully launched, in February 2010. This follows the launch of the telecommunications satellite in 2009 which has according to the head of the Iranian Aerospace Organization re-entered Earth's atmosphere. Further more ambitious programmes might be announced in the coming months after these victorious attempts.[398]

South Korea also tempts to take part in the space venture with more and less success after the Korea's Naro-1 satellite launcher blew up after lifting off from the Naro Space Center June 10, marking the second failure in as many tries for the vehicle, whose first stage was built by Russia. The Korean authorities have blamed the Russian part for the accident. Korea's prime objective is still to the use of Earth observation through two major programs, namely, the Kompsat and COMS-1 (Communication, Ocean and Meteorology) Two more satellites Kompsat-6 (radar) and Kompsat-7 (optical)are expected to be launched in 2016 and 2019.[399] South Korea should first of all fix its problem of launcher in order to fulfil its expectations concerning space activities.[400]

Israel launched with success its newest spy satellite in June which confirms the maturity of the Shavit rocket. After only three days the Ofeq-9 satellite began transmitting its first high-resolution imagery to military intelligence users. This satellite is going to be a key element of the intelligence network of Israel. The satellite would be capable of capturing black-and-white images at resolutions of 50 centimeters or better. A satellite's imaging resolution corresponds roughly to the size of ground objects or features it can distinguish. This reinforces the previous one TecSAR which had been previously launched. The successful launcher is due to be an integrant part of the nuclear capacity of the country.[401]

Turkey has also steadily increased its investments essentially destined to reinforce its Earth Observation capabilities. The country's space defense spending is gauged around $93 in 2010.[402] Turkey appears firmly decided to develop its own independent satellite system to protect national. This will symbolised by the GÖKTÜRK satellite. A €250 million contract had been already awarded in2009 to ThalesAleniaSpace and Telespazio preceded the construction kick-off in 2010 for the 80-cm resolution optical imaging satellite GÖKTÜRK satellite. The project is mainly due to provide image over borders nations. It is a quite critical topic given the tensions which has been recently developed with Israel, but also the development of the Kurd resistance. Launch of the satellite is scheduled for 2012/ 2013 timeframe.

4.9. Threats to the space environment

Volatile solar activity can seriously affect the functioning of satellite and others high tech materials. An insight of the effects was observed in 1989 while Quebec suffered an electrical power blackout due to a massive solar storm.[403] Numerous of studies are undergoing or planned to better undertand the composition and cycles of the Sun in order to anticipate the dramatic potential damages on Earth and its orbit. A longer description of this worldwide effort is provided in the chapter concerning Space exploration.

Space debris is a growing subject of concern while the most powerful nations seem to be helpless in face of this challenge, The chapter concerning new technologies describes the will of Russia and the U.S. to develop revolutionary systems to tackle this crucial issue[404]

The collision between an Iridium mobile communications satellite with a retired Russian spacecraft in February 2009, creating a new debris field in low Earth orbit, has accelerated talks on collaboration on space surveillance.[405] The new diplomatically strategy of the U.S. should be thus more axed on international cooperation, a cornerstone announced by the Obama's administration. A crucial step given that the U.S. maintains the world's most sophisticated space-surveillance network of ground-based sensors but it has been unclear in the past how willing the U.S. Air Force.[406] The European code of conduct might have oriented the protagonists in the right way to mitigate as much as possible the creation of debris.

The International Telecommunication Union (ITU) was seized by the french regulators to intervene with the Iranian government to persuade Tehran to stop jamming satellite signals from the BBC World Service's Persian-language broadcasts into Iran. The request come after that many complains had been already done to Iran to stop the jamming. The BBC Persian programming carried on the Eutelsat Hot Bird 6 satellite. The jamming had started last spring during Iran's elections and has continued intermittently.[407]

[1] International Monetary Fund. World Economic Outlook: Sustaining the Recovery. Washington D.C.: IMF, October 2009: 1.
[2] Ibid: xiv–xv+29.
[3] Ibid: 2–4.
[4] Ibid: 5–6.
[5] Ibid: 4+12+18+39.
[6] Ibid: 18+29–30.
[7] Ibid: 7+31–32.
[8] Ibid: 49.
[9] Ibid: 28.
[10] Ibid: 42–47.
[11] Ibid: 46–50.
[12] Ibid: 11+72. The World Bank. China Quarterly Update. Beijing: WB, Nov. 2009: 1–7.
[13] International Monetary Fund. World Economic Outlook. . . : 71–75.
[14] The World Bank. Global Economic Prospects 2010. WB: Washington D.C., 21 Jan. 2010: 17.
[15] International Monetary Fund. World Economic Outlook. . . : 79–82. The World Bank. Russian Economic Report No 20. Moscow: WB, Nov. 2009: 2+9.
[16] Ibid: 2–9.
[17] Bureau of Economic Analysis, U.S. Department of Commerce. Press Release: GDP 3rd Quarter 2009 Estimate. BEA: Washington D.C., 22 Dec. 2009.
[18] International Monetary Fund. World Economic Outlook. . . : 67–71.
[19] Ibid: 75–79.

[20] Ibid: 71–75.
[21] Cabinet Office of Japan, Economic Social and Industrial Research Office. GDP Quarterly Estimates. Tokyo: ESRI. 9 Dec. 2009. http://www.esri.cao.go.jp/jp/sna/qe093-2/main1.pdf.
[22] Cabinet Office of Japan. Monthly Economic Report. Tokyo, Jan. 2010. 21 May 2010. http://www5.cao.go.jp/keizai3/getsurei-e/2010jan.html.
[23] International Monetary Fund. World Economic Outlook. . . : 71–75.
[24] "iCasualties.org: Operation Enduring Freedom." 22 May 2010. http://www.icasualties.org/oef/.
[25] United Nations Secretary General. The Situation in Afghanistan and its Implications for International Peace and Security. Report to the U.N. General Assembly Security Council, New York: 28 Dec. 2009: 5–8.
[26] Schifrin, Nick. "Intel Official: Time is Running Out in Afghanistan." ABC News Report 27 Dec. 2009. http://abcnews.go.com/International/Afghanistan/intel-official-time-running-afghanistan/story?id=9429416.
[27] United Nations Secretary General. The Situation in Afghanistan . . . : 7+10.
[28] Ibid: 6.
[29] Ibid: 1–5.
[30] The White House. President Obama Statement: The Way Forward in Afghanistan. Washington D.C: 1 Dec. 2009. http://www.whitehouse.gov/issues/defense/afghanistan. Landay, Jonathan S. "Analysis: Focus on Withdrawal Could Jeopardise Afghan Mission." McClatchy 4 Mar. 2010. http://www.mcclatchydc.com/2009/12/01/79879/analysis-focus-on-withdrawal-could.html.
[31] The White House. White Paper of the Interagency Policy Group's Report on U.S. Policy Toward Afghanistan and Pakistan. Washington D.C:.
[32] International Atomic Energy Agency. Implementation of the NPT Safeguards Agreement and Relevant Provisions of Security Council Resolutions 1737 (2006), 1747 (2007), 1803 (2008), and 1835 (2008) in the Islamic Republic of Iran. Vienna: IAEA 18 Feb. 2010.
[33] Ibid: 8.
[34] United Nations Framework Convention on Climate Change-Secretariat. "Fact Sheet: Copenhagen-Why is a Deal so Important?" UNFCCC Fact Sheet Nov. 2009. 23 May 2010. http://unfccc.int/files/press/application/pdf/fact_sheet_copenhagen_deal.pdf.
[35] United Nations Framework Convention on Climate Change-Secretariat. "Historic United Nations Climate Change Conference Kicks Off in Copenhagen with Strong Commitment to Clinch Ambitious Climate Change Deal and Unprecedented Sense of Urgency to Act." UNFCCC Press Release 7 Dec. 2009. http://unfccc.int/files/press/releases/application/pdf/pr_cop15_20091207.pdf.
[36] Ibid. United Nations Framework Convention on Climate Change-Secretariat. "Copenhagen United Nations Climate Change Conference Ends with Political Agreement to Cap Temperature Rise Reduce Emissions and Raise Finance." UNFCCC Press Release 19 Dec. 2009. http://unfccc.int/files/press/news_room/press_releases_and_advisories/application/pdf/pr_cop15_20091219.pdf.
[37] United Nations Framework Convention on Climate Change-Secretariat. "Fact Sheet: Copenhagen . . . ."
[38] Ibid.
[39] United Nations Framework Convention on Climate Change-Secretariat. "Historic United Nations Climate Change Conference Kicks Off. . . ."
[40] United Nations Framework Convention on Climate Change-Secretariat. "Copenhagen United Nations Climate Change Conference Ends with Political Agreement . . . ."
[41] Ibid.
[42] World Meteorological Organisation. World Climate Conference 3-Conference Statement. Geneva: 8 Sept. 2009. http://www.wmo.int/wcc3/documents/WCC-3_Statement_07-09-09_mods.pdf. World Meteorological Organisation. WCC-3 High Level Declaration. Geneva: Sept. 2009. http://www.wmo.int/wcc3/documents/WCC3_declaration_en.pdf.
[43] Council of the European Union. "The Copenhagen Climate Change: Key EU Objectives." EU Council Memo/09/534. Brussels: 2 Dec. 2009. http://europa.eu/rapid/pressReleasesAction.do?reference=MEMO/09/534&format=HTML&aged=0&language=EN&guiLanguage=en.

[44] Swedish Presidency of the European Union. "Summary of the Work in the Environment Council of 22 Dec. 2009." EU Presidency: 23 Dec. 2009.
[45] EU Council. "EU Associates Itself with Copenhagen Accord and Submits Emissions Reduction Target." EU Council Press Release 5762/10. Brussels: 28 Jan. 2010.
[46] Swedish Presidency of the European Union. "Summary of the Work..." European Commission. "Options for an EU Vision and Target for Biodiversity Beyond 2010." European Commission, Brussels: 19 Jan. 2010.
[47] International Energy Agency. "Oil Market Report 2010" Highlights. Paris: IEA, 15 Jan. 2010. www.oilmarketreport.org.
[48] Organisation of the Petroleum Exporting Countries. Monthly Oil Market Report. Vienna: OPEC, 19 Jan. 2010: 1–9.
[49] Ibid.
[50] Ibid.
[51] Organisation of the Petroleum Exporting Countries. World Oil Outlook 2009. Vienna: OPEC, 8 Jul. 2009: 1–16.
[52] International Energy Agency. Natural Gas Market Review 2009-Executive Summary. Paris: IEA, 29 June 2009. http://www.iea.org/Textbase/npsum/gasmarket2009SUM.pdf.
[53] Ibid.
[54] Ibid.
[55] Ibid. International Energy Agency. World Energy Outlook 2009 – Executive Summary. IEA: Paris, 10 Nov. 2009: 3–7. http://www.iea.org/Textbase/npsum/weo2009sum.pdf.
[56] International Energy Agency. Natural Gas Market Review...: 15. International Energy Agency. World Energy Outlook...: 13.
[57] Ibid: 10–13.
[58] Ibid:.
[59] The International Bank for Reconstruction and Development. Global Economic Prospects 2010. Washington DC: World Bank, 21 Jan. 2010: 37.
[60] Ibid: 31–34.
[61] Ibid: 37.
[62] Ibid: 31–34.
[63] United Nations Conference on Trade and Development. Trade and Development Report, 2009. Geneva: UNCTD, 7 Sept. 2009: 10.
[64] The International Bank for Reconstruction and Development. Global Economic Prospects...: 34.
[65] United Nations Conference on Trade and Development. Trade and Development Report ...: 6–12.
[66] Ibid: 10.
[67] Ibid: 66–74.
[68] Ibid: 74.
[69] European Security Research and Innovation Forum. ESRIF Final Report. Brussels, Dec. 2009: 11.
[70] Ibid: 155–166.
[71] Ibid: 166–169.
[72] Ibid.
[73] United Nations Conference on Trade and Development. Review of Maritime Transport 2009. Geneva: UNCTD, 2009: 3. http://www.unctad.org/en/docs/rmt2009_en.pdf.
[74] Ibid: 5.
[75] International Maritime Organization, Maritime Knowledge Centre. International Shipping and World Trade Facts and Figures. London: IMO, Oct. 2009: 6–9. http://www.imo.org/includes/blastDataOnly.asp/data_id%3D26834/International Shipping and World Trade – facts and figures final 26 October 2009_.pdf.
[76] United Nations Conference on Trade and Development. Review of Maritime Transport.: 6.
[77] Ibid: 8–11.

[78] International Maritime Organization Assembly, 26th Session. Piracy and Armed Robbery Against Ships in Waters Off the Coast of Somalia (Res. A.1026 [26]). London: IMO, 2 Dec. 2009. http://www.imo.org/includes/blastDataOnly.asp/data_id%3D27087/1026.pdf.
[79] International Civil Aviation Organisation. Air Transport Records Worst-Ever Performance in 2009. Montreal: ICAO News Release, 18 Dec. 2009. http://www.flightglobal.com/assets/getasset.aspx?ItemID=32168.
[80] Ibid.
[81] Giovanni Bisignani. "State of the Air Transport Industry." International Air Transport Association Annual General Meeting. Kuala Lumpur, 8 Jun. 2009. http://www.iata.org/pressroom/speeches/2009-06-08-01.htm.
[82] "ERA an understanding area." 29 Jun. 2010 EU. http://ec.europa.eu/research/era/understanding/understanding_en.htm.
[83] Godin, Benoit. "Science, Accounting and statistics: The input-output framework" Research policy 36: (2007): 1388-1403.
[84] Commission of the European Communities. Innovation Union Competitiveness report 2011 : Part I: Investment and performance in R&D Investing in the future 2011. Analysis 17 jun. 2011. Brussel : European Union.
[85] Ibidem.
[86] "Crise financière 2007-2008 : les raisons du désordre mondial." 18 Apr. 2011 Ladocumentation-francaise. http://www.ladocumentationfrancaise.fr/dossiers/crise-financiere-2007-2008/index.shtml.
[87] "Budget 2010 en chiffres." 18 Apr. 2011 EU. http://ec.europa.eu/research/era/understanding/understanding_en.htm.
[88] Commission of the European Communities. Innovation Union Competitiveness report 2011 : Part I: Investment and performance in R&D Investing in the future 2011. Analysis 17 juin 2011. Brussel : European Union.
[89] Ibidem.
[90] Ibidem.
[91] Ibidem.
[92] Ibidem.
[93] Ibidem.
[94] "Number of Nations with Space Agencies is Rising." Space News 1 Mar. 2010: 9.
[95] United Nations General Assembly. Resolution adopted by the General Assembly on 2 Dec. 2009. UN Doc. A/RES/64/28 of 12 Jan. 2010. United Nations.
[96] United Nations General Assembly. Resolution adopted by the General Assembly on 2 Dec. 2009. UN Doc. A/RES/64/49 of 12 Jan. 2010. United Nations.
[97] United Nations General Assembly. Resolution adopted by the General Assembly on 10 Dec. 2009. UN Doc. A/RES/64/86 of 13 Jan. 2010. United Nations.
[98] United Nations General Assembly. Report of the Scientific and Technical Subcommittee on its forty-seventh session, held in Vienna from 8 to 19 Feb. 2010. UN Doc. A/AC.105/958 of 11 Mar. 2010. Vienna: United Nations.
[99] Ibid.
[100] ITU World Radiocommunication Seminar closes. "12 December 2008. International Telecommunication Union. 30 May 2010. http://www.itu.int/newsroom/press_releases/2008/NP08.html. World Radiocommunication Seminar 2010 (WRS-10)." 30 May 2010. http://www.itu.int/ITU-R/index.asp?category=conferences&rlink=wrs-10&lang=en.
[101] "United Nations Programme on Space Applications Activities Schedule: 2009." UNOOSA. 30 May 2010. http://www.oosa.unvienna.org/oosa/en/SAP/sched/2009.html.
[102] United Nations General Assembly. Committee on the Peaceful Uses of Outer Space. Fourth Meeting of the International Committee on Global Navigation Satellite Systems, held in Saint Petersburg, Russian Federation, from 14 to 18 Sep. 2009. UN Doc. A/AC.105/948 of 05 Nov. 2009. Vienna: United Nations.

[103] "UNGIWG tenth meeting." United Nations Geographic Information Working Group. 30 May 2010. http://www.ungiwg.org/meeting2009.htm.
[104] United Nations Conference on Disarmament. Report of the Conference on Disarmament to the General Assembly of the United Nations. UN Doc. CD/1879 of 17 Sep. 2009. Vienna: United Nations.
[105] "Past meetings and events." 27 June 2011. GEO. http://www.earthobservations.org/meetings/meetings.html.
[106] De Selding, Peter B. "ESA Freezes Spending At '09 Level for 2 Years." Space News 18 Jan. 2010: 13.
[107] De Selding, Peter B. "400 Million-Euro Shortfall Prompts Moratorium on Some ESA Contracts." Space News 30 Nov. 2009: 1.
[108] De Selding, Peter B. "ESA Spending Freeze Ends with Deals for Sentinel Satellites, Ariane 5 Upgrade." Space News 4 Jan. 2010: 6.
[109] "Galileo Cost Overrun Passes European Commission Audit." Space News 29 June 2009: 15.
[110] De Selding, Peter B. "EU Reps Express Support for big Investment in Exploration." Space News 26 Oct. 2009: 1.
[111] "Space Situational Awareness: Consolidated Activity Plan for SSA, GSTP and FP7." Commission of the European Union. 30 Apr. 2010. http://ec.europa.eu/enterprise/policies/space/files/research/esa_pb_ssa_consolidated_activity_plan_for_fp7_en.pdf.
[112] Official Journal of the European Union. Treaty of Lisbon.DOC C115/128 of 9 May 2008. Brussels: European Union. http://ec.europa.eu/enterprise/policies/space/files/policy/lisbon_treaty_space_en.pdf.
[113] Spanish Presidency of the European Union. Conference on Governance of European Space Programmes: Questionnaire and Background, 3–4 May 2010, Segovia, Spain.
[114] Council of the European Union. Outcome of Proceedings from the Competitiveness Council of 29 May 2009: Draft "Space Council" Orientations/Council Resolution. SEC (2009) 10500 of 29 May 2009. Brussels: European Union. http://ec.europa.eu/enterprise/policies/space/files/policy/6th_space_council_en.pdf.
[115] "Panel: Satellites Should be Part of European Stimulus." Space News 1 June 2009: 3.
[116] Council of the European Union. Outcome of Proceedings from the Competitiveness Council of 29 May 2009: Draft "Space Council" Orientations/Council Resolution. SEC (2009) 10500 of 29 May 2009. Brussels: European Union. http://ec.europa.eu/enterprise/policies/space/files/policy/6th_space_council_en.pdf.
[117] "European Commission Eyes Ocean Surveillance." Space News 26 Oct. 2009: 10.
[118] José Manuel Durão Baroso. "The Ambitions of Europe in Space." SPEECH/09/476. Conference on European Space Policy. Brussels, Belgium. 15 Oct. 2009. "EC President Eyes Space Surveillance Network." Space News 19 Oct. 2009: 3.
[119] De Selding, Peter B. "With Galileo Contracts at Stake, Disagreement Over Way Forward." Space News 22 June 2009: 4.
[120] "Last Minute Issues To Delay initial Galileo Deployment." Space News 12 Oct. 2009: 3.
[121] De Selding, Peter B. "First Eight Galileo Spacecraft Said to Go to Germany's OHB Technology." Space News 7 Dec. 2009: 1.
[122] EC-ESA-EDA Joint Task Force. European Non-Dependence on Critical Space technologies: List of Urgent Actions for 2009. 6 Mar. 2009. http://ec.europa.eu/enterprise/policies/space/files/research/jtf_critical_technologies_list_en.pdf.
[123] De Selding, Peter B. "European Defence Agency Promoting Long-term Satellite Leases." Space News 11 Jan. 2010: 16.
[124] "The FP7 Space 3rd Call." Commission of the European Union. 30 Apr. 2010. http://ec.europa.eu/enterprise/policies/space/files/research/fp7_space_3rd_call_en.pdf.
[125] De Selding, Peter B. "EUMETSAT Approves Jason-3." Space news 6 July 2009: 4.
[126] De Selding, Peter, B. "ESA Again Fails to Select a Prime Contractor for Meteosat." Space News 14 Dec. 2009: 6.
[127] "Thales Alenia, OHB Team Win $1.9 B Meteosat Deal." Space News 8 Feb. 2010:3.

[128] "French Intel Agency Eyes Outsourcing Data Production." Space News 15 June 2009: 13.
[129] "France Begins Work on New Surveillance Sats." Space News 22 June 2009: 15.
[130] De Selding, Peter B. "French Helios 2B Spy Sat Sends Back First Test Images." Space News 4 Jan. 2010: 16.
[131] "Astrium Inks Two-Satellite Deal with Kazakhstan." Space News 12 Oct. 2009: 9.
[132] "France and Germany to Split CHARME Cost." Space News 15 Feb. 2010: 3.
[133] "France, Germany To Build Methane-monitoring Craft." Space News 8 Feb. 2010: 3.
[134] De Selding, Peter B. "International Space Station Cutbacks Encounter German Resistance." Space News 1 Feb. 2010: 5.
[135] De Selding, Peter B. "Germany Eyes Teaming With Industry For Its Own Optical Satellite System." Space News 19 Oct. 2009: 1.
[136] De Selding, Peter B. "ESA and Germany Reach Agreement on Data Relay System." Space News 19 Oct. 2009: 6.
[137] De Selding, Peter B. "DLR Takes Step Toward In-orbit Servicing Demonstration." Space News 1 Mar. 2010: 14.
[138] De Selding, Peter B. "Italian Space Agency Expects Budget to Remain Flat for 2010." Space News 18 Jan. 2010: 6.
[139] Ibid.
[140] Ibid.
[141] De Selding, Peter B. "U.K. Eyes NASA-Style Agency." Space News 27 July 2009: 4.
[142] "Britain Replacing BNSC With New National Space Agency." Space News 14 Dec. 2009: 3.
[143] De Selding, Peter B. "Panel Urges Britain To Boost Space Spending, Support Exports." Space News 15 Feb. 2010: 15.
[144] Ibid.
[145] De Selding, Peter B. "New Government Space Agency To Centralise British Space Efforts." Space News 29 Mar. 2010: 11.
[146] "Briefing by Senior Administration Officials on the President's National Space Policy via Teleconference." 28 June 2010. U.S. Department of State 20 Dec. 2010. http://www.state.gov/r/pa/prs/ps/2010/06/143752.htm.
[147] "National Space Policy of the United States of America." 28 June 2010. The White House 20 Dec. 2010. http://www.whitehouse.gov/sites/default/files/national_space_policy_6-28-10.pdf.
[148] Ibid: 8. "Briefing by Senior Administration Officials . . . " See also: "Space Foundation Statement on New U.S. national Space Policy." 28 June 2010. Space Foundation 20 Dec.2010. http://www.spacefoundation.org/news/story.php?id=973.
[149] "Press Conference: Deputy Assistant Secretary Frank A. Rose -U.S. National Space Policy 2010." 13 July 2010. U.S. Mission to the U.N. in Geneva 20 Dec. 2010. http://geneva.usmission.gov/2010/07/13/rose-press-briefing/. See also: Robinson, Jana. The Role of Transparency and Confidence-Building Measures in Advancing Space Security. ESPI Report 28. Sept. 2008.
[150] See also: Remuss, Nina. Responsive Space for Europe. ESPI Report 22. Feb. 2010.
[151] "National Space Policy of the United States of America." 28 June 2010 . . . : 10.
[152] Ibid: 12.
[153] "Briefing by Senior Administration Officials . . . ."
[154] "U.S. National Space Policy." 31 Aug. 2006. The White House 20 Dec. 2010. http://www.whitehouse.gov/sites/default/files/microsites/ostp/national-space-policy-2006.pdf. "Fact Sheet: National Space Policy." 19 Sept. 1996. NASA 20 Dec. 2010. http://history.nasa.gov/appf2.pdf.
[155] "U.S. National Space Policy." 31 Aug. 2006: 7. "Fact Sheet: National Space Policy." 19 Sept. 1996: 5.
[156] "U.S. National Space Policy." 31 Aug. 2006: 9.
[157] "A Bold new Approach for Space Exploration and Discovery." 1 Feb. 2010. The White House 20 Dec. 2010. http://www.whitehouse.gov/files/documents/ostp/press_release_files/NASA%20OSTP%20Joint%20Fact%20Sheet%20FINAL%202020.pdf.
[158] "Briefing by Senior Administration Officials . . . ."

[159] "Obama's National Space Policy Authorizes Use of Foreign GNSS Services to Strengthen GPS." Inside GNSS July/Aug. 2010. 20 Dec. 2010. http://www.insidegnss.com/node/2150.
[160] "Briefing by Senior Administration Officials . . . ."
[161] "Deputy Assistant Secretary Frank Rose on U.S. National Space Policy." 13 July 2010. U.S. Mission to the U.N. in Geneva 20 Dec. 2010. http://geneva.usmission.gov/2010/07/13/das-frank-rose-space-policy.
[162] "National Space Policy of the United States of America." 28 June 2010: 14. "U.S. National Space Policy." 31 Aug. 2006: 4. "Fact Sheet: National Space Policy." 19 Sept. 1996: 4.
[163] "National Space Policy of the United States of America." 28 June 2010: 9. "Deputy Assistant Secretary Frank Rose . . . ."
[164] "Ask a Diplomat: The International Elements of the National Space Policy." 29 June 2010. U.S. Department of State 20 Dec. 2010. http://www.state.gov/t/avc/rls/143893.htm.
[165] Klamper, Amy. "House Intelligence Committee Weighs in on ITAR Reform." Space News 20 July 2009: 16.
[166] Klamper, Amy. "Obama ITAR Reform Could Move Satellites Back to Commerce." Space News 6 July 2009: 6.
[167] Ibid.
[168] Klamper, Amy. "House Intelligence Committee Weighs in on ITAR Reform." Space News 20 July 2009: 16.
[169] Klamper, Amy. "Official Reaffirms White House Support for ITAR Reform." Space News 14 Sept. 2009: 20.
[170] Muradian, Vago and Bennett, John D. "Pentagon Acquisition Chief Says Space Industrial Base May Warrant Protection." Space News 14 Sept. 2009: 18.
[171] Klamper, Amy. "AIA Presses Obama on Export-Control Reform." Space News 7 Dec. 2009: 13.
[172] Klamper, Amy. "Obama Repeats Call for Export Control Reform in State of Union." Space News 1 Feb. 2010: 12.
[173] "Major Provisions of the Federal Space Programme of the Russian Federation for 2006-2015" of 22 Oct. 2005. Russian Federal Space Agency (Roscosmos). 30 Apr. 2010. http://www.federalspace.ru/main.php?id=85.
[174] "The Outlines of the Russian Federation Policy in the Field of Space Activities for the Period Until 2020 and Further Perspective" Apr. 2008. Russian Federation Security Council. 30 Apr. 2010 (Available in Russian). http://www.scrf.gov.ru/documents/96.html.
[175] "Russia's Space Industry Output to Grow 18% in 2009-Putin." 30 Nov. 2009. RIANovosti 30 May 2010. http://en.rian.ru/russia/20091130/157042100.html.
[176] "Basic Plan for Space Policy" of 2 June 2009. Strategic Headquarters for Space Policy. 30 Apr. 2010. http://www.kantei.go.jp/jp/singi/utyuu/basic_plan.pdf.
[177] Ibid: 6–15.
[178] Ibid.
[179] Kallender-Umezu, Paul. "Amid Shift in Power, Japan Seeks Space Budget Hike." Space News 7 Sept. 2009: 14.
[180] Ibid.
[181] Kallender-Umezu, Paul. "Japan Urged To Break up JAXA and Establish New Space Agency." Space News 3 May 2010: 10.
[182] Ferster, Warren and Amy Klamper. "China, U.S. Put Spaceflight Cooperation Talks on Agenda." Space News 23 Nov. 2009: 12.
[183] Klamper, Amy. "Official Details 11-year Path to Developing China's Own Space Station." Space News 19 Apr. 2010: 6.
[184] Jayaraman, K.S."ISRO Budget Receives 40 Percent Increase from 2008." Space News 13 July 2009: 12.
[185] "Space Activities in Africa." 27 Dec. 2009. Space Issues. 25 May 2010. http://www.space-issues.com/blog/?p=316.

186 "Nigerian Space Agency signs MoU with Infoterra." 17 July 2009. Infoterra. 25 May 2010. http://www.infoterra.de/detailview/date/2009/07/17/nigerian-space-agency-signs-mou-with-infoterra-gmbh.html.
187 Wolstenholme, Robin. "Nigeria completes milestone in space." 07 Oct. 2010. Surrey. 25 May 2010. http://blog.sstl.co.uk/archives/254-Nigeria-completes-milestone-in-space.html.
188 Fairly, Peter. "South Africa's Polar-Orbiting Ploughshares – A National Space Agency could help it become a regional powerhouse in Earth observation." 18 Jan 2010. earthzine. 25 May 2010. http://www.earthzine.org/2010/01/18/south-africas-polar-orbiting-ploughsharesa-national-space-agency-could-help-it-become-a-regional-powerhouse-in-earth-observation.
189 "SA satellite finally lifts off." 18 Sep. 2009. News24. 25 May 2010. http://www.news24.com/SciTech/News/SA-satellite-finally-lifts-off-20090918.
190 "Declining Terminal Sales Harm Thaicom Revenue." Space News 1 Mar. 2010: 9.
191 De Selding, Peter B. "Thailand's Political Unrest has Thaicom Caught in the Middle." Space News 3 May 2010: 6.
192 De Selding, Peter B. "France Seeks ITU Help To Halt Satellite Signal Jamming by Iran." Space News 11 Jan. 2010: 4.
193 "Brazil, Belgium Sign Space Cooperation Agreement." 08 Oct. 2009. Space Mart. 28 May 2010. http://www.spacemart.com/reports/Brazil_Belgium_Sign_Space_Cooperation_Agreement_999.html.
194 "Brazil, China To Postpone Joint Satellite Launching To 2011." 16 Feb. 2010. Space Travel. 28 May 2010. http://www.space-travel.com/reports/Brazil_China_To_Postpone_Joint_Satellite_Launching_To_2011_999.html.
195 The Space Foundation. The Space Report 2010. The Space Foundation: Colorado Springs, 2010: 30.
196 Figures in this section are based on Euroconsult data.
197 The data used is the nominal GDP converted to current U.S. dollars using the official exchange rates as indicated by the International Monetary Fund.
198 "Gross domestic product, current prices." April 2010. International Monetary Fund. 04 May 2010. http://j.mp/aL2sTZ.
199 "World Population Prospects: The 2008 Revision." United Nations. Department of Economic and Social Affairs – Population Division. 04 May 2010. http://www.un.org/esa/population/publications/wpp2008/wpp2008_text_tables.pdf: 15.
200 "World Population Prospects: The 2008 Revision." United Nations. Department of Economic and Social Affairs – Population Division. 04 May 2010. http://www.un.org/esa/population/publications/wpp2008/wpp2008_text_tables.pdf: 15.
201 "World Population Prospects: The 2008 Revision." United Nations. Department of Economic and Social Affairs – Population Division. 04 May 2010. http://www.un.org/esa/population/publications/wpp2008/wpp2008_text_tables.pdf: 15.
202 De Selding, Peter B. "Eutelsat Fleet Expansion Fuels Double-Digit Sales Growth." Space News 9 Nov. 2009: 4.
203 "Astrium Sales on the Rise But Profit Margins Decline." Space News 23 Nov. 2009: 9.
204 De Selding, Peter B. "Astrium Revenue Soars, Margins Dip." Space News 3 Aug. 2009: 17.
205 De Selding, Peter B. "Eutelsat's 7.2 Percent Revenue Growth Exceeds Forecast." Space News 3 Aug. 2009: 6.
206 De Selding, Peter B. "SES Sticking with Growth Projections Despite Soft Spots." Space News 3 Aug. 2009: 6.
207 "Thales Orders Rose, Revenue Remained Flat." Space News 3 Aug.2009: 9.
208 "India Reports Big Rise in Satellite TV Subscribers." Space News 20 July 2009: 9.
209 De Selding, Peter B. "Avanti Finances Improve Inspite of Tardy Satellite." Space News 28 Sept. 2009: 12.
210 Werner, Debra. "FCC: Stimulus Not Enough for Broadband Plan." Space News 1 June 2009: 14.
211 De Selding, Peter B. "National Export-Credit Agencies Stepping up Satellite Financing." Space News 14 Sept. 2009: 10.

212 De Selding, Peter B. "SES, Eutelsat May be Rethinking Mobile S-Band Services Venture." Space News 10 Aug. 2009: 1.
213 De Selding, Peter B. "Solaris Stymied by Antenna." Space News 14 Sept. 2009: 10.
214 Klamper, Amy. "Report: Commercial Spaceflight Investment on the Rise." Space News 12 Oct. 2009: 14.
215 Brinton, Turner. "Low-Cost Imaging Satellites Encouraged in Defence Bill." Space News 13 July 2009: 7.
216 Brinton, Turner. "NGA to Seek Higher-Resolution Commercial Satellite Imagery." Space News 28 Sept. 2009: 4.
217 De Selding, Peter B. "U.S. Satellite Firms Vie for Broadband Stimulus Funds." Space News 5 Oct. 2009: 5.
218 De Selding, Peter B. "With WildBlue Acquisition, ViaSat Doubles Bet on Satellite Broadband." Space News 5 Oct. 2009: 1. De Selding, Peter B. "Design of Rival's Satellite Has ViaSat Asking Questions." Space News 5 Oct. 2009: 4.
219 Brinton, Turner. "U.S. To Consolidate, Modify Bandwidth Buying Programs." Space News 10 Aug. 2009: 4.
220 Brinton, Turner. "U.S. Loosens Restrictions on Commercial Radar Satellites." Space News 12 Oct. 2009: 4.
221 "Rosy Revenue Forecast for Mobile Satellite Services." Space News 12 Oct. 2008: 8.
222 De Selding, Peter B. "$738M Financing Package Gives Globalstar a New Lease on Life." Space News 6 July 2009: 1.
223 De Selding, Peter B. "105,000 Globalstar Telephone Subscribers Await Return of Full Service." Space News 10 May 2010: 5. De Selding, Peter B. "Globalstar Predicts Return to Growth Once New Satellites Launch." Space News 13 July 2009: 10.
224 "Thuraya Escalates Fight for Satellite Phone Customers." Space News 3 May 2010: 9.
225 The Space Foundation. The Space Report 2010. The Space Foundation: Colorado Springs, 2010: 31.
226 Federal Aviation Administration. Commercial Space Transportation: 2009 Year in Review. Washington DC: FAA, Jan. 2010. http://www.faa.gov/about/office_org/headquarters_offices/ast/media/year_in_review_2009.pdf.
227 "TomTom Annual Report and Accounts 2009." 11 May 2010. http://ar2009.tomtom.com/pdf/TomTom_AR09.pdf.
228 "Garmin Annual Report 2009." 11 May 2010. http://www8.garmin.com/aboutGarmin/invRelations/reports/2009_Annual_report.pdf.
229 "Recent High Profits Attracting New Underwriters." Space News 8 Mar. 2010: 15. "Ariane 5 Lofts Amazonas-2, COMSATBw-1 Satellites." Space News 5 Oct. 2009: 3. De Selding, Peter B. "2009 Shaping up as Profitable Year for Insurers." Space News 07 Sep. 2009: 10.
230 De Selding, Peter B. "Yahsat Buys Shariah-compliant Satellite Insurance." Space News 12 Apr. 2010: 29.
231 De Selding, Peter B. "Spot Commits to New Satellites, But Funding Questions Remain." Space News 15 June 2009: 1.
232 De Selding, Peter B. "Astrium Services Cleared To Buy New Spot Satellite." Space News 22 June 2009: 13.
233 De Selding, Peter B. "Italian Company Formed To Market Cosmo-SkyMed Imagery." Space News 13 July 2009: 6.
234 "Italy's e-Geos Lands GMES Radar, Optical Imagery Deal." Space News 23 Nov. 2009: 8.
235 De Selding, Peter B. "Thales Alenia To Build ESA's Expert Test Vehicle." Space News 27 July 2009: 11.
236 Singer, Jeremy. "NATO Readies Competition for Satcom Services." Space News 2 Nov. 2009: 12.
237 De Selding, Peter B. "SES Selects Astrium to Build Four Direct Broadcast Satellites." Space News 30 Nov. 2009: 4. De Selding, Peter B. "Four Satellite Order Bolsters SES's Direct Broadcast Presence." Space News 7 Dec. 2009: 6.

[238] Ibid.
[239] "Avanti Gets Export-Credit Backing for Ka-band Satellite." Space News 4 Jan. 2010: 8.
[240] "SES Taps Coface Financing For Four Astra Satellites." Space News 4 Jan. 2010: 9.
[241] "Astrium Positioned for Substantial Galileo Work." Space News 1 Feb. 2010: 8.
[242] De Selding, Peter B. "Thales Alenia Gets $384 Million CNES Contract for Athena-Fidus." Space News 15 Feb. 2010: 12.
[243] Brinton, Turner. "Firm Pitches Low-cost Replacements for Ikonos, QuickBird." Space News 15 Feb. 2010: 13.
[244] De Selding, Peter B. "Booming Eutelsat Raises 3-year Growth Forecast, Eyes Asia." Space News 22 Feb. 2010: 6.
[245] "Dutch Space Will Provide Solar Arrays for Sentinel." Space News 1 Mar. 2010: 8.
[246] "Thales Alenia Space Gets Jason-3 Satellite Contract." Space News 1 Mar. 2010: 8.
[247] De Selding, Peter B. "Delays Continue To Affect Vega, European Soyuz Programs." Space News 3 May 2010: 6.
[248] De Selding, Peter B. "Sea Launch Expects Dept Refinancing To Close in June." Space News 1 June 2009: 6.
[249] De Selding, Peter B. "$2Billion Dept Forces Sea Launch into Bankruptcy." Space News 29 June 2009: 6. De Selding, Peter B. "Failed Negotiations Over Arbitration Payment Was Last Straw for Sea Launch." Space News 29 June 2009: 6.
[250] De Selding, Peter B. "Sea Launch Bankruptcy Stokes Fears of Rising Launch Prices." Space News 29 June 2009: 1.
[251] De Selding, Peter B. "Sea Launch Promised Future Contracts." Space News 14 Sept. 2009: 12. Also see below, page.
[252] De Selding Peter B. "Boeing Seeks To Reduce Losses From Sea Launch Bankruptcy." Space News 2 Nov. 2009: 4.
[253] "U.S. Air Force Sows Seeds For Next-Gen Space Fence." Space News 15 June 2009: 3.
[254] Ibid.
[255] Matthews, William. "TacSat-3 Earth Observing Satellite Puts Hyperspectral Imagery Analysis in Space." Space News 22 June 2009: 16.
[256] Brinton, Turner. "Missile Tracking Demonstration Satellites Readied for Launch." Space News 22 June 2009: 12.
[257] De Selding, Peter B. "SES, Intelsat Asking Lawmakers to Rethink Launch Ban on China, India." Space News 3 Aug. 2009: 1.
[258] De Selding, Peter B. "Satellite Firms Tap Warner in Bid for Wider Access to Launchers." Space News 21 Sept. 2009: 5.
[259] "Raytheon Unveils Bigger, Better Infrared Detector." Space News 24 Aug. 2009: 11.
[260] "Ariane 5 Launcher Puts Terrestar-1 into Orbit." Space news 6 July 2009: 3.
[261] De Selding, Peter B. "TerreStar-1 Launch Slips as Insurers Request More Info." Space News, 15 June 2009: 4.
[262] "Ariane 5 Launcher Puts Terrestar-1 into Orbit." Space news 6 July 2009: 3.
[263] Klamper, Amy. "SpaceX: Dragon Prototype Will Fly on First Falcon 9." Space news 28 Sept. 2009: 7.
[264] Brinton, Turner. "U.S. Air Force Studying Wider Use of MDA Radars for Space Tracking." Space News 2 Nov. 2009: 16.
[265] Opall-Rome, Barbara. "Israeli Experts: Arrow-3 Could Be Adapted for Anti-Satellite Role." Space News 9 Nov. 2009: 16.
[266] Iannotta, Ben. "Pentagon Eyes Satellites Equipped with Joystick-controlled Video Cameras." Space News 23 Nov. 2009: 12.
[267] De Selding, Peter B. "Inmarsat Acquires Segovia to Boost Government Business." Space News 30 Nov. 2009: 10.
[268] Brinton, Turner. "Cisco Views IRIS Experiment as market Primer for Space Routers." Space news 30 Nov. 2009: 12.

[269] "SpaceShipTwo Rolls Out at Mojave Air and Space Port." Space News 14 Dec. 2009: 8.
[270] Brinton, Turner. "2010 Shaping Up as Busiest Year Yet for EELV." Space News 11 Jan. 2010: 15.
[271] De Selding, Peter B. "With U.S. Contracts Delayed, DigitalGlobe Looks Elsewhere for 2010 Growth." Space News 1 Mar. 2010: 10.
[272] Ibid.
[273] Ibid.
[274] "Russia Delays Phobos-Grunt Mars Mission Until 2011." Space News 28 Sept. 2009: 9.
[275] De Selding, Peter B. "National Export-Credit Agencies Stepping up Satellite Financing." Space News 14 Sept. 2009: 10.
[276] De Selding, Peter B. "Dnepr Launch Stalled by Rocket-Debris Cleanup Negotiations." Space News 26 Oct. 2009: 11.
[277] "Russian Proton Lofts Trio of Glonass Navigation Satellites." Space News 8 Mar. 2010: 8.
[278] Klamper, Amy. "Japan's HTV Delivered U.S. Navy Experiments to Station." Space News 28 Sept. 2009: 14.
[279] "Japanese Cargo Tug Burns Up Over Pacific Ocean." Space News 9 Nov. 2009: 8.
[280] "H-2A Rocket Lofts Japanese Reconnaissance Satellite." Space News 7 Dec. 2009: 8.
[281] De Selding, Peter B. "With High Launch Rate in Tow, China Great Wall Courts Western Business." Space News 19 Apr. 2010: 6.
[282] Federal Aviation Administration. Commercial Space Transportation: 2009 Year in Review. Washington DC: FAA, Jan. 2010. http://www.faa.gov/about/office_org/headquarters_offices/ast/media/year_in_ review_2009.pdf.
[283] Ibid.
[284] Ibid. Milbank Space Smart. Space Business Review. Los Angeles: Milbank, Dec. 2009. http://www.milbank. com/NR/rdonlyres/4429D43E-4A1A-4F8C-B129-067607E6530A/0/December_2009_Space_Business_ Review.pdf.
[285] Federal Aviation Administration. Commercial Space Transportation: 2009 Year in Review. Washington DC: FAA, Jan. 2010. http://www.faa.gov/about/office_org/headquarters_offices/ast/media/year_in_review_2009.pdf.
[286] Federal Aviation Administration. Commercial Space Transportation: 2009 Year in Review. Washington DC: FAA, Jan. 2010. http://www.faa.gov/about/office_org/headquarters_offices/ast/media/year_in_review_2009.pdf.
[287] Ibid.
[288] Federal Aviation Administration. Commercial Space Transportation: 2009 Year in Review. Washington DC: FAA, Jan. 2010. http://www.faa.gov/about/office_org/headquarters_offices/ast/media/year_in_review_2009.pdf.
[289] According to the mass classification of payloads of the FAA: Micro: 0 to 91 kg. (0 to 200 lbs.); Small: 92 to 907 kg. (201 to 2000 lbs.); Medium: 908 to 2268 kg. (2001 to 5000 lbs.); Intermediate: 2269 to 4536 kg. (5001 to 10000 lbs.): Large: 4537 to 9072 kg. (10001 to 20000 lbs.); Heavy: over 9072 kg. (20000 lbs.). Federal Aviation Administration. Semi-Annual Launch Report: Second Half of 2009. Washington DC: FAA, 2009. http://www.faa.gov/about/office_org/headquarters_offices/ast/media/10998.pdf.
[290] Ibid.
[291] Federal Aviation Administration. Commercial Space Transportation: 2009 Year in Review. Washington DC: FAA, Jan. 2010. http://www.faa.gov/about/office_org/headquarters_offices/ast/media/year_in_review_2009.pdf.
[292] Ibid.
[293] Arianespace. Company Profile. Paris: Arianespace, Jan. 2010. http://www.arianespace.com/Press-center/pdf/Company_Profile_EN_2010.pdf.
[294] "Launch Log" 18 Dec. 2009. Arianespace 19 Mar. 2010. http://www.arianespace.com/news-launch-logs/2000-2010.asp.
[295] De Selding, Peter B. "Commercial Launch Providers Looking Ahead to Busier 2009." Space News 5 Jan. 2009: 11.

[296] Arianespace. Company Profile. Paris: Arianespace, Jan. 2010. http://www.arianespace.com/Press-center/pdf/Company_Profile_EN_2010.pdf.

[297] Federal Aviation Administration. Commercial Space Transportation: 2009 Year in Review. Washington DC: FAA, Jan. 2010. http://www.faa.gov/about/office_org/headquarters_offices/ast/media/year_in_review_2009.pdf.

[298] Ibid.

[299] Ibid.

[300] Foust, Jeff. "Satellites, launches, and the recession" 30 Mar. 2009. The Space Review 19 Mar. 2010. http://www.thespacereview.com/article/1341/1.

[301] Federal Aviation Administration. Commercial Space Transportation: 2009 Year in Review. Washington DC: FAA, Jan. 2010. "Year in Review." Space News 14 Dec. 2009: 22.

[302] "Sea Launch Receives Final Approval for DIP Financing" 3 Dec. 2009. Sea Launch Company 19 Mar. 2010. http://www.boeing.com/special/sea-launch/news_releases/2009/nr_091203.html.

[303] Federal Aviation Administration. Commercial Space Transportation: 2009 Year in Review. Washington DC: FAA, Jan. 2010. http://www.faa.gov/about/office_org/headquarters_offices/ast/media/year_in_review_2009.pdf. "Launch Archives" 12 Feb. 2010. ILS Launch 19 Mar. 2010. http://www.ilslaunch.com/launch-archives/.

[304] Federal Aviation Administration. Commercial Space Transportation: 2009 Year in Review. Washington DC: FAA, Jan. 2010. http://www.faa.gov/about/office_org/headquarters_offices/ast/media/year_in_review_2009.pdf.

[305] Ibid. De Selding, Peter. "Arianespace 2009 Revenue Boosted by Higher Launch Rate" 5 Jan. 2010. Space News 19 Mar 2010. http://www.spacenews.com/launch/100105-arianespace-revenue-boosted-launch-prices.html.

[306] De Selding, Peter B. "With Sea Launch Sidelined by Debt, Satellite Operators Want More Choices." Space News 4 Jan. 2010: 11.

[307] Ibid.

[308] "Galileo IOV launch services contract signed" 16 June 2009. European Space Agency 19 Mar. 2010. http://www.esa.int/esaCP/SEMCUA3XTVF_index_0.html.

[309] Arianespace. Company Profile. Paris: Arianespace, Jan. 2010. http://www.arianespace.com/Press-center/pdf/Company_Profile_EN_2010.pdf.

[310] "Arianespace orders 35 Ariane 5 ECA launchers from Astrium" 2 Feb. 2009. Arianespace 19 Mar. 2010. http://www.arianespace.com/news-press-release/2009/02-02-09-launcher-order-from-astrium.asp.

[311] De Selding, Peter B. "With Sea Launch Sidelined by Debt, Satellite Operators Want More Choices." Space News 4 Jan. 2010: 11. Kaiser, Dustin. "The Impact of the Sea Launch Bankruptcy On The Launch Industry" Oct. 2009. Satmagazine 19 Mar. 2010. http://www.satmagazine.com/cgi-bin/display_article.cgi?number=1774104459. "Sea Launch Preparing for Bankruptcy Exit" 12 Mar. 2010. Space News 19 Mar. 2010. http://www.spacenews.com/launch/100312-sea-launch-preparing-bankruptcy-exit.html.

[312] De Selding, Peter B. "With Sea Launch Sidelined by Debt, Satellite Operators Want More Choices." Space News 4 Jan. 2010: 11.

[313] Ibid.

[314] de Selding, Peter. "Arianespace 2009 Revenue Boosted by Higher Launch Rate" 5 Jan. 2010. Space News 19 Mar. 2010. http://www.spacenews.com/launch/100105-arianespace-revenue-boosted-launch-prices.html. Foust, Jeff. "How competitive is commercial launch?" 19 Oct. 2009. The Space Review 19 Mar. 2009. http://www.thespacereview.com/article/1493/1.

[315] Payloads are assigned to the nation that commissioned them, not according to the nationality of the manufacturer. Federal Aviation Administration. Commercial Space Transportation: 2009 Year in Review. Washington DC: FAA, Jan. 2010. http://www.faa.gov/about/office_org/headquarters_offices/ast/media/year_in_review_2009.pdf.

[316] Futron Corporation. Satellite Manufacturing Report. Wisconsin: Futron, Jan. 2010. http://www.futron. com/pdf/friends_of_futron_reports/satellite_manufacturing_reports/FutronSM2010-01.pdf.

[317] Ibid.
[318] Ibid.
[319] De Selding, Peter B. "Export-Credit Financing Helps Boost 2009 Satellite Orders." Space News 4 Jan. 2010: 12. Futron Corporation. Satellite Orders Report. Wisconsin: Futron, Dec. 2009. http://www.futron.com/pdf/friends_of_futron_reports/satellite_manufacturing_reports/FutronSM2009-EOY.pdf.
[320] "OHB und Hispasat unterzeichnen Vertrag über Bau und Test des Small Geo-Satelliten 'Hispasat AG1'" 18 June 2009. OHB Technology AG 19 Mar. 2010. http://www.ohb-technology.de/media-relations/pressemitteilungen-detail/items/ohb-und-hispasat-unterzeichnen-vertrag-ueber-bau-und-test-des-small-geo-satelliten-hispasat-ag1.html.
[321] "OHB-System und ESA unterzeichnen Galileo-Vertrag" 26 Jan. 2010. OHB Technology AG 19 Mar. 2010. http://www.ohb-technology.de/media-relations/pressemitteilungen-detail/items/ohb-system-und-esa-unterzeichnen-galileo-vertrag.769.html.
[322] "MDA contracted to deliver the national communications satellite system of Ukraine" 15 Dec. 2009. MacDonald Dettweiler and Associates 19 Mar. 2010. http://www.mdacorporation.com/corporate/news/index.cfm?year=2009.
[323] "ISS Reshetnev Names Express AM5, AM6 Partners; RSCC Reveals Plans for New Fleet" 17 Aug. 2009. Satellite Today 19 Mar. 2010. http://www.satellitetoday.com/twitter/ISS-Reshetnev-Names-Express-AM5-AM6-Partners-RSCC-Reveals-Plans-for-New-Fleet_31855.html. "New international contract of the JSC 'ISS'" 5 Mar. 2009. Reshetnev 19 Mar. 2010. http://www.iss-reshetnev.com/?cid=news&nid=105. De Selding, Peter B. "Export-Credit Financing Helps Boost 2009 Satellite Orders." Space News 4 Jan. 2010: 12.
[324] "CGWIC Signs LaoSat-1 Communications Satellite Contract" 25 Feb. 2010. China Great Wall Industry Corporation 19 Mar. 2010. http://www.cgwic.com/news/2010/0225_LaoSat-1.html. Futron Corporation. Satellite Manufacturing Report. Wisconsin: Futron, Jan. 2010. http://www.futron.com/pdf/friends_of_futron_reports/satellite_manufacturing_reports/FutronSM2010-01.pdf.
[325] De Selding, Peter B. "Palapa-D to be Salvaged After Being Launched into Wrong Orbit." Space News 7 Sept. 2009: 10.
[326] De Selding, Peter B. "Export-Credit Financing Helps Boost 2009 Satellite Orders." Space News 4 Jan. 2010: 12.
[327] Ibid. Futron Corporation. Satellite Orders Report. Wisconsin: Futron, Dec. 2009. http://www.futron.com/pdf/friends_of_futron_reports/satellite_manufacturing_reports/FutronSM2009-EOY.pdf. "Boeing Signs 4-Satellite Contract With Intelsat" 16 July 2009. Boeing 19 Mar. 2010. http://boeing. mediaroom.com/index.php?s=43&item=748. "2009 – ein glänzendes Jahr für Astrium" 19 Jan. 2010. EADS 19 Mar. 2010. http://www.eads.net/1024/de/pressdb/pressdb/Astrium/20100119_astrium_2009.html.
[328] De Selding, Peter B. "Orbital Bullish on U.S. National Security Satellite Market" 19 Feb. 2010. Space News 19 Mar. 2009. http://www.spacenews.com/military/100219-orbital-bullish-national-security-satellite-market.html.
[329] De Selding, Peter B. "Export-Credit Financing Helps Boost 2009 Satellite Orders." Space News 4 Jan. 2010: 12.
[330] State of the Satellite Industry Report 2010. Washington: FUTRON,SIA. June 2010.
[331] De Selding, Peter. "The List: Top fixed Satellite Service Operators." Spacenews 5 July 2010.
[332] Ibidem.
[333] Figures are derived from Euroconsult data.
[334] De Selding, Peter B. "Europe Keeping Increasingly Capable Eye on Orbital Debris." Space News 26 Apr. 2010: 6.
[335] Ibid.
[336] "MUSIS Ground System Deal Teeters on Edge of Collapse." Space News 26 Apr. 2010:6.
[337] "OHB To Support Franco-German imagery Sharing." Space News 10 May 2010: 9.
[338] De Selding, Peter B. "France Seeks Military Space Investment Partners." Space News 30 Nov. 2009: 6.

[339] De Selding, Peter B. "British Military's Satellite Capacity Nearing Its Limit." Space News 23 Nov. 2009: 11.
[340] "Extension Prompts Paradigm To Order 4th Skynet 5." Space News 15 Mar. 2010: 8.
[341] De Selding, Peter B. "Thales Alenia Space, Telespazio To Build Sicral 2 Satellite." Space News 10 May 2010: 5.
[342] De Selding, Peter B. "Europe Keeping Increasingly Capable Eye on Orbital Debris." Space News 26 Apr. 2010: 6.
[343] De Selding, Peter B. "European Space, Defence Agencies Study New Unmanned Aerial Vehicle Service." Space News 22 Feb. 2010: 12. See also: Roma, Alfredo, Sanchez-Aranzamendi, Matxalen and Schrogl, Kai-Uwe. Opening Airspace for UAS. A Regulatory Framework to Introduce Unmanned Aircraft Systems into Civilian Airspace. ESPI Report 31. Vienna, March 2011.
[344] "ESA Eyes Military Addition To Civil Satellite Network." Space News 26 Apr. 2010: 6.
[345] Bennett, John T. "Small Satellites Attract Interest from Large U.S. Companies." Space News 27 July 2009: A4.
[346] Klamper, Amy. "NRO Embraces Cubesats for Testing Advanced Technologies." Space News 10 Aug. 2009: 11.
[347] "SMC Looks to Demonstrate Military Utility of Cubesats." Space News 23 Nov. 2009: 3.
[348] "Missile Defence Agency Eyes Small Tracking Sats." Space News 24 Aug. 2009: 3.
[349] Brinton, Turner. "Low-Cost Imaging Satellites Encouraged in Defence Bill." Space News 13 July 2009: 7.
[350] "Cancelled KEI Program Is Resurrected in U.S. House." Space News 3 Aug. 2009: 3.
[351] "War Game Shows Need for Better Space Surveillance." Space News 24 Aug. 2009: 3.
[352] Brinton, Turner. "NGA Solicits Proposals for Commercial Radar Imagery." Space News 14 Sept. 2009: 4.
[3523] Ibid.
[354] Brinton, Turner. "NRO Director Defends Plan for Electro-Optical Spy Satellites." Space News 12 Oct. 2009: 1.
[355] "DARPA Seeks Ideas for Orbital Debris Removal." Space News 21 Sept. 2009: 3.
[356] David, Leonard. "Orbital Debris Cleanup Takes Centre Stage." Space News 28 Sept. 2009: 15.
[357] Brinton, Turner. "Air Force Weighing Options For Space Weather Data." Space News 5 Oct. 2009: 16.
[358] Sunk-Ki, Jung. "U.S. Promises Missile Shield for South Korea." Space News 2 Nov. 2009: 14.
[359] "DARPA Plan Would Connect Satellites to the Internet." Space News 9 Nov. 2009: 8.
[360] United States Department of Commerce, Federal Business Opportunities. "Persistent Broadband Ground Connectivity for Spacecraft in Low Earth Orbit (LEO)." Sol. No DARPA-SN-10-28, 3 Feb. 2010. https://www.fbo.gov/index?s=opportunity&mode=form&id=08e21adc3b19b1a900dd2b8c3643cff9&tab=core&_cview=0.
[361] Ibid.
[362] See below.
[363] "Lockheed Martin Launches Secret Rocketplane Prototype." Space News 19 Oct. 2009: 8.
[364] Brinton, Turner. "Defence Spending Act Pushes Smarter Investments in Space." Space News 4 Jan. 2010: 4.
[365] Brinton, Turner. "NGA Awards Three Contracts for Radar Satellite Data." Space News 4 Jan. 2010: 5.
[366] Brinton, Turner. "U.S. Navy To Rely on Netted Iridium Service as Gap-Filler." Space News 11 Jan. 2010: 16.
[367] Ibid.
[368] Brinton, Turner. "Pentagon Urged To Budget for Commercial Satcom." Space News 11 Jan. 2010: 17.
[369] "Cisco Completes Test of Orbiting Internet Router." Space News 18 Jan. 2010: 9.
[370] Brinton, Turner. "Intelsat Nabs Big U.S. Navy Satcom Contract." Space News 1 Feb. 2010: 6.

[371] Brinton, Turner. "Testing Plans Uncertain for Missile Tracking Satellites." Space News 1 Feb. 2010: 12.
[372] Brinton, Turner. "U.S. Air Force Scales Back Missile Warning Technology Program." Space News 8 Feb. 2010: 12.
[373] Ibid.
[374] Brinton, Turner. "MDA Budget Request Reflects New European Strategy, More Emphasis on Testing." Space News 8 Feb. 2010: 12.
[375] Brinton, Turner. "Parts Problems Still Plaguing U.S. Space and Missile Hardware." Space News 15 Feb. 2010: 12.
[376] Brinton, Turner. "Airborne Laser Goes Two for Three in 1st Intercept Tests." Space News 15 Feb. 2010:14.
[377] Brinton, Turner. "First Operational ORS Satellite Readied for Payload Integration." Space News 22 Feb. 2010: 13.
[378] Brinton, Turner. "NRO Chief Aims To Restore Technology Development Funding." Space News 19 Apr. 2010: 7.
[379] Brinton, Turner. "U.S. Air Force Official Touts Space Plane Applications." Space News 26 Apr. 2010: 11.
[380] "Russia Approves New Defence Industry Policy." 19 Mar. 2010. RIANovosti 30 May 2010. http://en.rian.ru/russia/20100319/158253423.html.
[381] "Russia to Build Submarine-Detecting Satellite." 15 Apr. 2010. RIANovosti 30 May 2010. http://en.rian.ru/russia/20100415/158597419.html.
[382] "Russia Launches Military Satellite." 16 Apr. 2010. RIANovosti May 2010. http://en.rian.ru/russia/20100416/158616394.html. "Russia's Cosmos-3MSpace Carrier Orbits Military Satellite." 27 Apr. 2010. RIANovosti 30 May 2010. http://en.rian.ru/science/20100427/158762058.html.
[383] "Russian Military Lays Hope on New Angara Carrier Rockets." 15 Jan. 2010. RIANovosti 30 May 2010. http://en.rian.ru/russia/20100115/157550645.html.
[384] "Russia to Have Full Gamut of Air, Outer Space Defences by 2015." 2 Apr. 2010. RIANovosti 30 May 2010. http://en.rian.ru/mlitary_news/20100402/158414990.html. "Russia's Space Defences in Shambles-Experts." 13 may 2010. RIANovosti 30 May 2010. http://en.rian.ru/mlitary_news/20100513/159003853.html.
[385] "Russian Officer Says Developing New Weapon for Space Defence." 15 May 2010. RIANovosti 30 May 2010. http://en.rian.ru/russia/20100515/159029349.html.
[386] "Russia Has No Plans to Deploy Weapons in Space-Top Brass." 9 Apr. 2010. RIANovosti 30 May 2010. http://en.rian.ru/mlitary_news/20100409/158497786.html.
[387] "The Council on Security and Defence Capabilities Report." Aug. 2009. Cabinet Office of the Prime Minister of Japan (Kantei) 30 May 2010. http://www.kantei.go.jp/jp/singi/ampobouei2/090928houkoku_e.pdf.
[388] Sawako, Maeda. "Transformation of Japanese Space Policy: From the 'Peaceful Use of Space' to 'The Basic Law on Space'." The Asia-Pacific Journal: Japan Focus Vol. 44-1-09 2 Nov. 2009. 30 May 2010. http://www.japanfocus.org/-Maeda-Sawako/3243. Clark, Stephen. "Japan Launches Spy Satellite Under Veil of Secrecy." Space Flight Now 28 Nov. 2009. 30 May 2010. http://spaceflightnow.com/h2a/f16/.
[389] "Quasi-Zenith Satellite-1 Michibiki." Japan Aerospace Exploration Agency (JAXA) 30 May 2010. http://www.jaxa.jp/projects/sat/qzss/index_e.html.
[390] Sawako, Maeda. "Transformation of Japanese Space Policy: From the 'Peaceful Use of Space' to 'The Basic Law on Space'." The Asia-Pacific Journal: Japan Focus Vol. 44-1-09 2 Nov. 2009. 30 May 2010. http://www.japanfocus.org/-Maeda-Sawako/3243.
[391] "Japan Scraps GX Rocket Development Project." I StockAnalyst 16 Dec. 2009. 30 May 2010. http://www.istockanalyst.com/article/viewiStockNews/articleid/3716870.
[392] "China to Launch Module for Future Space Station in 2011." 3 Mar. 2010. RIANovosti 30 May 2010. http://en.rian.ru/world/20100303/158080728.html.
[393] Euroconsult data.

[394] "India Developing Means to Destroy Satellites." Space News 11 Jan. 2010: 9.
[395] Euroconsult Research Report, Government Space Markets, World Prospects to 2020. Paris: Euroconsult, 2010.
[396] http://www.spacenews.com/earth_observation/082610canadian-leader-endorses-radarsat-constellation.html.
[397] De Selding, Peter B. "Canadian Radarsat Constellation To Get $374 Million Cash Infusion." Spacenews 26 Aug. 2010. http://www.spacenews.com/earth_observation/082610canadian-leader-endorses-radarsat-constellation.html.
[398] De Selding, Peter B. "Iran Says Launch Puts Biological Capsule in Orbit." Spacenews 3 Feb. 2010. http://www.spacenews.com/earth_observation/082610canadian-leader-endorses-radarsat-constellation.html.
[399] Euroconsult Research Report, Government Space Markets, World Prospects to 2020. Paris: Euroconsult, 2010.
[400] De Selding, Peter B. "Korean Rocket Fails a Second Time." Spacenews 10 June 2010. http://www.spacenews.com/commentaries/100610-fromwires-korean-rocket-fails.html.
[401] Opall-Rome, Barbara. "Israel Declares Ofeq-9 Reconnaissance Satellite Operational." Spacenews 22 June 2010. http://www.spacenews.com/commentaries/100610-fromwires-korean-rocket-fails.html.
[402] Euroconsult Research Report, Government Space Markets, World Prospects to 2020. Paris: Euroconsult, 2010.
[403] ScienceDaily staff. "The Day The Sun Brought Darkness" Spacenews 19 Mar. 2009. http://www.sciencedaily.com/releases/2009/03/090316144521.htm.
[404] Werner, Debra."Israel Declares Ofeq-9 Reconnaissance Satellite Operational." Spacenews 9 Aug. 2010. http:ATK//www. Proposes Satellite To Fight Space Debris.
[405] De Selding, Peter B."U.S. Open to Ideas for Limiting Space Weapons." Spacenews 14 July 2010. http://www.spacenews.com/policy/100714-open-ideas-for-limiting-space-weapons.html.
[406] Ibidem.
[407] De Selding, Peter B."France Seeks ITU Help To Halt Satellite Signal Jamming by Iran." Spacenews 1 Sept. 2010. http://www.spacenews.com/satellite_telecom/100108-france-seeks-itu-signal-jamming-iran.html.

Developments in space policies, programmes and technologies throughout the world and in Europe

Spyros Pagkratis

1. Space policies and programmes

All major space policy developments worldwide were presented in the previous section of part one, in an attempt to clarify the principal space faring nations' strategies in 2009 and 2010. In the section bellow, there will be a biref discussion on developments in technology related areas, including access to space technologies and policies. The aim of this section is to clarify how the strategies already presented above interact with and influence specific space programmes and related research and development projects.

2. Space transportation

2.1. Europe

The most important development in European space transportation programmes in 2009 and 2010 was related to the deployment of GNSS, for which the European Commission opted for a two launcher scheme to lift the satellites: out of the 14 initial spacecrafts ten would be launched onboard 5 Soyuz rockets, whereas the last four would be carried to orbit on a single Ariane 5 ECA launch. Launching costs however were thought to be considerably higher than expected and their budget override could exceed €1 billion. A third €85 million contract for system support and validation was awarded to ThalesAleniaSpace. Further contracts on ground control and mission segments were anticipated in mid-2010.[408] Eventually, the first three contracts for the Galileo full operational capability system were signed in the premises of ESA's European Space Research and Technology Centre (ESTEC) in the Netherlands, on 27 January 2010.[409]

In a related development, the European Commission also announced on 7 January 2010 that the Galileo satellite navigation system's launching campaign was facing a 59% budget override, mostly due to the Soyuz rocket's unexpectedly high cost. In July 2008 the European Commission had announced that launching the 28 satellite constellation would cost €700 million, or approximately €25 million per spacecraft. The new figure announced in January for lifting only the first ten satellites to orbit mounted to €397 million, or €39.7 million per satellite (with each rocket carrying two spacecraft). This substantial budget increase was attributed by the European Space Agency's (ESA) Director General J.J. Dordain in a 14 January interview to higher Russian launchers' prices, as well as to initially underestimating the costs of adapting the Soyuz rocket for the Galileo mission.[410]

In conclusion, it can be noted the the deployment of the Galileo system is a major driver behind the development of European space transportation capabilities, making full use of both Ariane 5 and the European Soyuz rockets. Since the GNSS deployment campaign is expected to continue and intensify in the medium term, it can be stated that any future developments regarding these two vectors would have to take into account their ability to satisfy the system's operational needs. At the same time, ESA's small launcher VEGA continued its development, with a first flight expected in 2011. This rocket would complement operationally and commercially the heavier launchers, consequently improving the European launcher family's adaptability and affordability. It would also contribute to increasing European independent launch capabilities in the small payload category in an efficient way instead of relying on the use of Russian small launchers, which is the case today.

2.2. United States

The most significant change in U.S. Space transportation programme during the period in question was the adoption by the Obama Administration of a new NASA orientation that included the cancellation of the Constellation programme and the development of private Human space flight services providers as a possible long-term substitute. Indeed, on 1 February 2010, U.S. President Barack Obama surprised policy and industry officials alike with his NASA budget proposal for 2011. The new budget constituted in fact a major change in the U.S. civil space programme and a radical departure from the previous administration's NASA policy. The new policy's corner stone was the cancellation of the Constellation human spaceflight programme, including all of its components (namely the Orion spacecraft, the Ares heavy rocket and the Altair lunar lander). Instead, the new policy called for funding the development of radically new human space flight technologies that would enable NASA to venture not only to the moon, but to more distant destinations as well. While waiting for the development of these new technologies, NASA would increase funding to the private space flight industry, effectively outsourcing the entire U.S. programme for manned space flight to Low Earth Orbit. At the same time, a considerably increased budget was foreseen for the American participation to the International Space Station (ISS) so as to maintain the country's current LEO space flight capabilities. For the purposes of the new civil space programme, the White House requested a slightly increased budget for 2011 ($19 billion, 1.5% higher than in the previous year). Furthermore, total funding through 2015 was now estimated at $45.5 billion or $6 billion more than in the 2010 proposed medium term spending plan.[411]

The key plank of the new policy has been the change in NASA's orientation from a mission-focused organisation to a technology development one. According to this doctrine, NASA would be responsible for related research and development, bringing about ground braking innovations that would ultimately revolutionise human spaceflight. As far as specific missions were concerned, it would only maintain authority over ISS operations and it would continue to promote international cooperation in future manned space exploration projects. All other manned space flights to low Earth orbit short of the ISS would be outsourced to private companies. NASA would thus become the driving motor for the development of a flourishing commercial spaceflight industry in the U.S. The agency would support private sector endeavors in two ways. On the one hand, it would conduct the necessary R&D to develop future space transportation technologies that private companies would then be able to commercialise. On the other hand, NASA itself would be a client for commercial spaceflight services, for instance in order to secure independent access to the ISS. In this fashion, NASA would become the industry's partner and client, instead of its competitor.[412]

It should be noted that the new paradigm for developing human spaceflight in the U.S. bares significant resemblance to the way the country's Department of Defence has handled its relations with the commercial satellite industry. U.S. armed forces have been increasingly relying on a balanced mixture of developing their own space assets while punctually procuring commercial space applications products on an ad hoc basis. This dual approach has allowed for the simultaneous development of dedicated high performance military satellites on the one hand, and for outsourcing demand for less sophisticated products to the commercial satellite communications and Earth observation industry. With the proposed NASA space budget, a similar approach seems to be adopted in regard to the U.S. civil space flight programme as well. As we saw above, according to this new scheme NASA would be conducting the advanced R&D required for manned space exploration missions outside the Earth's orbit, while commercial companies would take over the technologically less demanding task of launching people to Earth orbit.

The proposed Obama space policy seems to have been developed exclusively within the White House Office of Science and Technology, and it was admittedly loosely inspired by the Augustine Report on the future of the Constellation programme. However, although it was endorsed by NASA's Administrator Charles Bolden, it met with considerable resistance from lawmakers and NASA officials alike, including some skepticism expressed from Bolden's predecessor Mike Griffin. Criticism on the proposed course of action focused mainly on three issues. First, the Constellation programme cancellation would nullify the $9 billion investment already made in the project and it would jeopardise the thousands of jobs depending on it. Second, outsourcing human space flight even to LEO could prove a lot more complicated than expected and in the end NASA might not be able to disengage itself from LEO spaceflight completely (for example, it would have to certify that commercial spacecraft are safe for humans). Third, changing the nature of NASA's objectives in relation to human spaceflight (from missions to R&D) and passing some of its mission areas to the private sector could jeopardise U.S. national capabilities in this field during the transition period from one policy model to the other.[413]

Finally, the announcement of the new policy was closely followed on 23 February by changes in NASA's administrative structure. More precisely, the agency's ten regional field centres and four mission directorates would be reporting directly to Administrator's C. Bolden office instead of the Associate Administrator's office, as it was the case previously. This decision clearly increased the grip of NASA's Administrator on the agency's day to day activities and made its decision-making process more top-heavy. Furthermore, the restructuring also put additional focus on the agency's R&D activities by reestablishing the offices of the agency's Chief Technologist and Chief Scientist. In fact one could argue that the changes clearly reflected NASA's new R&D oriented direction, while at the same time ensuring a pivotal role of the agency's head in managing these activities.[414]

Understandably, the principal contractors for the Constellation programme were the first to contest the President's new space policy. Representatives of Alliant Techsystems Inc. (ATK) of Minneapolis, the programme's principal rocket subcontractor, expressed their disappointment with the administration's decision and their hope that it would be overruled by Congress, which should have the final word in approving the project's termination. They also voiced concern over whether commercial spaceflight companies would be capable to develop man-rated spacecrafts within the foreseeable future. The company was expected to lose at least $650 million in backlog orders related to the Constellation programme. At the same time, it was unclear how much of the NASA's $2.5 billion budget slated for wrapping up Constellation-related work ATK would receive. The decision to terminate the project could weight heavily on the company's future, as it was already trying to cope with the phase-out of the Space Shuttle, as well as the cancellation of USAF's Minuteman 3 ballistic missile programme. Maryland based Lockheed Martin, the contractor for the building of Constellation's Orion capsule, also questioned the wisdom of the administration's new policy.[415]

Another concern voiced by many opponents of the new NASA policy was that it could corrode the U.S. solid rocket motor industrial base. In an interview on 11 February, NASA Deputy Administrator Lori Garver revealed that long high level discussions were held between NASA and Defence Department officials prior to the new policy's announcement. The Pentagon's concern was that after cancelling the Constellation programme there would be no solid fuel rocket development project left in the U.S. This situation could jeopardise the very existence of the only two companies that currently make these motors, Alliant Techsystems Inc. of Minneapolis and Aerojet of Sacramento. If these two companies seized their production, their important know-how would be irreversibly lost. Since the same motor type is also used in ballistic missiles, the U.S. could loose the industrial capability to replenish their strategic missile reserves. In order to tackle this issue, the U.S. military has set up a joint working group, mandated by Congress to deliver by June 2010 a strategic plan to preserve the U.S. industrial base in this field. In the mean time, the U.S. Air Force, Navy and Missile Defence Agency have decided to pool their solid motor rocket demands so as to sustain a very low rate production of the motors for as long as possible.[416]

Concern over loosing skilled workforce and critical industrial know-how as a result of the new policy was even expressed by officials who endorsed it, such as the president of the Aerospace Industries Association Marion Blakey. Although the potential of the new policy to create jobs in the private sector is acknowledged by the industry, there are also worries that a part of the skilled workforce currently employed in government programmes might loose its job. In that case, manufacturing know-how on several critical technologies could be irreversibly lost. On the other hand, NASA officials argued that commercial space had the potential to create a lot more jobs than those lost. Consequently, there is consensus in the U.S. that the space industrial base would benefit from the new policy on the long term. Nevertheless, coherent strategic industrial planning on a nation wide scale would be necessary to preserve critical manufacturing capabilities during the transition period. In an attempt to increase awareness of this problem, the NASA Administrator himself urged private space firms to make better use of existing labour force and industrial infrastructure.[417]

In an effort to prevent further reaction from industrial and policy officials, NASA Administrator Charles Bolden confirmed on 6 February that development work on heavy-lift rocket technology would not stop under the new U.S. space policy. In fact, he said that NASA would conduct an evaluation of all technologies related to Ares 5, the heavy-lift launcher of the Constellation programme, in order to choose the most promising ones for further development regardless of the programme's cancellation. The objective of this decision would be to maintain R&D relating to key heavy launcher technologies in order to start the manufacturing of such a rocket by 2020 at the latest. In this respect, the Obama administration space policy seemed to adhere to the Augustin Report's key policy recommendation known as the Flexible Path. Under this policy NASA was to perform only unmanned missions to the Moon and rather concentrate on Mars as the principal destination of future U.S. human exploration missions. Furthermore, NASA officials pointed out that the new policy would actually speed up the development of new space transportation technologies through its provisions for larger R&D budgets and increased international cooperation. As far as research was concerned, it was clear that the new policy called for an approximately $500 million annual budget for R&D on new space exploration technologies through 2016, whereas the Bush administration planning only foresaw $100 million.[418]

2.3. Russia

One of the major preoccupations of the Russian space programme has been to secure an independent space launch capability. In this regard, the construction of the new space centre will enable launches from national territory, providing Russia with the desired unrestricted access to space. Furthermore, Roscosmos has resumed the development of the Angara, its first entirely new post-soviet era rocket, which is expected to make its debut in 2011. The new rocket will use predominantly Russian made components and it should therefore decrease the country's dependence on foreign suppliers. At the same time, Soyuz launches from French Guiana are scheduled to begin in 2010 and its Russian manufacturers have already secured 14 orders from the system's operator Arianespace. This development should help Russian companies to expand their global market share and provide them with the necessary starting funds to develop their next generation of rockets.

In the framework of this dual strategy to reduce its dependence on third countries, Russia moved to secure the future use of the Baikonur Cosmodrome in Kazakhstan and to consolidate its plans to build a new spaceport in Vostochny, in the Russian Far East. Indeed, after years of disputes the Kazakh Parliament officially ratified on 9 April 2010 a 2004 agreement between the two countries, granting Russia access to the site until 2050. The agreement followed the general terms of the lease contract currently applied, under which Russia pays an annual fee of $165 million for the use of the Cosmodrome. After this development, the road opened for the joint construction by the two countries of a new space launch facility in Baikonur, known as the Baiterek, in order to accommodate the new Angara heavy rocket currently under development.[419]

Furthermore, during a meeting with Russian Prime Minister Vladimir Putin in July, the Head of Russia's Roskosmos space agency Anatoly Perminov promised that delays in constructing the Soyuz launch pad in Europe's Guiana Space Centre would be completed in time for its first launch in February 2010. Soyuz rockets launched from Guiana are scheduled to lift two spacecraft for Europe's Galileo satellite navigation system in 2010.[420]

On the other hand, Russia also continued its engagement in international cooperation in the field of Human space flight. In fact, on 28 May 2009 the Russian space agency Roskosmos announced a $360 million extension to its long-running space transportation contract with NASA. Under this contract modification, Roskosmos would launch an additional six U.S. astronauts on board four Soyuz capsule missions, scheduled for 2012. The same capsules will return their crews to Earth by spring 2013.[421]

At the same time, Russian officials confirmed that preliminary work for the creation of a new spaceport in Vostochny were proceeding according to plans and that its construction would begin in 2011. The new launching site would be operational by 2015 and certified for manned spacecraft missions by 2018. The project's total budget was expected to exceed 400 billion Rubles ($13.5 billion). Government officials accorded great importance to the new spaceport, which is hoped to boost the Russian space launch industry's overall competitiveness. In fact, the site is expected to become a space industry hub, generating more than 20,000 jobs on a long term basis.[422]

2.4. Japan

Key developments in the field of space transportation in Japan included the completion of the development of a new variant of the H-2 rocket family, capable of lifting the HTV-1 spacecraft to the ISS. But the period under examination also saw the development of Japan's future launcher system fall victim to its budget overrides. As far as the first one was concerned, on 11 July the Japan Aerospace Exploration Agency (JAXA) and Mitsubishi Heavy Industries Ltd announced the completion of ground tests for the H-2B rocket.[423] This is a more powerful version of the H-2A rocket, designed to lift the H-2 Transfer Vehicle 1 (HTV-1) cargo spacecraft to the International Space Station (ISS). HTV-1 is a solar-powered cylinder-shaped spacecraft approximately 10 metres long and 4.4 metres wide.[424]

It can carry 6 tons of pressurised and/or unpressurised cargo. Its development started in 1997 and cost about $680 million.[425]

At the same time however, the country's future launcher development came under scrutiny because of budgetary issues. On 17 November the Government Administration Reform Committee, an advisory committee to the Japanese government set up to eliminate wasteful budget spending, proposed the cancellation of the future GX rocket. GX has been under development since 2003 and was originally scheduled to fly in 2006. However, development of the rocket's second-stage liquid gas engine was not expected to be completed before 2011 and at a cost that was double the originally estimated budget (¥35 billion instead of an estimated ¥15 billion).[426] The final decision on the GX's fate was taken 16 December 2009, when the Japanese government opted to cancel the programme, but continue development work on its intended eliquified natural gas-powered engine. As it was announced, the principal reasons behind the programme's cancallation were its budget overrides and an increase in its expected launching cost that was bound to compromise any future commercial prospects.[427]

2.5. China

The Chinese space transportation prpgramme in 2009 and 2010 evolved around the improvement of the Long March rocket family's reliability and consequently also of its commercial prospects. Nevertheless, this effort suffered a set back when the Chinese Long March 3B rocket failed to place Indonesian commercial communications satellite Palapa-D into Geosynchronous transfer orbit on 31 August, due to an engine malfunction in its third stage. This was the launcher's second failure in 12 flights, the last one being during its maiden flight in 1996. Long March 3B is China's principal communications satellite launcher. The incident also grounded Long March 3A and 3 C rockets, as they all share the same engine. The launch services provider China Great Wall Industry Corp. declined to comment on the causes of the failure, but later in September it appointed an independent enquiry board to investigate them. Palapa-D was eventually salvaged, when satellite constructor ThalesAlenia was able to propel it into a geosynchronous orbit using its own propulsion module.[428]

On 19 November the enquiry board, comprising of officials from the China Academy of Launch Vehicle Technology (CALT) and the Academy of Aerospace Propulsion Technology, published its findings. It had reached the conclusion that the upper stage's engine had delivered 38% less thrust than expected for 43% of its total working time. The cause of this underperformance was determined to be a burn-through of its gas generator, caused by foreign matter or icing in the engine's liquid hydrogen injectors. To prevent any future similar accidents, the rocket's manufacturer China Great Wall Industry Corp. of Beijing decided to install a filter to the liquid hydrogen gas feed system. The company also stated it would flight test the new system before the end of 2009.[429]

2.6. India

Launch vehicle development absorbs the largest share of India's space budget, illustrating that rocket development is at the top of the list of the country's space programme priorities. As it was indicated in previous chapters, the country's future rocket development plans include the Mk 3 version of the Geosynchronous Satellite Launch Vehicle (GSLV), with a lift capability of four tons, which was expected to fly in 2010 or 2011, soon after a new indigenous cryogenic upper stage engine has been tested. However, its development has met serious problems, including a catastrophic failure at its very first launch that destroyed its GSAT-4 communications satellite payload on 15 April 2010. According to ISRO's first statements, the failure was due to two vernuer control motors igniting. However, it is not certain whether the cryogenic stage started to fire. While the cryogenic engine had been tested by ISRO and other experts, it was not tested in conditions that simulate high altitude, according to Nambi Narayanan. This failure also had repercussions on the timetable of the Chandrayaan-2 lunar orbiter mission in 2012 and launches of communication satellites.[430]

2.7. Emerging actors

The second half of 2009 and the first half of 2010 was a challenging and somewhat frustrating year for South Korea and its attempts to launch a satellite onboard the country's first national space launching vehicle. The first attempt, which was made on 25 August 2009, failed when the Korea Space Launch Vehicle (KSLV-1) did not put its payload (the meteorological satellite STSAT-2) into orbit. The vehicle, built with considerable help from the Russian rocket engine manufacturing company Khrunitsev, was presumed to have suffered an upper stage malfunction.[431] At the same time, Korea proceeded with its plans to acquire a completely independent launching capability by 2018.

In another development, on 3 February Iran unveiled the country's latest rocket, Simorgh. A full-scale mock-up of the vehicle was demonstrated to the public during the Iranian national Space Day ceremonies, which also included the launch of the Kavoshgar-3 rocket that carried small animals to space. According to the published pictures of the Simorgh, the vehicle seems to be 27 m long with a maximum diameter of at least two metres. It has two stages, with four engines for the first and one for the second one. Iranian officials claimed that it had the capacity to lift 100 kg into orbit, but western analysts claimed this could be upgraded in the medium term. The same analysts noted that if it was used as a ballistic missile, Simorgh could propel a one ton warhead to a range of 4,000 km. U.S. and Israeli experts were surprised to see an entirely new launcher developed by Iran, as they predicted Iran would focus on evolving its existing rockets, namely the liquid-fuelled Safir-2 and the solid-fuelled Sejil-2. It appears that Shehab 3B engines were used in the rocket's clustered first stage. Some observers also noticed similarities with the first stage of North Korea's Taepodong-2 three-stage launcher. In spite of its size, Simorgh apparently lacked a thrust-to-weight ratio capable of an intercontinental range. Its radically new first stage design however, could eventually be used in a future Iranian ICBM.[432]

Finally, Brazil has been increasingly focusing during our reporting period on the development of its national space rocket VLS-1, capable of carrying satellites to LEO. This project has been managed in close cooperation with Ukrainian and Russian companies. In April 2010, Brazil's space agency AEB announced a schedule for the programme, which foresaw the first launch of a version of the VLS-1 capable of lifting small satellites into LEO by 2014 at the latest. This mission would be the first since 2003, when an explosion of the rocket on its launching pad cost the lives of 21 people and put a halt to its further operational testing. Reportedly, this improved version has been developed with the assistance of Russian companies.[433]

3. Space sciences and exploration

Space exploration continues to attracting people imagination and governmental attention. In spite of the financial crisis which has entailed reassessment of certain projects or cooperation in this field, a lot of missions are ongoing or under development.

3.1. Human spaceflight activities

The human spaceflight activities are off courses largely dominated by the ISS events which are quite numerous between mid 2009 and mid 2010. The period covered has been mainly marked by the successive retirements of American shuttles while a new generations of commercial ones is under development and more described in the propulsion and launcher section. NASA has launched the first human-like robot to join the ISS team in 2010. A new fellow for the crew which could be particularly useful in the future especially for risky missions. In any case, "R-2" will be the base for development of next generations of robots which could be crucial to support human spaceflight activities. Its functioning within the particular gravitational environment of the ISS will be especially studied. The dexterous robot not only looks like a human but also is designed to work like one. With human-like hands and arms, R2 is able to use the same tools station crew members use. R2 will constantly updated by software to handle different kind of tasks and test its adaptability. It is therefore a crucial step in the ling term space exploration in which robots are due to play a significant role.[434]

The new tranquillity node attached to the ISS allow the astronauts to have a breathtaking view on earth and thus alleviating the sensation of confinement. The STS-130 astronauts delivered the two new space station pieces, the final components of the U.S. segment of the station, aboard space shuttle Endeavour during the first mission of the year. Throughout the mission, supplies and new equipment delivered by Endeavour were stowed, and work to outfit Tranquillity and cupola was the focus of the combined crews. The new segments added 2,600 cubic feet to the station's interior. An intense episode of the assembly of the ISS was thus achieved.[435]U.S. seems to look for another impulsion in its way to yield the advantages provided by the ISS after having invested so much effort. Several hundred leaders in space and science met in Cape Canaveral, Fla. on Nov. 16 and 17 2010 to explore ways to open the vast and exciting research capabilities of the space station to a wide array of uses. The setting for the discussion was the national conference of the American Astronautical Society. An occasion to present the capabilities of the achieved station and their hope for the upcoming decade.[436]

In April 2010 the STS-131 mission operated by the space shuttle Discovery to the International Space Station has been successfully fulfilled. Discovery delivered supplies and equipment to the station stowed inside the Italian-built multi-purpose logistics module Leonardo. The payload included new crew sleeping quarters, an ammonia tank, gyroscope and experiments.[437] It was the last flight for the shuttle Atlantis in May. The STS-132 delivered to the International Space Station the Russian Rassvet Mini-Research Module-1, only the second Russian module to ever be carried into space by a space shuttle. It's a fitting final payload for the orbiter that not only launched the first into space, but also was the first shuttle to dock to the Russian Space Station Mir. Atlantis was the shuttle behind seven of the 11 shuttle missions to Mir and this vehicle has visited

10 times the ISS. An historical page in spaceflight is thus turned while there are still a lot of questions unsettled about the news spacecrafts able to replace it.[438] Russia is therefore taking the lead to resupply the ISS affirming itself as an indisputable partner in this venture while the shuttle retirement is approaching. Numerous flights have been successfully undertaken by Russian to sustain the ISS as in april 2010 with the Progress 36 unpiloted spacecraft.[439] The station's 37th Progress unpiloted cargo craft docked also successfully to the International Space Station despite the failure of the automated rendezvous system, the cargo was manually flew to the dock by Russian crew resupplying successfully the ISS.[440] In July, the ISS Progress 38 cargo has also resupplied the ISS successfully without failure this time of the automated rendezvous system. It was also the last flight for Discovery. Among other noticeable tasks, it had delivered the Japanese Kibo laboratory to the International Space Station.[441] A new project of cooperation has been also settled between NASA and Roscosmos. A special lab may be installed in the International Space Station devoted to growing crystals for solar arrays in the nearest future. The lab to be initiated in 2013, is intended for growing up crystals of a brand new type. The lab is due to improve the crystals's properties thanks to the conditions not available on the ground. It is so expected to enhance their efficiency by around 60%. The project includes NASA and Roscosmos.[442]

The solution to assure the resupply of the ISS is coming up between Japan and Europe. Germany has committed to paying a 38 percent share of an estimated 3.8 billion euros that European governments will need to continue their work on the international space station (ISS) in the next 10 years. Germany confirms thus its strong support to the pursuit of the project until 2020. The discussions are rather fierce between the European partners concerning the possibility to develop an autonomous Automated Transfer Vehicle (ATV) to permit it to return station cargo to Earth. Indeed, France and Italy which currently pay about 27 percent and 19 percent of Europe's space station costs, respectively, have not come out in favor of the ATV upgrade. ESA has engaged talks with the Japan Aerospace Exploration Agency to determine whether Japan's plans to upgrade its current throwaway H-2 Transfer Vehicle cargo tug could be merged with ESA's vehicle into a single program. In order to develop a vehicle with payload-return capabilities, what could save money rather than developing alone a major hardware project such as an ATV. Two options are thus under studying and the European position should be cleared in 2012. But it is obvious that such a vehicle would be an asset that would serve Europe's future space-exploration plans beyond the space station.[443]

Indeed, Japan is currently developing an ambitious programme to enhance its role on the ISS, announcing its intentions to better utilize the Kibo laboratory and build a variant of the H-2 Transfer Vehicle (HTV) that would be capable of bringing cargo back to Earth. The project intends to add human capability after 2020 to assure Japan space flight operation in the future. A long term mission which has been started in the 1990s while it was seen as a necessity that Japan develop basic technologies for independent human space capabilities. The present-generation HTV, which was launched for the first time in September 2009 by Japan's heavy-lift H-2B rocket, is set to play a major role in keeping the ISS in service, ferrying roughly 6 tons of cargo on each of the six space station resupply missions planned between next January and 2015. Not less than 650 Japanese companies are involved in the space station and this project could be a crucial step for the Japan space programme which could become in the nearest future both more present in the international cooperation and autonomous trough its own exploration programmes.[444]

3.2. Lunar exploration

The moon still seems to spark international vocations. The perspective of finding water to establish a base and resource to eventually exploit attracts the international attention. This discovery makes utopian lunar bases closer to reality. Water provides the way to obtain propellant and oxygen supplies for the cosmonauts in the future bases.

In Europe a new step has been achieved for the ESA's first Moon lander. The mission foresees to land autonomously with pinpoint precision near the Moon's south pole, a region full of dangerous boulders and high ridges by 2018. The aim is to probe the moonscape's unknowns and test new technologies to prepare future human landings. A new contract has been signed between EADS-Astrium in Berlin, Germany. The region envisaged is particularly important because it concentrates continuous sunlight for power and potential access to vital resources such as water-ice. A milestone in the ESA's spaceflight programme to prepare a future human landing by accumulating data on this area as poorly understood as crucial for the upcoming lunar exploration.[445] Projects which could be helped by the recent discoveries of a new type of solar wind interaction with airless bodies in our solar system. Magnetized regions called magnetic anomalies, mostly on the far side of the Moon, were found to strongly deflect the solar wind, shielding the Moon's surface. This crucial element will help understand the solar wind behaviour near the lunar surface and how water may be generated in its upper layer. Evidences of this phenomenon have been presented at the European Planetary Science Congress in Rome, on 24 September 2010. The study of Atmosphere-less bodies being particularly interesting in their way of interact with the solar wind quite differently than the Earth. Indeed, surface of such bodies are exposed without any shielding by a dense atmosphere or magnetosphere forming a very rough and chaotic surface called "regolith". A significant flux of high energy particles was found to originate from the lunar surface, and the information collected will be particularly useful understand the role that solar wind can have as potential source of water on the Moon. Another studies presented by searchers from NASA suggest that the moon has cooled since its formation it has contracted and could still be tectonically active today. Indeed, the moon is smaller than it used to be and could continue shrinking.[446]

China has announced that lunar exploration would be its main priority with the aim of returning lunar samples by 2017. Without being an absolute priority for China, the space exploration sector has seen steadily increased its funding through several new scientific initiatives. Reflecting these new ambitions, China's space science budget is estimated to have increased substantially over the past five years, from about 800 million Yuan in 2006 (€69 million) to 1.88 billion Yuan ($194 million) in 2010. The first lunar orbiter probe was launched in 2007. The lunar program is divided in three phases consisting of landing on and returning samples from the lunar surface. China's ultimate objective is for a manned mission to the moon in the 2020s.[447] Besides the China's new Moon probe reached its destination Oct. 6 after firing braking thrusters to enter into lunar orbit. There are still two manoeuvres to execute in order to bring the probe to orbit desired allowing this one to map the Moon from an altitude of 100 kilometers.[448]

Two missions to the Moon are planned by Russia for 2013, one of them exclusively Russian and the other one devoted to studying the lunar resource on the poles is the result of a Russian-Indian agreement The Russian Luna-Glob will apply seismic methods instead of complicated penetrators to better figure out the composition of the lunar underground.[449] This new comes after the chunks of frozen water detected, within the moon's perennially shadowed polar craters. The data provided by the Indian probe Chandrayaan suggest a massive presence of water which could the result of a comet which would have smashed into the moon eons ago. It would have enough water to supply a human expedition. In addition to this important finding the probe has also mapped most of the area near the north pole of the moon.[450]

3.3. Mars exploration

Some questions about Mars still attract the international scientific awareness. This is particularly true concerning any form of life which could have or could exist on the Red planet. An obsessing question which explains the numerous missions undertaken to respond to this quest. Another matter often raised concerns the atmosphere which would have been disappeared. The ghosts of Mars seem to all the more haunt the Terrian inspirations as that could somehow be the destiny of our planet.

ESA and NASA have selected the scientific instruments to install for their first joint Mars mission scheduled for 2016.[451] This mission will study chemical makeup of the Martian atmosphere included previously methane important for life. An unprecedented alliance which is only due to begin because the ExoMars Trace Gas Orbiter is the first in a planned series of joint missions leading to the return of a sample from the surface of Mars. Scientists worldwide were invited to propose the spacecraft's instruments. Among its objectives, one is to characterise the planet's atmosphere, and in particular search for trace gases like methane. Discoveries could bring about crucial information about the eternal question, namely has Mars ever hosted any form of life. The selected materials are an infrared spectrometer to detect very low concentrations of molecular constituents of the atmosphere, An infrared spectrometer to detect trace constituents in the atmosphere and to map their location on the surface, An infrared radiometer to provide daily global measurements of dust, water vapour and chemical species in the atmosphere to aid the analysis of the spectrometer data, A camera to provide 4-colour stereo imaging over an 8.5 km swathe and finally A wide-angle multi-spectral camera to provide global images in support of the other instruments. U.S., ESA and a lot of European national agencies and Canada are coopering under the present agreement to develop jointly these equipments.[452]

Since July 2010, the German Aerospace Centre (Deutsches Zentrum für Luft- und Raumfahrt DLR) is currently studying the existence of liquid salt solutions on Mars which could host a form of life in spite of the extremely low temperatures. The previous modelling has showed that the conditions required were met. Triggered by the findings of NASA's Phoenix Mars mission these study could bring crucial information on this fluid medium that supports life because there is no water. This liquid solution could allow flow processes, known as rheological processes, on the Martian surface. In the context of possible biological processes, this could also be a life-sustaining transport of nutrition and waste. The characteristics of this fluid are therefore rather close to water and so favourable to form of life. This is all the more true since U.S. scientists have found that a form of life was partially composed of arsenic.[453] Recent observations made by Mars Express could help German scientists. Indeed, A small crater called Schiaparelli embedded in the north western rim of the Schiaparelli impact attracts the attention of scientists. The images provided by ESA's Mars Express would show evidences for past water and the great Martian winds that periodically blow. However, other astronomers thought he meant canals, meaning artificial irrigation and transportation routes, which led to a few astronomers, and a large number of the general public, believing that they had been created by intelligent Martians. No doubt that the future missions will enlighten this enigma.[454]

The NASA's Mars Reconnaissance Orbiter resumed observing Mars after having recovered from an unplanned reboot which has plunged the probe in a safe mode waiting for plenty recovery. The spacecraft appeared to finally operate normally, making science observations and returning data. It has already completed its primary science phase of operations in November 2008.[455] It continues to observe Mars both for science and was used by NASA to try to reach Phoenix but NASA has finally abandoned hopes to recontact the NASA's Mars Lander[456] which had studied by digging up during six months the Martian soil. Indeed, Mars Reconnaissance Orbiter shows signs of severe ice damage to the lander's solar panels and has tried unsuccessfully to catch any radio signal. However there are still possibilities to Phoenix to survive and it could try later to recontact the Orbiter. During its mission, Phoenix confirmed and examined patches of the widespread deposits of underground water ice detected by Odyssey and identified a mineral called calcium carbonate that suggested occasional presence of thawed water. The lander also found soil chemistry with significant implications for life and observed falling snow. The mission's biggest surprise was the discovery of perchlorate, an oxidizing chemical on Earth that is food for some microbes and potentially toxic for others.[457] In the same way, The NASA's Spirit Mars Rover had hold evidence of a wet, non-acidic ancient environment that may have been favorable for life. However additional information put into perspectives these discoveries because the rover has also found that the environment may have been acidic and so much less favourable to any form of life. The core of the questions arisen by these discoveries is to determine where most of the carbon dioxide went which could bring crucial element to respond why the Martian's atmosphere had disappeared.[458] A new project concerning Mars is on the verge of being carried out since NASA gave the green light Oct. 4 to what is expected to be the last of its Mars Scout missions, a $438 million probe that could help one more time scientists to understand how the red planet lost much of its atmosphere. The 2,500-kilogram spacecraft is due to be launched by 2013. The yearlong mission is designed to sample the planet's upper atmosphere in an effort to understand a dramatic climate change that left Mars unable to support the presence of liquid water on its surface. Earlier this year, however, NASA decided to discontinue the Scout program and join the European Space Agency in 2016 on the first of several proposed joint Mars missions.[459]

In a statement in October 2010 The Chinese Academy of Space Technology has expressed its plan to independently conduct a Mars orbiting exploration by as early as the year 2013. The objective is to use the same platform as the unmanned lunar orbiter Chang'e I successfully launched in 2007. Meanwhile, China's first Mars probe "Yinghuo-1" is expected to be launched on a Russian carrier rocket in November 2011. The Chinese Mars probe, designed for a two-year life span, will try to discover why water disappeared from the planet and explain other environmental changes on Mars.[460]

3.4. Saturn exploration

The ESA and NASA's Cassini spacecraft has successfully returned images of Enceladus and the nearby moon Dione.[461] The Cassini-Huygens mission is a cooperative project of NASA, the European Space Agency and the Italian Space Agency. Several pictures show Enceladus backlit, with the dark outline of the moon crowned by glowing jets from the south polar region. The images show several separate jets, or sets of jets, emanating from the fissures known as "tiger stripes". Scientists will use the images to pinpoint the jet source locations on the surface and learn more about their shape and variability. The Enceladus flyby took Cassini within about 48 kilometers of the moon's northern hemisphere. Cassini's fields and particles instruments worked on searching for particles that may form a tenuous atmosphere around Enceladus. They also hope to learn whether those particles may be similar to the faint oxygen and carbon-dioxide atmosphere detected recently around Rhea, another Saturnian moon. The scientists were particularly interested in the Enceladus environment away from the jets emanating from the south polar region. Scientists also hope this flyby will help them understand the rate of micrometeoroid bombardment in the Saturn system and get at the age of Saturn's main rings.[462] An experiment also designed to probe the moon's interior composition. The instruments are designed to measure the gravitational pull of Enceladus against the steady radio link to NASA's Deep Space Network on Earth. Detecting any wiggle will help scientists understand what is under the famous "tiger stripe" fractures that spew water vapor and organic particles from the south polar region.

3.5. Venus exploration

Europe has taken the lead in the Venus exploration with its orbiter. ESA's Venus Express which has returned the clearest indications proving that Venus is still geologically active. Indeed, relatively young lava flows have been identified by the way they emit infrared radiation. The finding suggests the planet remains capable of volcanic eruptions. The data were collected by the Visible and Infrared Thermal Imaging Spectrometer (VIRTIS) on ESA's Venus Express spacecraft, which has been orbiting the planet since April 2006. Venus still keeps jealously certain of her secret as explained Dr Smrekar : "There are some intriguing models of how Venus could have completely covered itself in kilometres of volcanic lava in a short time, but they require that the interior of Venus behaves very differently from Earth. If volcanism is more gradual, this implies that the interior may behave more like Earth, though without plate tectonics."[463] The orbiter has also permitted to detect high density of Sulfur dioxide has been detected as well by special equipment developed by the French space agency (CNES).[464] Moreover, Venus express has successfully studied the Venus's atmosphere. The polar atmosphere of Venus is thinner than expected, Venus Express has flown through the upper reaches of its poisonous atmosphere. The orbiter went diving into the alien atmosphere during a series of low passes in July August 2008, October 2009, and February and April 2010 which was unprecedented. The results suggest that unanticipated natural processes be at work in the atmosphere as 10 measurements so far and shown that the atmosphere high above the poles is a surprising 60% thinner than predicted. Unfortunately, Venus Express had not been foreseen to sample atmosphere and use radio tracking stations on Earth watch for the drag on the spacecraft as it dips into the atmosphere and is decelerated by the Venusian equivalent of air resistance. The resistance and longevity of the orbiter are however remarkable and promises other interesting data.[465]

From their side, Russian are preparing a mission to Venus planned to 2016. The main objectives are to study the planet's turbulent atmosphere and surface, and find out why it has no water. A leading Russian firm specializing in automated probes would be in charge of designing the hardware but no contract has been signed so far. The orbiter should reach its target by 2017 and will be carried by a heavy Proton-M or Angara-A5 rocket. The lander and probes will work for up to several days before harsh corrosion damages them. The orbiter's life is expected to be much longer.[466] It has been also announced that France should participate significantly to the project[467] Japan's Akatsuki spacecraft has unfortunately failed to enter into Venus orbit but mission planners hope to attempt orbital insertion again in six years. Whereas it was about to be injected into the Venus orbit, the Venus Climate Orbiter switched itself into safe mode for unknown reasons. A painful failure which recalls the Japan's Nozomi Mars mission. This one suffered in 1998 a thruster failure that delayed the orbiter's arrival at the red planet by more than four years. The Orbiter will be back in position between December 2016 and January 2017 to make a second attempt to enter orbit around Venus. It is however not clear whether the spacecraft will have enough battery and fuel at this date.[468]

3.6. Neptune exploration

New measurements provided by a European space telescope suggest that a comet may have crashed into the gas giant about 200 years ago in July 2010. Composition of Neptune's atmosphere is currently analyzed from data furnished by the Herschel space observatory. The scientists examined the atmosphere of Neptune, which mainly consists of hydrogen and helium with traces of water, carbon dioxide and carbon monoxide. They detected an unusual distribution of carbon monoxide in Neptune's atmosphere, with much higher concentrations in the upper layer, called the stratosphere, compared to the troposphere layer beneath. They also found higher concentration of Methane than expected which operates in a similar ways to water vapor on Earth. The European Space Agency launched the Herschel infrared space telescope in May 2009 and is the largest and most powerful infrared telescope in orbit today.[469]

3.7. Jupiter exploration

In June 2010 the NASA/ESA Hubble Space Telescope has provided insights of two recent events on Jupiter: the mysterious flash of light on 3 June and the disappearance of the planet's dark Southern Equatorial Belt. Another flash has been noticed by an amateur astronomer. However there is no sign of debris above Jupiter's cloud tops. The flash is thought to have come from a giant meteor burning up high above Jupiter's cloud tops. An interesting occasion to observe how Jupiter's atmosphere is going to react to such extremely violent event.[470]

3.8. Solar observation

The Sun vitally linked to our liveable environment on earth is subject to many studies to better understand its nature and functioning. This all the more true as it can show unforeseen phenomenon which can entail huge consequences on Earth concerning as much as the climate or technological aspect particularly vulnerable. In Europe, ESA has begun to exploit the probe Proba-2. Launched in the end of 2009, Proba-2 is a small but innovative experimental technologies. In June 2010, after only eight months of life, it has already returned more than 90 000 images of the Sun. It embodies this new generation of miniaturised science instruments, focused on the Sun and space weather, as well as 17 state-of-the-art technology payloads. Moreover Proba-2 constitutes a technological stepping stone to future missions which will be used onboard of BepiColombo mission to Mercury.

An ideal Platform of test for material such as credit-card-sized magnetometer or among others experimental solar panel produced in Belgium. The scientific payload of Proba-2 is equally innovative with a Sun camera (SWAP) that is the first solar physics instrument with Active Pixel Sensor (APS) detectors. The data provided are satisfactory and allow operational application related to the risk entailed by space weather on satellite.[471] In July 2010 the French satellite Picard has provided its first images from the Sun. An opportunity for the ground teams to operate the last adjustments before the beginning of the mission. The functioning of the satellite launched has been successfully achieved and it can now start its scientific mission. The satellite is to stay several years to study in detail the Sun's activity by taking one photo each minute and measuring regularly its power during several years.[472] These experiences take place at a cornerstone of the Sun understanding. A lot of questions are raised all around the world such as a recent study published by researchers from Stanford and Purdue University that shows that the decay rates of radioactive elements are changing. An explanation put forward would explain this phenomenon. The celestial body would emit a previously unknown particle that is meddling with the decay rates of matter. The issue appears to be particularly tricky for the searchers and could even bring us to rethink our way of understanding the true nature of these subatomic particles.[473] As herein explained, the solar activity has been provoking great concern among the scientist community given its erratic and sometimes volatile behaviour with potential heavy damages. For this reason, representatives from more than 25 of the world's most technologically-advanced nations were gathered in July 2010 for the International Living with a Star (ILWS) meeting in Bremen, Germany, to discuss the importance of developing better methods for forecasting space weather. Indeed, Streams of charged particles that fly off the sun can interfere with electronics on Earth and satellites orbiting our planet. The dependence of technology on earth being more and more important it appears as a necessity to face the risk entailed by the unforeseeable nature of the Sun.[474]

On the other side of the Atlantic, many efforts are deployed in the same way. In February 2010, the space-based Solar Dynamics Observatory (SDO) was launched successfully from Cape Canaveral on an Atlas V. A project carried out by the DLR supporting NASA-led mission with the SDO Data Center at the Max-Planck-Institute for Solar System Research (MPS) in Katlenburg-Lindau. The data provided by the observatory are due to improve our ability to forecast space weather significantly. From its geosynchronous orbit (36,000 kilometres) SDO should observe the Sun during five years. It is particularly expected from this mission to better understand the negative effects of the Sun. These adverse factors are caused by massive explosions in the Sun's atmosphere, known as "flares" which expel billions of tons of solar material into interplanetary space.

Some of these electrically-charged particles can travel at the speed of light as well as solar radiation. The areas most affected by these harmful solar outputs are near-Earth space and Earth's polar regions. The system will improve the forecasting of solar radiation also known as "space weather". Foreseen more efficiently the threats from space enable satellite operators, to switch their equipment to a secure mode whenever danger threatens, thereby protecting sensitive devices from damaging overloads and surges. When danger looms, astronauts in the ISS take shelter in a specially-protected room. A significant step in the understanding of our Sun achieved to better adjust our behaviour to him.[475] NASA is envisaging devising a Solar Probe intended to plunge directly into the Sun as well.The probe must be launched by 2018 and is due to bring crucial information about solar physics. The announcement means that researchers can begin building sensors for unprecedented in situ measurements of the solar system's innermost frontier. A very stimulating experience in an unexplored territory which will demand cutting-edge technologies given the extreme conditions that the probe is due to encounter.[476] The awareness of the potential damages entailed by the Sun activity has been spreading among the politicians in the U.S., especially the changes in the solar magnetic field which can have a number of effects on Earth, some of which can be disruptive to critical technologies such as GPS and the electric power grid. The congress has subsequently decided to add $5 million to NASA's 2010 budget to refurbish and launch the Deep Space Climate Observatory, which had been shelved since 2001, to replace ACE. NASA and the National Oceanic and Atmospheric Administration will be responsible for getting the satellite ready for launch, and the Air Force will provide the launch vehicle in 2013. The system should dramatically improve the quality of space weather forecasts.[477]

3.9. Outer solar system exploration and observation

A couple of projects are ongoing in Europe, one of the most excitant technology speaking is surely the ESA's Rosetta probe which has returned the first close-up images of the asteroid Lutecia defining it as most probably a primitive survivor from the violent birth of the Solar System. The pictures taken are provided by Rosetta's OSIRIS instrument. Rosetta has successfully completed its flyby and transmitted the data corresponding to Earth. The Lutecia asteroid is due to be a fragment of the cores of much larger objects. The next step for the probe is to meet its primary target, namely the comet Churyumov-Gerasimenko scheduled in 2014. A successful and informative first step in its mission which announces others in the future. An event that also symbolizes a real achievement in term of know-how for the European industry and scientist community.[478] The French's space telescope Corot has discovered an alien planet orbiting another star and which could host potentially water in its atmosphere. The suspected temperate nature of the planet whose surface temperature is between minus-20 degrees and 160 degrees Celsius could mean that it could harbour liquid water. But this water would not be in the form of Earth-like oceans; more likely it would be only in the form of clouds with water droplets reports scientists. The study of the planet's composition is ongoing and should enlighten our knowledge of similar ones already observed.[479] From its part, the powerful Herschel space telescope enlarges our horizons thanks to the effects called "gravitational lensing" or "cosmic magnifying glass". The European satellite Herschel has been able to detect and characterize galaxies how they are used to be 10 milliards years ago. Herschel is particularly adapted to take advantage of this kind of phenomenon. Indeed, a massive object tends to amplify the visibility of a target located behind it. That allows to observe much more farther than using only its own capacities providing an insight of galaxies never studied before.[480]

A great achievement has been also reached with the return of the Hayabusa capsule in Australia on 13 June 2010. Launched in 2003 the capsule has successfully delivered its sample from the near-Earth asteroid, Itokawa. JAXA concludes with this success a remarkable step in exploration of celestial bodies in partnership with their US's fellows. It was the occasion for Japanese engineers to deepen their knowledge in information on electrical propulsion and autonomous navigation and obtain useful asteroid samples. It is worthy to note that was only the second time in history a spacecraft descended to the surface of an asteroid. A preliminary cataloguing and analysis of the capsule's contents is ongoing by scientist from JAXA in cooperation with Australian ones. The sample should be then distributed to scientists worldwide for more detailed analysis. This return concludes thus a successful cooperation between NASA and JAXA which will be surely pursued in other projects.[481] A short time after JAXA has announced to find minute particles in the capsule of the space probe Hayabusa which returned to Earth. Scientists hope that could help them to better understand the solar system's origin.[482]

In the U.S., the period has been particularly busy in outerspace discoveries. The first is the detection of the youngest nearby black hole by the NASA's Chandra X-ray observatory. The 30-year-old black hole provides a unique opportunity to watch this type of object develop from infancy. The subject would be a remnant of a supernova in the galaxy M100 approximately 50 million light years from Earth. That could constitute an decisive opportunity to enlarge our understanding about mysterious and fascinating black holes.[483] Secondly, a team of planet hunters from the University of California (UC) Santa Cruz, and the Carnegie Institution of Washington has announced the discovery of a planet with three times the mass of Earth orbiting a nearby star at a distance that places it squarely in the middle of the star's "habitable zone". The W. M. Keck Observatory in Hawaii, one of the world's largest optical telescopes has been used to detect the body. It could host form of life and constitute the most Earth-like exoplanet yet discovered representing the first strong case for a potentially habitable one.[484] Then, NASA has made another discovery thank to its Kepler spacecraft with the first confirmed planetary system with more than one planet crossing in front of, or transiting the same star. The Kepler's ultra-precise camera measures tiny decreases in the stars' brightness that occur when a planet transits them. The size of the planet can be derived from these temporary dips. This discovery would be the first clear detection of significant changes in the intervals from one planetary transit to the next, what we call transit timing variations according to the scientists in charge of the project. There is thus evidence of gravitational interaction between the two planets as seen by the Kepler spacecraft.[485] NASA has also discovered a crucial element which could answer the long-standing question of how massive stars are. Kraus' team whom is known to be at the origin of the discovery used the Very Large Telescope Interferometer of the European Southern Observatory in Chile to observe of a massive disk of dust and gas encircling the giant young star. The presence of the disk is strong evidence that even the very largest stars in the galaxy form by the same process as smaller ones growing out of the dense accumulation of vast quantities of gas and dust, rather than the merging of smaller stars, as had been previously suggested by some scientists. This could be a major step towards a better understanding of Space and development of celestial bodies such as Earth around this kind of stars.[486] Still in the U.S scientists from the California Institute of Technology and UCLA have discovered evidence of "universal ubiquitous magnetic fields" that have permeated deep space between galaxies since the time of the Big Bang. The images provide an insight of the most powerful objects in the universe supermassive black holes that emit high-energy radiation obtained by NASA's Fermi Gamma-ray Space Telescope. There would be signs of primordial magnetic fields in deep space between galaxies. The universal magnetic fields may have formed in the early universe shortly after the Big Bang, long before stars and galaxies formed report scientists in charge of the programme.[487] Finally, Wide-field Infrared Survey Explorer, or WISE has completed its first survey of the entire sky on July 2010. The mission has generated more than one million images so far, of everything from asteroids to distant galaxies. This infrared view highlights the region's expansive dust cloud, through which the Seven Sisters and other stars in the cluster are passing. Infrared light also reveals the smaller and cooler stars of the family.[488]

3.10. International cooperation in space exploration

Rather more than concluding new agreements, NASA has reinforced the previous ones between 2009 and 2010. More explained in the part concerning Mars the cooperation between NASA and ESA has reached another steptstone with the selection of the five science instruments for the first mission. The ExoMars Trace Gas Orbiter, scheduled to launch in 2016 and will be one of the joined robotic mission among in series planned. The second one scheduled for 2018 consists of a European rover with a drilling capability, a NASA rover capable of caching selected samples for potential future return to Earth, a NASA landing system, and a NASA launch vehicle. A crucial experience might constitute the base for further common space exploration mission.[489] Another cutting edge project concerns the common telescope whose new components are under development for the Webb Space Telescope destined to that focus the attention of the infrared camera on specific targets to the exclusion of others. They can therefore concentrate themselves on objects like very distant stars and galaxies. The microshutters will enable scientists to block unwanted light from objects closer to the camera in space, like light from stars in our Galaxy, letting the light from faraway objects shine through. This is a technical challenge and a big step in the cooperation between ESA and NASA to assembly the future most powerful space telescope. Constructed in the U.S. the pieces are installed by the European Space Agency.[490] With national agencies, NASA and the DLR concluded a framework agreement for bilateral cooperation in Washington D.C. on December 2010. The partners have also agreed to cooperate on lunar research, through the Lunar Science Institute Agreement. The NASA-DLR framework agreement encompasses cooperation in all relevant aspects of aerospace research. In terms of space, the emphasis will be on Earth observation and conducting research in the space environment, as well as space operations and planetary research. In addition, the agreement also covers the exchange of research staff and scientific data. There will also be even closer cooperation in encouraging the development of young researchers.[491] NASA and SLR have also concluded an agreement on June 2010 to extent the Gravity Recovery and Climate Experiment (GRACE) mission through the end of its on-orbit life, expected in 2015.[492] NASA and the Israel Space Agency have also signed a joint statement of intent on August 2010 to expand the agencies' cooperation in civil space activities. Several key sectors of cooperation have been identified such as space science, life sciences, space exploration and other areas of mutual interest. An agreement might announce several projects in common in the upcoming years.[493]

Meanwhile, Russia is strengthening its links with Europe and a good many of others countries. An agreement has been indeed concluded on September 2010 between ESA and the Russian Ministry of Foreign Affairs in order to facilitate continuation and evolution of fruitful cooperation between Roscosmos and ESA.[494] The cooperation with ESA member's states is also developing. A complex series of plasmas physics experiments have been carried out by a Russian-German team in January 2010 on the ISS. The German DLR has funded both the development of the experimental equipment and the research itself. The development of the equipment for space and the experiments themselves has resulted in a surprising spin-off in the field of medicine, known as a "cold plasma torch". This is a small medical tool that uses cold plasma to sterilise chronic, antibiotic-resistant wounds. It has already been used with success in clinical trials on more than 100 subjects. A successful cooperation leaded by the DLR which shows just once will not hurt that ISS can provide useful result.[495] Common Scientifics experiences which follow a frame agreement between Roscosmos and the DLR signed in Berlin in June 2010. The agreement which covers long-term cooperation on scientific microgravity research in Photon-M and Bion-M spacecraft. The document defines the framework and general principles of the long-term cooperation program.[496]

Vladimir Putin has emphasized mutual advantages that could obtain Russia and France by uniting their space technological potentials at the meeting of the Russian-French Bilateral Cooperation in December 2010. According to him "Speaking about aviation and space, France and Russia have cumulated essential heritage her". A declaration which follows a clear will to strengthen their relation in the future.[497] Meanwhile, Roscosmos and NASA have begun talks about the possibility of signing a protocol in order to define and implement different space programs. The protocol could include missions to asteroids and the moon.[498] In July 2010 an important agreement was concluded between Russia and the UK to enhance and develop their mutual cooperation in particular concerning space exploration, the Memorandum of Understanding highlights the new step in relations between the UK Space Agency and Roscosmos.[499]

Prime Minister Vladimir Putin has also reiterated its commitment to strengthening Russia and Ukraine cooperation in space exploration. The two countries are devoted to modernizing their economy by creating high-paying jobs and new technologies clusters. The areas concerned are nuclear power industry but also aircraft industry and space exploration. [500] Russia has also been sustaining a dialog with India especially to develop several project related to space exploration such as the successful lunar exploration projects (Chandrayan-2) which could be the fisrt to a serie of common project.[501]

Egypt and Saudi Arabia have signed a memorandum of understanding on remote sensing and space sciences cooperation. The agreement is aimed at promoting joint action in the domain of space sciences and remote sensing research and exchanging expertise and information on this score, he added.[502]

4. Satellite applications

4.1. Space-based communications

Apart from development in government programmes and industrial activities presented above, one of the most prssing issues that preoccupied global satellite communicatios was the issue of interference. On 27-29 October the Satellite Users Interference Reduction Group (SUIRG) met in Cannes, France. The Group has been trying for years to raise industry awareness of the importance of interference for commercial communication satellites operators. During the meeting, Intelsat, SES and Inmarsat decided to create a voluntary satellite database named Space Data Association (SDA) and based on the Isle of Man. Its mission will be to collect voluntary data contributions from commercial operators on their respective satellites, such as satellite location, broadcast frequencies and power, signal polarisation and coverage areas. Using this database should shorten the time needed to localise interference sources. Other commercial operators were expected to join in the effort, but questions remained on whether they would be willing to distribute such sensitive information. Furthermore, although this initiative is expected to contribute to limiting unintentional interference, it will still have to overcome the absence of a legal framework that would oblige operators to cease activities that create interference.[503]

In November, SDA begun to take form as its three founding companies issued a Request for Proposals (RfP) for a contract to design and build the database. The RfP called for the creation of a central Space Data Centre on the Isle of Man, as well as two more backup data storage servers on different continents. These would probably operate from within SDA member companies. The centre would provide accurate and timely information on commercial satellites and would also act as a registry of interference incidents and sources. SDA's members also signalled their future intention to link the database to the U.S. Space Surveillance Network and to invite operators from Russia, China and India to participate as well.[504]

4.2. Space-based positioning, navigation and timing systems

The European GNSS programme has withnessed considerable progress during the period in question. Apart from its progrees in contracting its first spacecraft mentioned above, the programme has also matured in its purpose, ellevated to a strategic asset fo Europe. For example, in March 2010 the European Commission announced that it was considering the removal of all non-European built components from the Galileo satellites currently under final assembly. This decision, mainly affecting Chinese-built parts of the spacecrafts for which European-made substitutes would have to be found, reflected the EC's view of the system as a strategic asset for Europe, in which a cetain degree of manufacturing independence should be achieved. However, if this this decision were to materialise it might also impede the procurement of the Canadian-built search and rescue terminals currently envisaged for the system. This significant change in the programme's procurement policy could also be seen as a result of its change from a private-public partnership to a 100% public financed project.[505]

In any case, this decision was reinforced by China's own step towards the development of its own national satellite navigation system called Beidou (Compass). The system was originally planned to provide only regional coverage, with China relying on Galileo for global wide use. However, this policy is now changed and China is developing Beidou as a global system, directly competitive to Galileo. This development has also created friction between European and Chinese authorities over radio frequencies reserved for each system's government-only use, known in Europe as the Public Regulated Service (PRS) and in China as the Authorised Service. Although these systems should in principal emit in different frequencies, negotiations over allocating them between Galileo and Beidou have been fruitless up to now. On the contrary, a similar agreement between Europe and the U.S. was signed already in 2004 and similar negotiations with Russia on cooperation with its Glonass system also saw progress in 2010.[506]

Another important developmnet concerning Galileo was the preliminary agreement reached among EU membern states in March 2010 on the service use policy of the PRS. Negotiations were expected to conclude by the end of 2010, but it appeared that EU governments would be granted direct and unconstrained access to Galileo's encrypted military signal. This arrangement would imply that EU member-states would all create their own national points of contact with the system's two Galileo Security Monitoring Centres (GSMC) in France and the UK and they would be solely responsible for the service's use by their authorities. Furthermore, officials from the Galileo Supervisory Authority (GSA) disclosed that all member-states would have unrestrained access to the service and no prior approval by EU institutions or authorities would be required for its military use. However, according to the same sources, the question of whether the PRS would be completely free of charge like its GPS counterpart or if its use would entail a certain fee was not yet decided. Needless to say that any charge related to PRS could damage Galileo's competitiveness to its U.S. counterpart.[507]

Finally, in a related development the European GPS Navigation Overlay Service Egnos was declared ready for use for its freely accessible service on

1 October 2010. Higher reliability versions, as well as an Egnos Commercial Service were expected to become operational in 2010. The system uses two Inmarsat and ESA's Artemis satellites to augment GPS performance. The programme's budget was €350 million. A French-based consortium of seven European air-navigation agencies, called ESSP, is under contract to the European Commission to manage Egnos until 2013.[508]

Another country that paid particular attention to the development of its satellite navigation capabilities in 2009 and 2010 was Russia. The Glonass constellation is expected to be completed by the end of 2010, with the launch of 7 spacecraft, bringing their total number to 28 satellites, of which 23 fully operational.[509] Nevertheless, the constellation would have to include at least 24 spacecraft to provide global coverage.[510] The development of a new generation of satellites (Glonass-K) has already matured and the first spacecraft are expected to fly by 2011. The programme's budget reached 2.5 billion Rubles in 2009 and it was expected to rise to 3.7 billion Rubles through 2011.[511] The first three of the Glonass satellites scheduled for 2010 were put into orbit on 2 March onboard a Proton-M rocket launched from the Baikonur Cosmodrome.[512]

Further steps for the programme include improving its performance and accuracy, as well as giving it an international dimension by providing services outside Russia. In this respect the Russian government has launched a high political level effort to convince neighbouring countries and emerging space powers to subscribe to the system's services. The issue was discussed during the Russian Prime Minister's visit to India in March. According to Russian officials, the two countries would establish a joint venture in India to produce Glonass-compatible navigation equipment. According to the same sources Indian authorities would use the system's civilian signal at first, but negotiations were underway to allow them to access its more accurate military signal as well. All navigation equipment made in India would also be GPS compatible.[513] Finally, in order to improve the system's commercial prospects Russian authorities announced that the next generation satellites' civilian use signal would be compatible with the GPS and Galileo systems.

In a similar fasion, Russian officials discussed with their Ukrainian counterparts the possibility of creating a joint venture for the supply of Glonass services to that country as well. Although Ukraine has previously opted for using the GPS satellite positioning system, the recently elected government seemed to prefer the simultaneous use of both the U.S. and Russian systems, but no final decision had been made yet. Finally, broader negotiations between the two countries on possible cooperative ventures in the field of communications, including Satcom's, were expected to start in the second half of 2010.[514]

4.3. Space-based Earth observation

In the field of Earth observation, 2009 and 2010 saw considerable advances in the European GMES EO programme. The European Commission also had a number of decisions to make during the period in question, including defining a data-access policy and securing future funding. GMES would consist of three dedicated Sentinel surveillance satellites and two payloads on other satellites, for which the European Commission has secured a €2.3 billion budget while most of the hardware will be paid for by ESA's 18 Member States. In addition to this, the programme will utilise national satellites built by Member States. However, funding of the future generation Sentinels after 2013 still remained unclear. Although European Commission officials were previously leaning towards a user-defined/user-paid approach, press reports indicated that longer operating cost recoupment methods were also considered.[515]

The question of the future funding of the GMES space segment preoccupied the European Commission and Member States throughout the year, as the European Commission budget line for the programme would expire in 2013 and further funding would be needed to develop the next generation of its Sentinel satellites. For the time being, six Sentinel satellites and two dedicated payloads onboard EUMETSAT satellites are planned. However, with an expected satellite operational life-span of seven years, the next generation of Sentinels will have to enter into development soon, in order to sustain operations after 2020.[516]

As far as future funding is concerned, all possible options were investigated by the European Commission. The conclusion was that only an adequate long-term budget for GMES operations after 2013 would guarantee a profitable return for the investment already made on the programme. In other words, backing away form the project now would defeat its declared objective of a sustainable Earth observation capability for Europe. Furthermore, without additional future funding the programme would not reach its full research and development potential and it would not produce any significant technological returns for the European space industry.[517] An independent ESA long-term analysis of the programme also reached the same conclusion. According to this, a €600 million annual budget between 2014 and 2020 would be required, including €470 million for operational activities and €170 million for future research and development, (R&D).[518]

Several other issues concerning GMES also remained open, including ownership of the system, data policy, procurement policy and governance arrangements. The most probable scenario contemplated would be for the transfer of the Sentinels' ownership from ESA to the European Commission (EC). Such an arrangement would make the EC the operator of the system's space segment, thus assuring a free and open access policy to its data. It would also normalise the project's governance, with the EC acting as the programme manager deciding on system upgrades, ESA working as the development and procurement agent on behalf of the EC, and EUMETSAT operating oceanographic and meteorological components onboard its satellites.[519] Finally, another milestone for the programme's governance was accomplished on 5 February 2010, with the European Commission's decision to set up the GMES Partners Board. The 27 member board (one from each EU member country) would function as a panel of experts, monitoring the GMES project's implementation and providing strategic guidance for its future development.[520]

ESA, on the other hand, launched on 1 November 2009 its Soil Moisture and Ocean Salinity (SMOS) Earth observation satellite and the Proba-2 technology demonstration spacecraft, aboard a Russian Rockot vehicle operated from the Plesetsk Cosmodrome. The 658 kg SMOS was put into a near-polar sun-synchronous orbit at an altitude of 760 km. It is equipped with a Spanish Microwave Imaging Radiometer using Aperture Synthesis (MIRAS) instrument, the first major satellite instrument built for ESA in that country. The device is using L-wave microwave frequencies to derive information about soil moisture and ocean salinity levels based on their reflection. Proba-2 is an ESA spacecraft that is set to test future satellite systems and instruments currently under development. Total costs of the programme reached €333 million.[521]

5. Technology developments

5.1. Propulsion

The period 2010–2010 has seen the emergence of a lot of new technologies and improvements while questions arise concerning the technological challenges to take up, brought by the prospective of Mars exploration. The main cutting-edge evolutions ongoing come predominantly from Europe, the U.S., Russia and Japan. It is worthy to note that almost each country or region have its own programme in this field particularly strategic. This is not astonishing because it affects directly the weight that the launcher can lift and so the profitability of the system. Europe has made significant progress especially concerning its IXV (Intermediate eXperimental Vehicle). It is now in a transitory period of its FLPP (Future Launchers Preparatory Programme) between the step two (2009–2012) which includes Completion of systems studies on expendable launch configurations, progression through ground demonstrators, in particular for high thrust engine, in-flight experiments and cryogenic upper stage technologies, and the step three (2012–2015) with the flight of the IXV on an ESA Vega Launcher. These tests are necessary to prepare the development of next generation launcher and begin to design the future vehicle.[522] This movement was particularly noticeable with the signature of an agreement between ThalesAleniaSpace and ESA for the development of the IXV.[523] A new step was reached with the workshop hold in Paris in September 2010 which came within the scope of developing a European re-entry flight module. The main industrials involved in the development of the next ESA's Intermediate experimental Vehicle had the opportunity to discuss and suggest their hardware development and issues. The event was the occasion to share technical achievement and to envisage the short and long term perspectives concerning this crucial project. The IXV project objectives are the design, development, manufacture, ground and flight verification of an autonomous European lifting and aerodynamically controlled re-entry system. Three sides of the project have been particularly studied namely, advanced instrumentation for aerodynamics and aerothermodynamics, thermal protection and hot-structure solutions, guidance, navigation and flight control through a combination of thrusters and aerodynamic flaps: It is foreseen that the IXV be launched in 2013 on Vega Europe's new small launcher, as part of the "VERTA" (Vega Research and Technology Accompaniment) programme. After re-entering Earth's atmosphere and being slowed by friction from 7.5 km/s, it will descend by parachute and land in the Pacific Ocean to await recovery and analysis.

A High-thrust engine demonstrator industrial day has been also hosted in Germany during February. It has confirmed the commitment to develop a liquid propulsion system for first stage propulsion providing flexibility and efficiency. The demonstrator is due to be definitely chosen by mid 2010 and will undergo firing test around 2015. The activities were presented by the Joint Propulsion Team, a contractor consortium composed of Astrium GmbH (D), Avio SpA (I) and SNECMA (F). It is a critical endeavour in the will to equip Europe with a new generation of launcher able to continue the "success-story" of our continent in this strategic domain.[524]

Concerning propellant itself a great step has been achieved to overtake hydrazine. This one is a high-performing storable propellant, given its characteristics and remains the main source of fuel for satellite. Unfortunately hydrazine is also highly toxic and ground personnel have to wear protective gear in all procedures dealing with this substance. This situation is due to change as ESA and a Swedish company called ECAPS, part of the Swedish Space Corporation Group develop a new propellant officially called LMP-103 S, this new fuel is a blend of ammonium dinitramide (ADN) with water, methanol and ammonia. The AND is not only due to give better performance (around 30%) but it will be much less toxic for humans and their environment as well. That would facilitate and reduce the cost of satellite handling as they would not have to be fuelled at the last moment as this is done now due to safety reasons.[525]

Concerning engine more particularly, the development of the new Vulcain equipping currently Ariane 5 has known a new step with the test of a new nozzle design (codenamed NE-X) for the first time. The nozzle is built by the Swedish company Volvo. The innovative design consists of a sandwich system construction "an external metal cone is laser-welded to an inner one, doing away with the numerous cooling ducts of the current design. The result is a construction that is considerably cheaper to produce than the existing Vulcain nozzles, as well as lighter than more powerful, thus increasing the payload capacity and cost-efficiency of the Ariane 5. Numerous of tests have been carried out so far to gauge the viability of the system and adapt it in order to enhance the global performances of the engine and so Ariane 5 itself.[526]

In the other side of the ocean Atlantic is not outdone with a couple of significant innovations in this field. The NASA's ion-propelled Dawn spacecraft has eclipsed the record for velocity change produced by a spacecraft's engines, deep in the heart of the asteroid belt, on its way to the first of the belt's two most massive inhabitants. The Dawn mission to Vesta and Ceres is managed by JPL, a division of the California Institute of Technology in Pasadena, for NASA's Science Mission Directorate, Washington[527] Then, NASA has achieved a milestone in its preparation for a third major rocket engine test project concerning the next generation J-2X rocket engine.[528] The transition work from the space shuttle main engine project to the J-2X test project included structural, electrical and plumbing modifications to accommodate the different geometry of the J-2X engine, and included the installation of a new J-2X engine start system. Liquid oxygen and liquid hydrogen transfer lines that dated back to the 1960s also were replaced, as was other piping on the stand. Control systems also were upgraded on the stand. The J-2X engine is being developed by Pratt & Whitney Rocketdyne for NASA as a next-generation engine that can carry humans beyond low-Earth orbit to deep space. These crucial improvements announce thus future and maybe more daring missions.

Another field explored by the U.S. agency consists of a new kind of manoeuvre without any fuel depending only on the power of Earth's magnetic field to move satellite and spacecraft in orbit.[529] This particular force could be employed to compensate the degradation of their orbits due to friction from colliding with atmospheric particles that have escaped into space. This technology would considerably reduce the need of propellant and could prolong the satellite's life expectancy. This would be useful as well to make them re-entry into the atmosphere when they will have accomplished their mission.

Japan has also undertaken research to develop a new kind of solid propellant for its rockets. The main objectives are to facilitate the operations in the grounds facilities and save about one fourth of the time previously needed. The new system expected should improve the liquid propellant consumption and simplify the procedures making them meanwhile safer.[530]

Another major Japan's space propulsion project is to launch a "space yacht" propelled by solar particles that bounce off its kite-shaped sails. The ikaros is a prototype of interplanetary Kite-craft accelerated by radiation of the Sun, the system is a "space yacht" that gets propulsion from the pressure of sunlight particles bouncing off its sail. The flexible sails, which are thinner than a human hair, are also equipped with thin-film solar cells to generate electricity to create an hybrid technology of electricity and pressure.

Nuclear energy seems to particularly interest the Russian which would like to take advantage from their know-how in this field. A lot of events have underpinned this will during 2010–2011. For instance, Russian President Dmitry Medvedev defined Keldysh R&D Center as a sole designer of the megawatt nuclear propulsion system.[531] This event took place before the statement of Roscosmos Head. According to him, attempts to improve parameters of the existing rocket propulsion systems are unreasonable. Indeed, the improvement of actual fuel propellant would be only measurebable in fraction of percentage, far from the power necessary to furnish reliable propulsion to reach Mars. Nuclear would impose itself as sole realistic alternative especially for large scale manned mission.[532]

This affirmation has been backed by Russian scientists from the Moscow Physical institute[533] which currently study the opportunity offered by powerful plasma engines in nuclear propulsion systems which could provide according to prof. Oleg Gorshkov the necessary technology to build a new generation of spacecraft. One more time this conclusion supports the idea that nuclear energy would be much more effective for long term and human mission, these kind of Ionic engines are featured by 5-year life-time. By this way the Russian authorities could prepare their participation to an eventual future mission towards Mars which would probably necessitate a nuclear propulsion system much more light and powerful than classic propellant.

5.2. Information technology

An important step has been made with the ESA's SMOS satellite which includes the use of fibre optic in the micro satellite.[534] It represents thus a historic step forward "photonics" in space. Proba-2's experimental payload includes a fibre-optic sensor system which monitors its propulsion, while SMOS's triple-armed MIRAS (Microwave Imaging Radiometer using Aperture Synthesis) is entirely reliant on an optical fibre-based communication harness. This represents a historical turning point for the use of lightwave technologies in space, there are more than 500 metres of optical fibres embedded within the MIRAS instrument. Through this project ESA extends the use and network of optical fiber to Space. This system allows not to produce electromagnetic radiation to transmit information which could be polluted among other electrical signals, indeed Any electrical noise leaking out from the electronic boxes in the arms can influence the correlation measured and blur the resulting images and optical fibres relay information with light pulses rather than electrical signals, so they do not produce any. But MIRAS's optical fibre harness turned out to bring other advantages. Their lower weight meant that the instrument's long arms could be unfolded in orbit using relatively light spring-loaded motors, and the fibres' mechanical flexibility meant their performance was unaffected by this movement. This is therefore a crucial step ahead which will probably lead to other applications on other future satellites. Another important advance has been done with Hylas-1, the first satellite created specifically to deliver broadband access to European consumers, is very much a commercial undertaking. It is also a significant technological achievement, encapsulating a decade of research and development by ESA and European industry. This satellite has also the specificity to be a successful public private partnership through UK operator Avanti Communications and ESA. The satellite is particularly adaptable and is due to enter in service very quickly contrary to its counterparts. It has also the capability to reuse its given allocation of radio frequencies between spot beams. The operator can fine-tune how much bandwidth and power to put in each beam so it is possible to match changes in data demand as they happen in order to be more reactive to the market.

The satellite is the result of an intense cooperation between ESA and Atrium for the development of a Modular Microwave Hybrid Technology' (MMHT). A successful relation which has produced this Generic Flexible Payload making the Hylas satellite so efficient. The flexibility that the GFP delivers to Hylas-1 is increased by another pioneering piece of technology, developed with ESA support by Tesat-Spacecom in Germany: the In-Orbit Adjustable Microwave Power Module (IOA-MPM) allows the transmit power signal to be adjusted to match demand while maintaining near-constant efficiency, preventing power being wasted in the form of heat. One last innovation is the most visible: Hylas-1's larger, double-sized antenna, which had to be carefully optimised for high-frequency Ka-band operations, the responsibility of EADS Casa Espacio in Spain. Hyas project is thus the vibrant example that we can associate in a PPP the needs of profitability and science.

5.3. Spacecraft operations and design

This subject has brought an intense reflexion in various areas such as Europe, Russia, U.S. and Japan to find solution or arise the future challenge that the future Spacecrafts will have to take on.

It is such a defy to fly multiple satellites in formation, separate multi-million-euro pieces of hardware, each one moving through space at several kilometres per second through an hostile environment. Lose the control even momentarily can entail dramatic consequences that a cutting-edge software try to avoid. To respond to this challenge ESA and industrial have developed a new software to get to grips with multiple-satellite missions to come. This one is a generic simulator for formation flying mission no matter how many spacecraft are concerned. Of course this system needs several CPU in order to calculate the different trajectories so the software is distributed among several machines without affecting the users of them. This new technology will be used to address crucial operational factors for formation flying, including mission and vehicle management, guidance navigation and control, fault detection, isolation and recovery and inter-satellite links.[535] A test is ongoing on Proba-3 and already announces itself has a great step in Spacecraft management.

In Russia, a constellation of nano-satellites is to be launched by Russian scientists in the coming two years.[536] It is another development of nano-tehnology in Russia which started 5 years ago when Russia launched its first nano-satellite. Russia is firmly committed to investing in this new branch particularly crucial in space application. The only national micro-satellite was launched manually from the ISS by cosmonauts Salidzhan Sharipov. Another ground of reflexion in Russia is symbolised by the CIS programme called Cosmos-NT which is aimed at reducing duration and expenses on development of new space technologies in order to achieve the goals of space programs for the benefits of Russia and Belorussia.[537]The program will include a multifunctional space system which is an integrated part of Russian and Belorussian ground and orbital systems. The system will provide space monitoring data to the users, maintain acquisition of navigation data and remote education on the basis of communication and relay satellite systems. Cosmos-NT includes development and verification of space data display systems, design of a unified micro-satellite platform, development of advanced propulsion systems and power supplies, etc, BELTA informs.

Russia has also undertaken the development of a new vehicle which will be capable of docking to the orbital station on the day of its launch.[538] The decision has not been definitively made yet and the convenience for the crew and the cost-effectiveness of this solution must be verified.

This vehicle would be multifunctional with several types of mission such as to LEO and LLO, spacecraft orbital maintenance, deorbiting big space debris and it could be also used to fly space tourists. The configuration expected for the vehicule is a capability to fly autonomously for 5 days, and in the lunar modification 4 cosmonauts shall fly in the vehicle for 14 days.

Japan is pressing forward with ambitious plans to enhance its role on the international space station (ISS), announcing its intentions to better utilize the Kibo laboratory and build a variant of the H-2 Transfer Vehicle (HTV) that would be capable of bringing cargo back to Earth.[539] JAXA officials have presented two enhancements which would allow the HTV to return cargo from the space station before 2020 and put the automated cargo supply ship on the path to eventually carrying humans. The Japanese government is particularly committed to improving the current HTV since a statement in which he has firmly expressed its will to procure to Japan its own human space capabilities. The main aim is to provide an alternative way to resupply the ISS while that would also allow to maintain the nation's space industrial base, and also its participation in the ISS which constitutes an important part of Japan's soft diplomacy and power projection.

5.4. Suborbital activities

The suborbital news are mainly dominated by the U.S. which extends important investments to foster the private activity in this field. However a significant English technology could soon reestablish the balance between Europe and the U.S. in term of reusable and suborbital spacecraft.

2010 is a milestone for suborbital activity, Masten Space Systems of Mojave has signed a letter of intent with space florida to explore conducting demonstration launches of Masten's suborbital reusable launch vehicle from Cape Canaveral.[540] A crucial step in space tourism has been also reached by virgingalactic. Though SpaceShipTwo did not reach space, the flight was a major step for the private suborbital spacecraft, which flew in glide mode for 11 minutes after being released from its WhiteKnightTwo carrier plane at an altitude of 13,700 meters. Accomplished on the 13 October 2010 this successful flight allows Virgin Galactic to go ahead in this non explored field.[541] Virgin Galactic is resolutely on the roll. Indeed, a new law has been enacted in New Mexico and saluted by the company. This legislation reduces significantly the risk that space tourism operators will face crippling lawsuits brought by surviving family members of a participant injured or killed during flight.[542] The performances targeted are quite similar in term of thrust and efficiency, but the real objective is to develop rather more an engine that could be dropped into existing stages. The real challenge is thus to make it much cheaper and adapt the existing engine to the XCOR's spacecraft Lynx. The suborbital spacecraft is designed to carry tourists along with research payloads. Those recent developments bring about analyst to foresee a global fall of the suborbital tickets for the next years.[543] Indeed hundreds or even thousands or flights could be soon operated once the technical aspects have been solved. The question of the future of this market is however a crucial question given the huge investments which have been made so far. It is expected that the markets growth as the price by seat drop but that does not mean necessarily that could become a sustainable activity. An important topic particularly treated during July 2010 at the Space Frontier Foundation's annual conference in Sunnyvale.

The development of commercial suborbital flights is kindly watched over by NASA which tries hard to encourage this activity through programmes such as the Commercial Reusable Suborbital Research (CRuSR) under which the agency proposes to spend $75 million over the next five years to make use of commercial suborbital vehicles. According to the U.S. agency those flights would constitute a real opportunity to carry out there scientific experimentations at a lower price.[544] Other scientific and independent entities have decided to invest this field by investing or booking pre-ticket. A lot of disciplines are concerned such as astronomy, life sciences, and microgravity physics. There are no less than five companies which are due to propose suborbital flights, namely These firms, led in some cases by technology industry pioneers, include Virgin Galactic, Armadillo Aerospace, Blue Origin, Masten Space Systems and XCOR Aerospace. That could be a crucial asset in order to rend the market profitable and competitive enough by maintaining several operators in course allowing thus lower prices.

The race for the NASA awards concerning the development of commercial launchers and spacecraft that would transport astronauts to and from low Earth orbit is particularly fierce. While Sierra Nevada Corp is the big winner in NASA's first round of Commercial Crew Development awards, Orbital Team could take the lead by proposing a new lifting-body spacecraft capable of carrying at least four passengers to orbit by 2015.[545]

The spacecraft, designed to launch atop an Atlas 5 rocket and dock with the international space station, could be ready for test flights as early as 2014. The remotely piloted spacecraft would be able to carry four passengers initially, including three astronauts and one paying ticketholder.

JAXA undertook scientific research with several releases of balloons from the Taiki Aerospace Research Field.[546] The release experiment, called BS10-06, aimed at testing the high-altitude thin film balloon flight performance and observing the ozone and atmospheric gravitational waves. The balloon that was expanded to its full capacity of 60,000 m^3 was made of a thin film for high altitude with a thickness of 3.4 micrometers, and it ascended about 300 meters per minute. In this experiment, we achieved the scheduled objectives of flight verification for a thin-film high-altitude balloon using the polyethylene thin film that is wider than conventional films, and the balloon's tearing mechanism for a thin-film balloon. At the same time, the balloon carried out precise observations on the ozone, wind velocity, temperature and air pressure using two kinds of ozone measurement devices, namely an optical type, and electrochemical type.

5.5. Other technologies

The 2009–2010 period has been particularly rich in technological and science progresses and advances in the current major programmes. First of all, Astrophysics and space observation have experienced significant progresses.

Dark matter has interested scientists for many years, two researches carried out could improve our knowledge about this crucial topic. Astronomers from NASA have created one of the most accurate map of dark matter in the universe thank to Hubble space Telescope.[547] An important step to better understand this crucial element of the universe functioning. The researches are particularly focused on the massive Galaxy Abell 1689 which could explain the particular role of this substance. Astronomers have devised a new method for measuring the dark matter through cosmic lens for this purpose.[548] The AMS (Alpha Magnetic Spectrometer) launched by endeavour on 16 April 2011, it should provide other crucial information about dark matter and black holes. It has been assembled in the CERN (European Center of Nuclear Research) based in Geneva and is the result of cooperation between ESA, NASA and the CERN.[549]

In Europe, the magnetic tests are ongoing to check if the spacecraft LISA Pathfinder is magnetically clean,[550] The project among the most difficult challenge treated by ESA is due to be launched in 2013 and concerns gravitational wave detection. To succeed, the mission has extremely stringent requirements to limit any disturbance of the test masses by magnetic materials or effects. Having first characterised these effects, designers sought to minimise them. Any magnetic disturbances could condemn the entire mission to failure.

Astrium is due to supervise the development of an atomic clock using two new-generation atomic clocks to be operated aboard the international space station under a contract with the European Space Agency (ESA).[551] The system will be launched by 2013 aboard a Japanese HTV. The first, called Pharao, was developed by the French space agency, CNES. It is a laser-cooled cesium clock designed specifically for use in a microgravity environment. The second, developed by the Observatory of Neuchatel, Switzerland, is a hydrogen maser clock. Ground based atomic clocks through a microwave network will be used as comparison with the two ACES clocks to measure the differences.

The ESA's Herschel infrared space observatory has discovered the key ingredient for making water in space-ultraviolet starlight.[552] A crucial finding which explains why a dying star is surrounded by a gigantic cloud of hot water vapour. Herschel's PACS and SPIRE instruments have revealed that the secret ingredient is ultraviolet light, because the water is too hot to have come from the destruction of icy celestial bodies. The Herschel water detection made the astronomers realise that ultraviolet light from surrounding stars could reach deep into the envelope between the clumps and break up molecules such as carbon monoxide and silicon monoxide, releasing oxygen atoms. The oxygen atoms then attach themselves to hydrogen molecules, forming water.

Concerning rather more life in space and medicine. In 2010 NASA had announced a capital discovery concerning other form of life,[553] if a lot of people have been finally disappointed by the new it was however an important step. Indeed, NASA astrobiology research has changed the fundamental knowledge about what comprises all known life on Earth. The first micro organism on earth capable to use Arsenic in its reproduction process has been observed. These researches allow us to rethink what we had previously considered as the necessary basis of all form of life. In the future this kind of discoveries could allow us to detect more easily and better understand extraterrestrial form of life.

ESA coordinates a project made of French companies MEDES Magellium and the CNES designed to use Ultrasound as medical technique.[554] This system is particularly useful for remote operation by distant specialists which could be crucial in case of operation aboard a spacecraft. ESA tests currently this new robotic ultrasound system within a project expected to last two years.

Medicine is a domain in which ESA has already shown its capacity to transform a scientific discovery into a successful innovative and commercial project as it has been recently illustrated with the telemedicine system telemedicine system adapted to answer to a potential onboard medical emergency.[555]

In a more practical view numbers of important programmes have been carried out. Experiment concerning nanotechnologies have been engaged to better understand the creation process of carbon nanotube,[556] and especially how carbon gets recycled in the regions of space that spawn stars and planets. Lead in cooperation between NASA and a Japanese university, this crucial study could bring about some cutting edge progress which could find application in various sectors and even enhance our understanding of certain supernova for instance. The first human-like robot Initially developed by NASA and General Motor the so called R2 is due to join the ISS crew and become a permanent resident of the ISS.[557] Its operational functions will be being studied on weightlessness.

It announces future enhancement in order to make it sufficiently autonomous to move inside and outside the complex executing tasks more and more complex. The current model is not yet equipped with system resisting to the extreme temperature of outer space but it is just a matter of time.

NASA experiences new tank dome technology[558] in partnership with Lockheed Martin Space Systems and MT Aerospace in Augsburg, Germany. This new development should reduce the weight of future liquid propellant tanks by 25 percent, compared to current tank designs that use a lower-strength aluminum alloy that weighs more. The concave net shape forming process patented by MT Aerospace simplifies the manufacturing and reduces thus considerably the costs Beyond the technological feat it is a good example of international cooperation between Europe and the U.S:. This project has been funded by the Exploration Technology Development Program for NASA's Exploration Systems Mission.

The study of earth and particularly its weather provokes concerns and thus a lot of innovative solutions to deal with the upheavals brought about by the climate change.

NASA has carried out an experiment using a new prototype designed to surveying the impacts of aerosols and clouds on global climate change.[559] The successful test is the first step to equip weather and scientific satellite with this instrument. NASA has also revealed its new tool for weather forecast called iSWA[560] (Integrated Space Weather Analysis) which gathers information from spacecraft including the National Oceanic and Atmospheric Administration's (NOAA) Geostationary Operational Environmental Satellites (GOES), NASA's Solar Terrestrial Relations Observatory (STEREO), the joint European Space Agency and NASA mission Solar and Heliospheric Observatory (SOHO), and NASA's Advanced Composition Explorer (ACE). It is expected that this new system in constant evolution allows to improve weather forecast and global understanding of it by making information more available to the scientific community.

Earthcare a satellite jointly developed by Japan and ESA is undergone vacuum test[561] before its launch scheduled for 2013. Equipped with four sensors (Cloud Profiling Radar, Backscatter Lidar, Multi-Spectral Imager and Broadband Radiometer) it will permit to improve the accuracy of climate change predictions. It is know that the effects from clouds and aerosols make sometimes current predictions unreliable. This phenomenon between radiation in interaction with clouds and aerosols can be studied and solved enhancing the reliability of the data collected to study the weather.

ESA is working on a largely ground-based project: designing the Agency's Space Situational Awareness infrastructure, which will allow Europe to track potential hazards in space. This system will permit Europe to predict, detect and assess the risk to life, property and in particular space assets due to natural or man-made space hazards. A contract was recently awarded to ThalesAleniaSpace (France) supported by Spanish, French, German and Belgian companies.[562] The network of computers organizes will allow to constitute an efficient and coordinate tools to detect any risky situation from space. This allows for iterative design at a fast pace, with customer and designers agreeing requirements and taking decisions in real time to ensure the best design for the right cost and an acceptable risk. New navsat sensors developed within the scope of the ESA Business Incubation Centre in the Netherlands allow to improve electricity production from a hydroelectric plant on Lake Laja in Chile by using navigation satellite signals to measure water levels and wave heights in real time.[563] The device provides a very accurate of the water level reducing drastically the maintenance needed to check the water sensors.

The burning topic of space debris has brought about some interesting development. Especially in Russia where tracking space debris has been becoming a real concern for Russian authorities which have planned to tackle this crucial issue for the benefit of the worldwide space activities. We have seen a great involvement and significant progress to develop adapted technologies. To achieve this purpose, a group of scientists from Lebedev Physical Institute (FIAN) have developed a unique special tracker and SW which can be used to search for small space debris from 1 to 10 cm.[564] The tracker could be installed on any spacecraft especially concerning those located in risky orbit such as GEO. A more ambitious programme in Russia again is devoted to launching a special orbital pod that would sweep up satellite debris.[565] The system whose the cost is estimated around ($1.9 billion) could help to clean up busy orbit such as LEO and GEO by collecting or sinking them into ocean. The cleaning satellite would work on nuclear power and would be capable to work up to 15 years, he said. The company Energia that would be in charge of this task announce a complete assembly by 2020 and test the device no later than in 2023. In the same way, Alliant Techsystems (ATK) is currently developing a new system specially designed to tackle debris too small to be tracked by ground-based telescopes but large enough to penetrate satellite shielding.[566] The plan is to launch a spherical spacecraft enclosed in multiple layers of a lightweight material. The spacecraft would operate in low Earth orbit as a sweeper or shield, breaking up debris particles and reducing their velocity.

5.6. Innovation policy

The U.S. space strategy seems to take a more practical path while President Obama makes a no binding promise of developing by 2025 "a new spacecraft designed for long journeys to allow us to begin the first-ever crewed missions beyond the moon into deep space" [567] and furthermore by the mid-2030s to send human on Mars. The devotement to favour new technologies can be seen through the budget which shows a sharp engagement in research and development programs[568] Obama is asking the U.S. Congress for $19 billion for NASA for the year ahead, a 1.5 percent increase over the agency's 2010 budget. The spending includes particularly new technology programmes such as robotic missions and propulsion research. The main rise concerns primarily science Earth observation that focuses the U.S. efforts. The agency intends to use prize competitions, to encourage public-private partnerships and other approaches to develop next-generation technologies. The great new of the Obama space program was the cancellation of the Constellation programme. Indeed, that has released funds particularly devoted to develop new initiatives such as commercial crew program, heavy Lift and Propulsion or in orbit refueling.

Russia shows a strong will to cope with international competition in space. A determination noticeable in several occasions. Russia is expected to launch a scientific mission after a long absence in this area.[569] The astrophysical observatory Spektr-R is to fly in May-June 2011 and should study interplanetary magnetic field and black holes. Interplanetary station to Mars' moon Phobos-Grunt is to be launched in late 2011. This one is to deliver soil from Mars' moon and to study both Mars and Phobos. Satellites such as Loutch-5A spacecraft whose the launch is scheduled for 2013 is intended to reinforce the Russian position in space by deploying a multi track and effective relay satellite capable of transmitting data from any other spacecraft. It will be effective for agriculture, military, weather forecast and other services.[570] Russian government includes the space sector in its plan of modernization and technological evolution by providing extra money to Roscosmos the national agency.[571]

ESA has chosen its main scientific missions in 2010. The topics selected are Dark energy which is currently studied by project such as the AMS, habitable planets around other stars, and the nature of our own Sun.[572] The Euclid mission will be particularly devoted to addressing key questions relevant to fundamental physics and cosmology and particularly concerning dark energy and dark matter which are known to be crucial in the functioning of the universe. The PLATO mission will concern rather more the frequency of planets around other stars and especially the fascinating possibility to find another habitable planet. Finally, Solar Orbiter will be designed to studying more closely the activity of our Sun especially concerning solar far side when it is not visible from Earth. All three missions present challenges that will have to be resolved at the definition phase, the decision is thus not definitely made. Participation to the Japanese SPICA infrared telescope programme is also envisaged. Beyond its scientific choices, ESA as its U.S.

counterpart is broadening its effort to encourage space technology spin-offs through investments in the creation of a new fund specially dedicated to this purpose with the Open Sky Technologies Fund specialised in arising from space technologies and satellite applications for terrestrial industries is expected to reach not less than €100 million in 2011.[573] This help will take the form of aid to small companies and protecting patents from ESA work which particularly crucial for modest size companies considering their deterrent cost especially when they must be taken in the U.S. ESA is also strongly engaged to find concrete application on earth of space technologies, we can observe a result for instance with a spin-off company supported by ESA which develops software that uses conventional satnav signals to obtain accurate positioning with centimetre precision. Based on ESA satellite-control software, it has already attracted customers in the oil and gas industry.[574]

---

[408] De Selding, Peter B. "European Commission Picks Underdog To Build 14 Galileo Navigation Satellites." Space News 11 Jan. 2010: 1.
[409] "Major Galileo Contracts Signed." 27 Jan. 2010. European Space Agency Galileo Navigation Website. 30 Mar. 2010. http://www.esa.int/esaNA/SEMPX0SJR4G_galileo_0.html.
[410] De Selding, Peter B. "Galileo Program Feels Sharp Rise in Russian Rocket Prices." Space News 18 Jan. 2010: 4.
[411] Klamber, Amy and Brian Berger. "Obama's 'Game Changing' NASA Plan Folds Constellation, Bets Commercial." Space News 8 Feb. 2010: 1.
[412] Ibid.
[413] Ibid.
[414] Berger, Brian. "Bolden Overhauls NASA Organisation, Centralises Chain of Command." Space News 1 Mar. 2010: 12.
[415] De Selding, Peter B. "Constellation's Demise Will Cost Ares-1 Builder ATK $650 Million in Backlog." Space News 8 Feb. 2010: 5.
[416] Klamper, Amy. "Obama's Move to End Constellation Prompts Industrial Base Questions." Space News 15 Feb. 2010: 1.
[417] Ibid.
[418] Klamper, Amy. "Bolden: Heavy-lift Development To Persist Despite Ares 5 Demise." Space News 15 Feb. 2010: 16.
[419] "Kazakhstan Finally Ratifies Baikonur Rental Deal With Russia." 9 Apr. 2010. RIANovosti 30 May 2010. http://en.rian.ru/world/20100409/158494207.html.
[420] De Selding, Peter B. "Perminov Briefs Putin on Glonass, Other Russian Space Programs." Space News 20 July 2009: 6.
[421] "NASA, Roskosmos Sign $306 Million Soyuz Deal." Space News 1 June 2009: 3.
[422] "Russia to Spend $13.5 Bln on New Space Centre in Far East." 15 Oct. 2009. RIANovosti 30 May 2010. http://en.rian.ru/russia/20091015/156475196.html. "Russia to Boost Share in Global Space Market." 26 Mar. 2010. RIANovosti 30 May 2010 http://en.rian.ru/russia/20100326/158322158.html.
[423] "Japan Conducts Ground Test of H-2B Rocket." Space News 20 July 2009: 8.
[424] Malik, Tariq. "Japan's First Space Cargo Ship ready to Fly." Space News 7 Sept. 2009: 14.
[425] Malik, Tariq. "Japan's HTV-1 Bound for Station." Space News 14 Sept. 2009: 22.
[426] "Japan's GX Rocket Targeted for Cancellation in 2010." Space News 23 Nov. 2009: 8.

[427] "Japan Scraps GX Rocket Development Project." I StockAnalyst 16 Dec. 2009: 30 May 2010. http://www.istockanalyst.com/article/viewiStockNews/articleid/3716870.
[428] De Selding, Peter B. "Palapa-D to be Salvaged After Being Launched into Wrong Orbit." Space News 7 Sept. 2009: 10. "Long March Mishap Report Due by Mid-Nov." Space News 14 Sept. 2009: 16.
[429] De Selding, Peter B. "Burn-through Blamed in China Long March Mishap." Space News 23 Nov. 2009: 4.
[430] "ISRO's Cryogenic Stage Fails in Maiden Flight." Space News 19 Apr. 2010: 3.
[431] Parks, Clinton. "S. Korean Officials Investigate Launch Failure." Space News 26 Aug. 2009. http://www.spacenews.com/launch/korean-officials-investigate-launch-failure.html.
[432] Opall, Barbara. "Israeli Missile Experts: Simorgh Sets Iran on Path to ICBM." Space News 15 Feb. 2010: 14.
[433] "Brazil To Develop Carrier Rocket By 2014." 06 Apr. 2010. Space Travel. 28 May 2010. http://www.space-travel.com/reports/Brazil_To_Develop_Carrier_Rocket_By_2014_999.html.
[434] "NASA to Launch Human-Like Robot to Join Space Station Crew." 15 Apr. 2010. NASA http://www.nasa.gov/topics/technology/features/robonaut.html.
[435] "Station Gains Unparalleled Views." 25 Feb. 2010. NASA. http://www.nasa.gov/mission_pages/shuttle/shuttlemissions/sts130/launch/130_overview.html.
[436] "Preparing for the Research Decade." 19 Nov. 2010. NASA. http://www.nasa.gov/mission_pages/station/behindscenes/astronautics_conference.html.
[437] "STS-131: Teamwork Overcomes Mission's Challenges." 23 Apr. 2010. NASA. http://www.nasa.gov/mission_pages/shuttle/shuttlemissions/sts131/launch/131mission_overview.html.
[438] "Atlantis' Final Mission." 14 May 2010. NASA. http://www.nasa.gov/mission_pages/shuttle/shuttlemissions/sts132/atlantis_final.html.
[439] "Progress 36 Docks to Station." 4 Feb. 2010. NASA. http://www.nasa.gov/mission_pages/station/expeditions/expedition22/p36_dock.html.
[440] "New Russian Cargo Vehicle Docks to Station." 1 May 2010. NASA. http://www.nasa.gov/mission_pages/station/expeditions/expedition23/p37_dock.html.
[441] "STS-133: Final Flight of Discovery." 2 Nov. 2010. NASA. http://www.nasa.gov/mission_pages/shuttle/shuttlemissions/sts133/discovery_final_flight.html.
[442] "Crystal Growth Space Lab May be Installed in the ISS." 30 Aug. 2010. Roscosmos. http://www.federalspace.ru/main.php?id=2&nid=10216.
[443] De Selding, Peter B. "Member To Commit to Station Extension? Germany is First ESA." Spacenews 30 Sept. 2010. http://www.spacenews.com/civil/100930-germany-commit-iss-extension.html.
[444] Kallender-Umezu, Paul. "Member Japan Aims To Boost Its ISS Role with Cargo Return Vehicle." Spacenews 23 Aug. 2010. http://www.spacenews.com/civil/1100823-japan-aims-boost-iss-role.html.
[445] "Next step for ESA's first Moon lander." 16 Sept. 2010. European Space Agency. http://www.esa.int/esaCP/SEMUV2KOXDG_index_0.html.
[446] "Moon may Be Shrinking!." 22 Aug. 2010. Roscosmos. http://www.federalspace.ru/main.php?id=2&nid=10146.
[447] Euroconsult Research Report, Government Space Markets, World Prospects to 2020. Paris: Euroconsult, 2010.
[448] Space News Staff. "China's Chang'e 2 Probe Enters Orbit Around Moon." Spacenews 11 Oct. 2010. http://www.spacenews.com/civil/101011-change-orbit-moon.html.
[449] "Russian Spacecraft to Study Lunar Ice." 7 May 2010. Roscosmos. http://www.federalspace.ru/main.php?id=2&nid=9803.
[450] "Lunar Ice Comes in Blocks." 22 Apr. 2010. Roscosmos. http://www.federalspace.ru/main.php?id=2&nid=9175.
[451] "Tracing the Big Picture of Mars' Atmosphere." 26 Aug. 2010. NASA. http://www.nasa.gov/mission_pages/mars/news/exo20100826.html.

[452] "Instruments selected for Mars." 2 Aug. 2010. European Space Agency. http://www.esa.int/esaSC/SEM6XEZNZBG_index_0.html.
[453] "DLR investigates the existence of liquid salt solutions on Mars." 23 July 2010. DLR. http://www.dlr.de/en/DesktopDefault.aspx/tabid-6215/10210_read-25885/10210_page-4/.
[454] "Wind and water have shaped Schiaparelli on Mars." 10 Dec. 2010. European Space Agency. http://www.esa.int/esaSC/SEMEEYOR9HG_index_0.html.
[455] "Orbiter Resumes Science Observations." 20 Sept. 2010. NASA. http://www.nasa.gov/mission_pages/MRO/news/mro20100920.html.
[456] Space News Staff. "NASA's Odyssey Probe To Listen for Radio Signal from Frozen Mars Lander." Spacenews 18 Jan. 2010. http://www.spacenews.com/civil/100118-odyssey-listen-radio-signal-mars-lander.html.
[457] "Phoenix Mars Lander is Silent, New Image Shows Damage." 24 May 2010. NASA. http://www.nasa.gov/mission_pages/phoenix/news/phx20100524.html.
[458] "NASA Rover Finds Clue to Mars' Past And Environment for Life." 3 June 2010. NASA. http://www.nasa.gov/mission_pages/mer/news/mer20100603.html.
[459] Klamper, Amy. "NASA Approves MAVEN Mars Scout Mission for 2013 Launch." Spacenews 11 Oct. 2010. http://www.spacenews.com/civil/101011-nasa-approves-maven-2013-launch.html.
[460] "China Eyes Independent Mars Exploration." 30 Oct. 2010. Roscosmos. http://www.federalspace.ru/main.php?id=2&nid=10701.
[461] "Cassini Measures Tug of Enceladus." 3 June 2010. NASA. http://www.nasa.gov/mission_pages/cassini/whycassini/cassini20100426.html.
[462] "Cassini Finishes Sleigh Ride by Icy Moons." 21 Dec. 2010. NASA. http://www.nasa.gov/mission_pages/cassini/whycassini/enceladus20101221.html.
[463] "Venus is alive – geologically speaking." 8 Apr. 2010. European Space Agency. http://www.esa.int/esaSC/SEMUKVZNK7G_index_0.html.
[464] "Du dioxyde de soufre dans la haute atmosphère de Vénus." 23 Nov. 2010. CNES. http://www.cnes.fr/web/CNES-fr/8928-gp-du-dioxyde-de-soufre-dans-la-haute-atmosphere-de-venus.php.
[465] "Venus Express finds planetary atmospheres such a drag." 7 Oct. 2010. European Space Agency. http://www.esa.int/esaSC/SEMVIWSOREG_index_0.html.
[466] "Russia Eyes Scientific Mission to Venus." 26 Jul. 2010. Roscosmos. http://www.federalspace.ru/main.php?id=2&nid=9948.
[467] "Russian Scientists Plan to Launch an Interplanetary Station to Venus." 7 July 2010. Roscosmos. http://www.federalspace.ru/main.php?id=2&nid=9822.
[468] Kallender-Umezu, Paul. "NASA's Japanese Akatsuki Probe Fails To Enter Orbit Around Venus." Spacenews 13 Dec. 2010. http://www.spacenews.com/civil/101213-akatsuki-fails-orbit.html.
[469] "Comet Smacked Neptune 200 Years Ago, Data Suggests." 27 july 2010. Roscosmos. http://www.federalspace.ru/main.php?id=2&nid=9953.
[470] "Jupiter's mysterious flashes and missing cloud belts." 17 Jun. 2010. European Space Agency. http://www.esa.int/esaSC/SEMKY9RVEAG_index_0.html.
[471] "Proba-2 tracks Sun surging into space." 30 June 2010. European Space Agency. http://www.esa.int/SPECIALS/Technology/SEM9XUOZVAG_0.html.
[472] "Picard fournit sa 1ere image du Soleil." 28 July 2010. DLR. http://www.cnes.fr/web/CNES-fr/8637-gp-picard-fournit-sa-1ere-image-du-soleil.php.
[473] "Is the Sun Emitting a Mystery Particle?." 29 Aug. 2010. Roscosmos. http://www.federalspace.ru/main.php?id=2&nid=10198.
[474] "Solar Flare Activity Continues to Increase." 21 Jul. 2010. Roscosmos. http://www.federalspace.ru/main.php?id=2&nid=9935.
[475] "On the trail of space weather: SDO solar observatory launched successfully." 11 Feb. 2010. DLR. http://www.dlr.de/en/DesktopDefault.aspx/tabid-6215/10210_read-22471/10210_page-6/.
[476] "Solar Probe Plus To Plunge Directly Into Sun's Atmosphere." 6 Sept. 2010. Roscosmos. http://www.federalspace.ru/main.php?id=2&nid=10266.

477 Brinton, Turner. “Solar Upswing Presents New Research Opportunities.” Spacenews 14 June 2010. http://www.spacenews.com/civil/solar-upswing-presents-new-research-opportunities.html.
478 “Rosetta triumphs at asteroid Lutetia.” 10 July 2010. European Space Agency. http://www.esa.int/esaSC/SEM44DZOFBG_index_0.html.
479 Space News Staff. “Newly Found Exoplanet May Have Water-laden Clouds.” Spacenews 29 Mar. 2010. http://www.spacenews.com/civil/100329-newly-found-exoplanet-water-laden-clouds.html.
480 “On Herschel détecte 5 nouvelles galaxies grâce aux loupes cosmiques.” 4 Nov. 2010. DLR. http://www.cnes.fr/web/CNES-fr/8853-gp-herschel-detecte-5-nouvelles-galaxies-grace-aux-loupes-cosmiques.php.
481 “Hayabusa Asteroid Mission Comes Home.” 14 June 2010. NASA. http://www.nasa.gov/topics/solarsystem/features/hayabusa20100609-revised.html.
482 “Japanese Lab Finds ‘Minute Particles’ in Asteroid Pod.” 7 May 2010. Roscosmos. http://www.federalspace.ru/main.php?id=2&nid=9809.
483 “NASA’s Chandra Finds Youngest Nearby Black Hole.” 15 Nov. 2010. NASA. http://www.nasa.gov/mission_pages/chandra/news/H-10-299.html.
484 “NASA and NSF-Funded Research Finds First Potentially Habitable Exoplanet.” 29 Oct. 2010. NASA. http://www.nasa.gov/topics/universe/features/gliese_581_feature.html.
485 “NASA’s Kepler Mission Discovers Two Planets Transiting the Same Star.” 26 Sept. 2010. NASA. http://www.nasa.gov/mission_pages/kepler/news/two_planet_orbit.html.
486 “Meet the Titans: Dust Disk Found Around Massive Star.” 7 Aug. 2010. NASA. http://www.nasa.gov/mission_pages/spitzer/news/spitzer20100714.html.
487 “Primordial Magnetic Fields Discovered Across The Universe.” 24 Sept. 2010. Roscosmos. http://www.federalspace.ru/main.php?id=2&nid=10403.
488 “NASA’s WISE Mission to Complete Extensive Sky Survey.” 16 July 2010. NASA. http://www.nasa.gov/mission_pages/WISE/news/wise20100716.html.
489 “NASA and ESA’s First Joint Mission to Mars Selects Instruments.” 2 Aug. 2010. NASA. http://www.nasa.gov/mission_pages/mars/news/exo20100802.html.
490 “ESA To Set Tiny Hair-Like Webb Telescope Microshutters.” 29 June 2010. European Space Agency. http://www.nasa.gov/topics/technology/features/microshutters.html.
491 “DLR and NASA conclude bilateral framework agreement.” 8 Dec. 2010. DLR. http://www.dlr.de/en/DesktopDefault.aspx/tabid-6215/10210_read-22471/10210_page-6/.
492 “NASA and DLR sign agreement to continue the GRACE mission through 2015.” 10 June 2010. NASA. http://www.dlr.de/en/DesktopDefault.aspx/tabid-6215/10210_read-24882/10210_page-5/.
493 “NASA and Israel Space Agency Sign Statement of Intent for Future Cooperation.” 12 Aug. 2010. Roscosmos. http://www.federalspace.ru/main.php?id=2&nid=10072.
494 “Russia and ESA Sign Intergovernmental Agreement Enclosure Protocol.” 24 Sept. 2010. Roscosmos. http://www.federalspace.ru/main.php?id=2&nid=10409.
495 “Twenty-fifth series of German-Russian plasma physics experiments.” 27 Jan. 2010. DLR. http://www.dlr.de/en/DesktopDefault.aspx/tabid-6215/10210_read-22207/10210_page-6/.
496 “Roscosmos and DLR Sign Frame Agreement.” 9 June 2010. Roscosmos. http://www.federalspace.ru/main.php?id=2&nid=9625.
497 “Russia and France to Obtain Significant Competitive Advantages Through Uniting Space Technological Potentials – Chairman of the Russian Government Vladimir Putin.” 10 Dec. 2010. Roscosmos. http://www.roscosmos.ru/main.php?id=2&nid=11007.
498 “Roscosmos and NASA to Sign Joint Protocol to Cover Different Space Programs.” 3 Dec. 2010. Roscosmos. http://www.roscosmos.ru/main.php?id=2&nid=10954.
499 “Historic Space Deal Between UK and Russia.” 21 July 2010. Roscosmos http://www.federalspace.ru/main.php?id=2&nid=9941.
500 “Russia and Ukraine will Cooperate in Space Exploration – Russian Prime Minister Vladimir Putin.” 28 Oct. 2010. Roscosmos. http://www.federalspace.ru/main.php?id=2&nid=10684.
501 “Space Cooperation between Russia and India is Prospective.” 2 Aug. 2010. Roscosmos. http://www.federalspace.ru/main.php?id=2&nid=10000.

502 "Saudi Arabia, Egypt Sign Agreement on Space Science." 1 Aug. 2010. Roscosmos. http://www.federalspace.ru/main.php?id=2&nid=9987&hl=egypt.
503 De Selding, Peter B. "Satellite Operators To Create Database To Combat Interference." Space News. 2 Nov. 2009: 20. De Selding, Peter B. "Execs Say Interference is Mostly Unintentional, but Often Hard to Resolve." Space News 2 Nov. 2009: 20.
504 De Selding, Peter B. "Satellite Operators Solicit Bids To Create Orbital Database." Space News 23 Nov. 2009: 6.
505 De Selding, Peter B. "European Officials Poised to Remove Chinese Payloads From Galileo Sats." Space News. 15 Mar. 2010: 1.
506 Ibid.
507 "Individual European Nations To Determine Uses for PRS." Space News. 15 Mar. 2010: 16.
508 "European GPS Overlay Ready for Operations." Space News. 5 Oct. 2009: 3.
509 "Russia to Orbit 7 New Glonass Satellites in 2010." 6 Apr. 2010. RIANovosti 30 May 2010. http://en.rian.ru/russia/20100406/158456450.html.
510 "Russian Spacecraft Succesfully Orbits 3 Glonass Satellites." 2 Mar. 2010. RIANovosti 30 May 2010. http://en.rian.ru/science/20100302/158060090.html.
511 "Russia to Orbit 7 New Glonass Satellites in 2010." 6 Apr. 2010. RIANovosti 30 May 2010. http://en.rian.ru/russia/20100406/158456450.html.
512 "Russian Spacecraft Succesfully Orbits 3 Glonass Satellites." 2 Mar. 2010. RIANovosti 30 May 2010. http://en.rian.ru/science/20100302/158060090.html.
513 "Russia, India May Jointly Make Glonass, GPS Navigation Devices." 9 Mar. 2010. RIANovosti 30 May 2010. http://en.rian.ru/russia/20100309/158137947.html.
514 "Kiev Says Russian Rival to GPS Good for Ukraine's Security." 19 May 2010. RIANovosti 30 May 2010. http://en.rian.ru/world/20100519/159078178.html.
515 De Selding, Peter B. "Uncertainty Besets GMES Data Policy and Satellite Contributions." Space News 26 Oct. 2009: 10.
516 Commission of the European Communities. Global Monitoring for Environment and Security (GMES): Challenges and next steps for the Space Component. COM (2009) 589 final of 28 Oct. 2009. Brussels: European Union.
517 Commission of the European Communities. Summary of the (GMES) Impact Assessment. SEC (2009) 1441 of 28 Oct. 2009. Brussels: European Union.
518 Commission of the European Communities. Global Monitoring for Environment and Security (GMES): Challenges and next steps for the Space Component. COM (2009) 589 final of 28 Oct. 2009. Brussels: European Union.
519 Ibid.
520 Commission of the European Union. Decision of 5 February 2010 Setting Up the GMES Partners Board (2010/67/EU). Official Journal of the European Union doc. L35/23 of 6 Feb. 2010. Brussels: European Union.
521 De Selding, Peter B. "European SMOS and Proba-2 Successfully Launched by Rockot." Space News 9 Nov. 2009: 6.
522 "New milestone in IXV development." European Space Agency 15 Sept. 2010. http://www.esa.int/SPECIALS/Launchers_Home/SEMOU1KOXDG_0.html.
523 "ESA and Thales Alenia Space establish agreement for development of Intermediate eXperimental Vehicle (IXV)" 16 June 2009 European Space Agency. http://www.esa.int/SPECIALS/Launchers_Home/SEM52E3XTVF_0.html.
524 "High-thrust engine demonstrator industrial day." 15 Feb. 2010 European Space Agency. http://www.esa.int/SPECIALS/Launchers_Home/SEMJPZMEG5G_0.html.
525 "Green' satellite fuel designed to make space safer."16 Mar. 2010 European Space Agency. http://www.esa.int/SPECIALS/Technology/SEMPJQ9KF6G_0.html.
526 "DLR tests new nozzle for Ariane 5 main engine." 29 Jan. 2010 DLR. http://www.dlr.de/en/DesktopDefault.aspx/tabid-6215/10210_read-22300/10210_page-6/.

[527] "NASA's Dawn Spacecraft Fires Past Record for Speed Change." 6 Mar. 2010 NASA. http://www.nasa.gov/mission_pages/dawn/news/dawn20100607.html.
[528] "NASA's Dawn Spacecraft Fires Past Record for Speed Change." 6 Mar. 2010 NASA. http://www.nasa.gov/topics/technology/features/A2TestStand-033111.html.
[529] "Tether Maneuvers Spacecraft Without Fuel." 5 sept. 2010. Roscosmos. http://www.federalspace.ru/main.php?id=2&nid=10256.
[530] "Lowering the hurdles to space." 20 May 2010. JAXA. http://www.jaxa.jp/projects/rockets/epsilon/index_e.html.
[531] "Russian President Defines Prime Contractor for Development of National Nuclear Propulsion System." 28 Aug. 2010. Roscosmos. http://www.federalspace.ru/main.php?id=2&nid=9771.
[532] "New Advanced Propulsion Systems to Make Mission to Mars in 2-3 Months Feasible." 30 Sept. 2010. Roscosmos http://www.federalspace.ru/main.php?id=2&nid=10461.
[533] "New Plasma Engine for Missions to the Other Planets." 11 June 2010. Roscosmos. http://www.federalspace.ru/main.php?id=2&nid=9636.
[534] "ESA missions extend fibre optics to orbit." 28 Apr. 2010. European Space Agency. http://www.esa.int/SPECIALS/Space_Engineering/SEMNB5BKE8G_0.html.
[535] "New testbed simulates space's formation flying future." 9 Sept. 2010. European Space Agency. http://www.esa.int/SPECIALS/Space_Engineering/SEMD8VDODDG_0.html.
[536] "Russian Scientists to Orbit Nano-Satellite Constellation in the Next Two Years." 12 Oct. 2010. Roscosmos. http://www.federalspace.ru/main.php?id=2&nid=10556.
[537] "Russian Cosmos-NT Program to Reduce Expenses and Duration of Spacecraft Design and Development." 31 Aug. 2010. Roscosmos. http://www.federalspace.ru/main.php?id=2&nid=10214.
[538] "Russian Advanced Crew Vehicle Developed in Russia." 20 Aug. 2010. Roscosmos. http://www.federalspace.ru/main.php?id=2&nid=10137.
[539] Kallender-Umezu, Paul. "Japan aims To boost Its ISS role with Cargo Return Vehicle." Spacenews 23 Aug. 2010. http://www.spacenews.com/civil/1100823-japan-aims-boost-iss-role.html.
[540] Space News Staff. "Masten Eyeing Canaveral For Suborbital Launches." Spacenews 26 Nov. 2010. http://www.spacenews.com/venture_space/masten-eyeing-canaveral.html.
[541] David, Leonard. "Virgin Galactic's Private Spaceship Makes First Solo Glide Flight." Spacenews 18 Oct. 2010. http://www.spacenews.com/venture_space/101018-virgin-galactic-spaceship-makes-glide-flight.html.
[542] http://www.spacenews.com/venture_space/100303-virgin-galactic-applauds-legal-hurdle-lawsuits.html.
[543] Space News Staff. "Suborbital Price Drop Foreseen by 2014." Spacenews 23 July 2010. http://www.spacenews.com/venture_space/100723-suborbital-price-drop-foreseen.html.
[544] David, Leonard. "Suborbital Ventures Eye NASA Business." Spacenews 26 Fev. 2010. http://www.spacenews.com/venture_space/100226-suborbital-ventures-eye-nasa-business.html.
[545] Svitak, Amy. "Orbital Teams with Virgin Galactic in CCDev 2 Bid." Spacenews 13 Dec. 2010. http://www.spacenews.com/civil/101213-orbital-virgin-ccdev2-bid.html.
[546] "Atmospheric balloon experiment BS10-06 completed." 9 Sept. 2010. JAXA. http://www.jaxa.jp/projects/sas/balloon/index_e.html.
[547] "Detailed Dark Matter Map Yields Clues to Galaxy Cluster Growth." 12 Nov. 2010. NASA. http://www.nasa.gov/mission_pages/hubble/science/dark-matter-map.html.
[548] "Cosmic Lens Used to Probe Dark Energy for First Time." 19 Sept. 2010. NASA. http://www.nasa.gov/mission_pages/hubble/news/hubble20100819.html.
[549] "L'expérience AMS en route pour le Centre spatial Kennedy." 18 Aug. 2010. CERN. http://public.web.cern.ch/press/pressreleases/Releases2010/PR16.10F.html.
[550] "Testing time to keep LISA Pathfinder clear of magnetism." 24 Sept. 2010. European Space Agency. http://www.esa.int/SPECIALS/Space_Engineering/SEMGYU7OTCG_0.html.
[551] "Astrium To Manage Atomic Clock Demo on Station." 21 July 2010. Roscosmos. http://www.federalspace.ru/main.php?id=2&nid=9933.

[552] "Ultraviolet Starlight Key to Making Water in Space." 9 May 2010. Roscosmos. http://www.federalspace.ru/main.php?id=2&nid=10255.
[553] "NASA-Funded Research Discovers Life Built With Toxic Chemical." 12 Feb. 2010. NASA. http://www.nasa.gov/topics/universe/features/astrobiology_toxic_chemical.html.
[554] "Long-distance ultrasound exams controlled by joystick." 8 Dec. 2010. European Space Agency. http://www.esa.int/esaCP/SEMP3OOWXGG_index_0.html.
[555] "ESA emergency telemedicine system soars to commercial success." 21 Sept. 2010. European Space Agency. http://www.esa.int/SPECIALS/Technology/SEM6OOMO7EG_0.html.
[556] "A Stellar, Metal-Free Way to Make Carbon Nanotubes." 22 Feb. 2010. NASA. http://www.nasa.gov/topics/technology/features/metal-free-nanotubules.html.
[557] "NASA to Launch Human-Like Robot to Join Space Station Crew." 15 Apr. 2010. NASA. http://www.nasa.gov/topics/technology/features/robonaut.html.
[558] "NASA Uses Twin Processes to Develop New Tank Dome Technology." 2 Dec. 2009. NASA. http://www.nasa.gov/topics/technology/twin_dome.html.
[559] "Prototype NASA Earth Camera Goes for Test Flight." 19 Oct. 2010. NASA. http://www.nasa.gov/topics/earth/features/misr20101019.html.
[560] "NASA Unveils New Space-Weather Science Tool." 23 Feb. 2010. NASA. http://www.nasa.gov/topics/technology/features/iswa-program.html.
[561] "CPR EM heat balance vacuum test." 25 Mar. 2010. JAXA. http://www.jaxa.jp/projects/sat/earthcare/index_e.html.
[562] "High-tech design for ESA's space hazards awareness system." 17 Nov. 2010. European Space Agency. http://www.esa.int/SPECIALS/Space_Engineering/SEMV5456JGG_0.html.
[563] "New navsat sensor improves water monitoring." 23 Apr. 2010. European Space Agency. http://www.esa.int/SPECIALS/Technology/SEM65WF098G_0.html.
[564] "Russian Physicists Developed Space Debris Tracker." 9 Apr. 2010. Roscosmos. http://www.federalspace.ru/main.php?id=2&nid=10250.
[565] "Russia to Clean Space around the Earth." 29 Nov. 2010. Roscosmos. http://www.roscosmos.ru/main.php?id=2&nid=10923.
[566] Werner, Debra. "ATK Proposes Satellite To Fight Space Debris." Spacenews 9 Aug. 2010. http://www.spacenews.com/civil/100809-atk-satellite-fight-space-debris.html.
[567] "President Outlines Exploration Goals, Promise." 15 Apr. 2010. NASA. http://www.nasa.gov/about/obamaspeechfeature.html.
[568] Berger, Brian and Klamper, Amy. "Budget Beneficiaries: Science, Research and Technology? NASA." Spacenews 5 Feb. 2010. http://www.spacenews.com/policy/100205-nasa-budget-beneficiaries-science.html.
[569] "Russia to Restart Science in Space." 12 Sept. 2010. Roscosmos. http://www.roscosmos.ru/main.php?id=2&nid=10988.
[570] "Relay Satellites Produced by Reshetnev ISS to Strengthen Russian Positions in Space." 20 Oct. 2010. Roscosmos. http://www.federalspace.ru/main.php?id=2&nid=10617.
[571] "Russian Government Allocates Budgeting of Roscosmos Efforts Aimed at Economy's Modernization." 6 July 2010. Roscosmos. http://www.federalspace.ru/main.php?id=2&nid=9817.
[572] "ESA chooses three scientific missions for further study." 19 Feb. 2010. European Space Agency. http://www.esa.int/esaSC/SEMSHM7CS5G_index_0.html.
[573] "New investment fund backs space technologies finding uses on Earth." 5 Mar. 2010. European Space Agency. http://www.esa.int/SPECIALS/Technology/SEMRE27K56G_0.html.
[574] "Satellite-control software helps the oil and gas industry." 15 Oct. 2010. European Space Agency. http://www.esa.int/SPECIALS/Technology/SEMSW3WO1FG_0.html.

PART 2

VIEWS AND INSIGHTS

1. Space in the financial and economic crisis

Christophe Venet

1.1. Introduction

The purpose of this article is to analyse the impact of the financial and economic crisis on the space sector from a political point of view. Indeed, it seems technically difficult and methodologically dangerous to make econometric predictions regarding future economic developments. This is particularly valid for the global economic situation, the best example being the regular revision and correction of the economic outlooks of the World Bank and International Monetary Fund (IMF). In the particular case of the space sector, this uncertainty is lessened by some features that are specific to the space economy (such as the relatively predictable manufacturing cycles in the upstream sector) and by the existence of several consulting firms specialised in providing space-specific economic forecasts.[575] This contribution however, does not intend either to make a market forecast or to predict when and how the space sector will overcome the crisis. It will rather propose a political analysis based on the thorough observation of the space sector in the last two years, enabling the identification of some political implications and prospects offered by the crisis. Consequently, the focus will be on the institutional space sector rather than on commercial space activities. In addition, Europe will be placed at the centre of the analysis and worldwide trends and developments will always be considered as structural constraints or enablers for the European space sector. Two interrelated central ideas underlie the present article. On the one hand, the crisis can be seen as a test case for the strategic nature of space. On the other hand, it also represents an opportunity to strengthen this strategic aspect of space, in the sense that it could anchor space even further to the European socio-economic framework, in the long term. In order to develop these ideas, a brief overview of the crisis itself will first be given, identifying the factual elements directly relevant to the space sector. In a second step, an analysis of what has happened in the space economy in the last two years will be conducted. This will finally enable the third step that is to specifically highlight the political implications and future prospects for the space sector that have been induced by the crisis.

1.2. The financial and economic crisis

After a brief presentation of some basic facts and figures on the crisis, national and international responses and reactions to it will be sketched. Finally, national and regional differences in the extent and gravity of the crisis will be highlighted, focusing on the major space faring nations. The overall purpose of this first section is to give an overview of the global background, against which an analysis of the space sector's behaviour during the crisis can be made.

1.2.1. Basic facts and figures

The crisis begun with the subprime mortgage collapse in the U.S. in August 2007 and reached its peak in September 2008, when the U.S. investment bank Lehman Brothers declared bankruptcy. What started as turmoil in the financial sector rapidly spread to the real economy in the second half of 2008 and in 2009, as investment and consumption dropped sharply, due to the diminished confidence of households. While commodity prices – in particular oil and gas prices – reached historical high levels in the first half of 2008, weakening global demand had a contracting effect on commodity prices in the second half of the year. This trend somewhat lightened the burden for advanced economies importing commodities, but it had adverse consequences for exporting emerging economies. Another negative consequence of the surge in food and oil prices was very high inflation rates. As a whole, the worldwide economic and financial situation at the end of 2008 was exceptionally uncertain and prone to major risks.[576] While world output grew by 5.1% in 2006 and 5% in 2007,[577] the figures for 2008 (+3%) and 2009 (–0.6%)[578] mirrored the rapidly deteriorating situation. Global recession continued to spread in the first half of 2009, but the first signs of recovery could be observed by the middle of that year. However, the stabilisation pace remained uneven, slow and uncertain.[579] By the middle of 2010, positive macroeconomic developments indicated a steady recovery, despite renewed financial turbulences in the beginning of the year. Overall, the IMF expects the world output to grow by 4.6% in 2010 and 4.3% in 2011.[580]

1.2.2. National and international reactions to the crisis

Two central features could be observed in the national and international reactions to the crisis. The first is that state intervention was widely used as a regulating tool, reflecting the growing criticism of the neoliberal economic model and responding to pressure from the IMF. This trend started from the very beginning of the crisis, with six central banks injecting 180 billion U.S. dollars into the monetary markets in a concerted action in October 2008.[581] Almost all the large economies adopted national stimulus packages to reboost their internal demand. The U.S. Congress approved President Obama's 787 billion U.S. dollars economic stimulus package in February 2009 while the European Council on 11 and 12 December 2008 approved a European Economic Recovery Plan, equivalent to about 1.5% of the EU's GDP, which represents around 200 billion Euros. Similar measures were implemented in the United Kingdom, France, Japan, Russia and China. The second point is the increased international consultation to reform the global financial and banking system. Discussions mostly took place within specialised international institutions, such as the World Bank and the IMF, and in the framework of the G7 and G20 summits. While a series of globally shared principles to fight the crisis could be agreed upon (such as the rejection of protectionism), national dissensions over the extent and the content of a reformed international financial system have prevented any concrete steps until now. However, these massive public macroeconomic interventions accelerated the pace of recovery.[582]

1.2.3. The uneven impact of the crisis

A final element of the crisis that needs to be highlighted concerns the regional disparities in its impact. Indeed, not all areas of the world were hit with the same intensity by the crisis, and a distinction should be made in particular between developed economies which suffered the most and emerging nations which were hardly affected. Looking specifically at spacefaring nations, established players such as the U.S., Europe and Japan faced a drastic slowdown of their economies.[583] While the U.S. experienced a moderate negative GDP growth in 2009 (−2.4%), Europe had to cope with a stronger contraction during the same period (−4.1% in the Euro area). The situation was even worse in Japan, as the recession had already started in 2008 (−1.2%) and worsened in 2009 (−5.2%). Russia remained relatively untouched by the crisis in the beginning (+5.6% GDP growth in 2008), but the country's economy experienced a significant setback in 2009 (−7.9%), mainly due to the fall in energy prices and the curtailed access to external funding. At the other end of the spectrum, the two fast emerging space powers India and China saw their GDP growth continuing during the crisis at almost the same pace (+5.7% for India and +9.1% for China in 2009).

To summarise, the financial crisis that started in the middle of 2007 spread to the real economy throughout 2008 and 2009, and recovery is still slow and uncertain. While the crisis hit developed economies more harshly than emerging countries, the globally coordinated response helped to mitigate its effects. Although it is difficult to establish a direct correlation between the crisis and developments in the space sector, its manifestations provide the background against which analysis should be conducted.

1.3. Space in the crisis

Keeping the main features of the crisis in mind, a specific focus will be put on both the commercial and the institutional space sector during the crisis. However, in order to better understand the impact of the crisis on the space sector, it seems first necessary to define the space economy.

1.3.1. The space economy

For many years, space was not considered as an economic sector as such, but rather as a limited scientific and technological domain. Governments played a central role for decades, as national security concerns dominated space activities during the Cold War. In addition, the huge investments needed in space and the associated economic risks were obstacles for the development of private endeavours.[584] As a consequence, noteworthy commercial space activities started only in the 1980's.[585] The end of the Cold War however, paved the way for the expansion of the space economy, as a conjunction of favourable structural factors emerged. The first of these enabling trends was the new impetus given to worldwide financial liberalisation and privatisation, symbolised in the space sector by the transformation of the two main intergovernmental organisations providing worldwide satellite telecommunications services (INMARSAT and INTELSAT) into privately held, profit oriented organisations.[586] This liberalising wave also touched the two other main space applications, navigation and Earth Observation (EO).[587] A second factor explaining the rise of private space activities in the 1990's was the reduction of national space budgets, especially concerning military spending. Although of a limited duration, these budget cuts forced private firms to seek new markets.[588] Finally, and perhaps most importantly, the emergence of new satellite applications with a high socio-political impact in all three areas (EO, navigation and telecommunications) offered large commercial opportunities for space activities.

The last point in particular, led to considering the space economy as an economic sector by itself, with high growth potential. Another related consequence was that the space economy became both broader and more difficult to define.[589] In light of these developments, the OECD launched a research project focusing on space in 2003. Its objectives were to identify the economic challenges and opportunities for the space sector, and more generally to gain a better understanding of the issues at stake. As a tangible result of this work, the OECD proposed a broad definition of the space economy:

> *All public and private actors involved in developing and providing space-enabled products and services. It comprises a long value-added chain, starting with research and development actors and manufacturers of space hardware (e.g. launch vehicles, satellites, ground stations) and ending with the providers of space-enabled products (e.g. navigation equipment, satellite phones) and services (e.g. satellite-based meteorological services or direct-to-home video services) to final users.*[590]

This definition reflects both the developments in space in recent years (as it comprises not only the traditional space industry but also space services) and the multiplicity of actors involved in the space sector (public and private actors, space and non-space actors, demand and supply side).

Due to their structural specificities, space activities are not comparable to other market sectors. These specificities need to be briefly presented, as they were often exacerbated by the crisis. The first and most characterising feature of space is the strong involvement of governments. Indeed, most of the R&D work in space is public, space agencies remain the largest buyers of space services and products and governments also set up the framework conditions for private space activities.[591] The two main reasons for this are, on the one hand the strategic relevance of space (both in the narrow military and in the broader, socio-political sense) and, on the other, the high economic risks associated with space endeavours, which can not be usually faced by private companies alone. In addition to the central role of governments, the high costs of access to space and the long development cycles of relevant technologies constitute further determinants of space activities. Finally, only low economies of scale are possible in the upstream sector, leading to concentration, while the downstream sector is characterised by high economies of scale, enabling the development of large and viable markets.[592]

To summarise, two simple facts need to be pointed out to better understand the effects of the crisis on the space sector. The first is that because of rapid developments since the 1990's, space has become a significant sector of activity: the Space Foundation estimated the global size of the space economy in 2009 at 261.61 billion U.S. dollars.[593] The second point is the close interrelation and mutual dependency between the different actors involved in space. Besides the obvious link between the demand and the supply side, strong relationships also exist between the private and the public sector, as well as between the upstream and the downstream sector.

1.3.2. The commercial space sector in the crisis

Looking only at revenues in the last two years, the crisis seems to have had very little impact on the commercial space sector. Indeed, revenues in 2008 and 2009 increased in virtually all the sectors. In the upstream sector first, global satellite manufacturing revenues have grown steadily in the last couple of years despite some year-to-year variations. While a slight decrease could be observed between 2007 and 2008 (11.6 billion U.S. dollars and 10.5 billion U.S. dollars respectively), global manufacturing revenues grew again in 2009 to reach 13.5 billion U.S. dollars.[594] Even the diminished results in 2008 should not be attributed to the crisis: as manufacturing cycles are long, satellites launched in 2008 were ordered many years before the crisis started.[595] To reinforce this picture of a resilient sector of activity, the number of GEO commercial satellites ordered in 2009 reached 41, almost twice as many as in 2008.[596] The launch sector also remained very dynamic in 2009, both in terms of the number of launches (68 in 2008 and 78 in 2009)[597] and in terms of revenue (3.9 billion U.S. dollars in 2008 and 4.5 billion U.S. dollars in 2009).[598] Finally, ground equipment revenue grew by 8% between 2008 and 2009. Despite this positive sign, this is a significantly lower rate than the 34% expansion recorded between 2007 and 2008.[599] This is mainly due to the fact that the ground equipment market is structurally dependant on activities that were hard hit by the crisis, such as the transportation sector that uses GPS receivers. As for the downstream sector, overall satellite services revenues expanded by 11% in 2009, although at a lower rate than in 2007 (+17%) and 2008 (+16%).[600] At first sight, the crisis itself doesn't seem to have a significant impact on the commercial space sector. This could be ascribed at least partially to the strong growth of innovative applications, such as High Definition TV (HDTV) and to the very resilient demand for satcom applications in certain regions of the world, such as the Middle East and India. Another trend that could be observed during the crisis was the tendency for space companies to consolidate their market positions through mergers, takeovers, alliances and restructuring. Although this is a structural feature of the space industry, the crisis may have contributed to accelerating its pace. Examples from the upstream sector include the takeover of Saab Space by RUAG in July 2008 and the acquisition of a majority stake at SSTL by EADS Astrium in January 2009.[601] This trend also expanded to the downstream sector, as was manifested by the merger between the two major satellite radio providers Sirius and XM in July 2008[602] and the joint venture between satellite broadband providers EchoStar and ViaSat to launch a dedicated broadband satellite by 2012.[603]

One of the most striking features of the last two years however, has been the strong government support provided to the commercial space sector. While strong government involvement in space is a structural constant, the crisis reinforced this trend. The first factor mirroring this was the increased institutional demand for satellite procurement in 2009. Indeed, commercially procured satellites for government missions were the primary driver of growth in this sector. While the value of such satellites amounted to 5.3 billion U.S. dollars in 2008, it reached 9.8 billion U.S. dollars in 2009.[604] Another way for governments to support the space industry during the crisis was the increased use of national export-credit agencies to finance industrial projects. As access to credit is one of the most crucial prerequisite for space entrepreneurs and as the crisis had a very negative impact on the availability of credit, space companies often turned to national export-credit agencies to finance their projects. In France for example, the national credit-export agency granted a loan guarantee to Gazprom Space Systems for the acquisition of two Yamal 400 satcoms to be built by Thales Alenia Space in September 2009. The U.S. Export-Import Bank implemented similar schemes for U.S. satellite manufacturers.[605] Similarly, a study on the future of UK space policy, released in February 2010, recommended that the Export Credit Guarantee Department should back the country's space exports.[606] As a whole, these trends seem to indicate that governments understood the strategic nature of space activities, as they backed the space sector in the crisis, both politically and financially.

1.3.3. The institutional space sector in the crisis

A first and important point to raise regarding the institutional space sector is that there is no direct correlation between the evolution of a country's GDP and the evolution of its public space expenditures.[607] In Europe for example, public space budgets during the last twenty years accounted for a stable percentage of the GDP, around 0.6–0.7%.[608] This fact was verified again during the crisis, as no space budget cuts could be observed despite the global recession. To the contrary, institutional space budgets continued to rise, although at a significantly different pace from country to country. Established spacefaring nations experienced moderate budget increases. The U.S. witnessed a 5% increase of its overall public spending on space between 2008 and 2009, reaching 48.794 billion U.S. dollars. The space expenditures of Japan increased by 2.17% (3.012 billion U.S. dollars in 2009) but in France the net increase was only 0.6% (2.712 billion U.S. dollars in 2009).[609] In other instances, space budgets are expected to remain flat for the next few years. ESA's budget spending for example will remain at the same levels in 2010 and 2011.[610] This will also be the case in Italy, although ASI's President announced that the crisis would not have any effect on the 2010 and 2011 budgets.[611] By contrast, emerging spacefaring nations seemed even less affected by the crisis. Although it is difficult to assess the exact size of the Chinese space budget, the country pursued its ambitious space programmes throughout the crisis, unveiling its plans for a future space station, continuing its lunar exploration programme and further developing its space infrastructure. India announced in July 2009 that the 2010 ISRO budget would reach 1.04 billion U.S. dollars, which represents a 40% increase from 2009.[612] The case of Russia finally, is particularly interesting: while the country was severely hit by the crisis (−5.2% GDP decrease in 2009), its space budget grew by almost 100% between 2008 and 2009, to reach 2.837 billion U.S. dollars.[613]

As a whole, the positive figures of global space spending during the crisis seem to indicate a strong public willingness to commit resources to the space sector in the long run. Two additional elements reinforce this impression. The first is the growing number of countries engaging in space activities and/or setting up national space agencies. In recent years, countries as different as South Africa, Australia, Venezuela, South Korea, Turkey, Kazakhstan, Nigeria and Vietnam have started space projects or programmes. The fact that such a high number of nations, including developing countries, have decided to invest in space despite the crisis is a strong indication of the strategic nature of space. A second point concerns established spacefaring nations more specifically. Most of these countries (such as

Fig. 1: *The G20 Summit on global economic recovery and financial markets (2009) (source: BBC.com).*

the U.S., France or Russia) launched economic stimulus packages to fight the crisis and channelled some of these funds to the space sector.

Combining all the elements related to the attitude of governments towards the space sector during the crisis, such as space budget increases, parts of stimulus packages poured into the space sector, support for the industry through credit-export agencies and enhanced institutional orders, the strategic importance of space seems undoubtedly verified. At a first glance, the space sector seems to have passed the test of the crisis. By looking more closely however, it becomes obvious that most public responses during the crisis were short-term oriented (e.g. providing stimulus packages and facilitating access to credit through credit-export agencies) and mainly for the benefit of the upstream sector (e.g. the satellite manufacturers). The last point in particular, corresponds to a rather traditional understanding of the strategic character of space. According to this reasoning, safeguarding the space industrial base (both as a technological asset and as a job provider) and guaranteeing a certain degree of autonomy and independence in space asset manufacturing should constitute the core goals. While this is an important component of the strategic character of space, it doesn't seem sufficient by itself. Indeed, the crisis should also be seen as an opportunity to unleash the economic potential of space on a long-term perspective, with a particular focus on the downstream sector (space applications). The strategic nature of space also rests with the huge socio-economic potential of space applications. For this reason, it seems necessary to define more accurately the political implications of the crisis and especially what the truly strategic aspects of space activities are.

1.4. The political implications of the crisis for the space sector

A broader understanding of the strategic nature of space is laid down in all space-related European official documents. Furthermore, Europe tried during the crisis to translate this conceptual framework into concrete policy initiatives. It seems however that a real paradigm shift necessary to enjoy the full long term strategic advantages of space has yet to be implemented.

1.4.1. The strategic relevance of space

The definition of the strategic dimension of space in Europe was the result of a long and progressive policy process, which culminated in the adoption of the

European Space Policy in 2007. The starting point for this process was the increasing consciousness of decision-makers that new challenges and opportunities associated with the rise of innovative space applications had to be tackled. While Europe was quite successful in space in the commercial and scientific fields, it lacked a comprehensive space policy. The necessity to fully exploit the political, social and economic potential of space was thus recognised quite early, as demonstrated by the first communications on space issued by the European Commission (EC) in 1988, 1992 and 1996.[614] The 1992 Communication in particular, made that point evidently clear:

> *In the earlier phase of Europe's space effort, the space agencies had an essential, almost exclusive role, since the main aim was to establish a technological and industrial capability (technology-push). Europe must now move progressively towards a demand-pull approach in order to integrate space activities into the broader socio-economic fabric of Europe. Space applications programmes should be oriented according to objectives defined outside the space sector [...]*[615]

In sum, the strategic dimension of space for Europe consists in its effective contribution to a wide variety of policy areas. This goes beyond mere support to the "strategic industrial basis" of space activities (e.g. the upstream sector), as it also encompasses all the possibilities and potential offered by downstream services. This redefinition of the strategic nature of space in Europe has to be placed in the broader framework of a paradigm change after the end of the Cold War. While space was mostly dominated by research and science and by national security considerations during the Cold War, the emergence of innovative space applications coupled with the new perspectives offered by global political and economic liberalisation in the 1990's induced a perception change. Consequently, the political relevance of space grew, as states needed to adapt to this new environment. In general, it is not only political preoccupations that influence space programmes, such as during the Cold War, but it is also space activities that can shape policies to a certain extent. This dual and reciprocal link between space and politics constitutes the central feature of the post-Cold War period.

In this regard, the crisis could represent an opportunity to fully exploit the potential of this paradigm change. It could serve as a strong political impetus to anchor space to the broader European socio-economic framework in the long term, focusing on space applications. Indeed, space is more than a high-technology innovative domain that can be occasionally helpful to boost Europe out of a crisis. In fact, European decision-makers took several decisions towards this direction during the crisis.

1.4.2. Space in the framework of a European policy against the crisis

A series of recent official European documents acknowledge the economic potential of space along two major axes. First, space is seen as a potential contributor to the efforts to overcome the crisis (European Council of 11-12 December 2008, 6th Space Council of 29 May 2009). Second, the long-term perspective was put forward by laying down the role space can play within the Lisbon strategy[616] (5th Space Council of 26 September 2008, ESA Ministerial Council of 26 November 2008). The overall focus was put on the development of new markets based on space applications, mainly in the framework of the GMES and Galileo programmes.

First, the European Council of December 2008 was dedicated to setting the overall European response to the crisis. This led to the adoption of a European Economic Recovery Plan (EERP) and to a further call for launching a European innovation plan and for developing a European Research Area (ERA). Space was explicitly mentioned as a part of these initiatives.[617] Secondly, the 6th Space Council, which took place on 29 May 2009, focused on the concrete contribution of space to innovation, competitiveness and economic recovery.[618] It called again for the inclusion of space in the EERP and the Lead Market Initiative (LMI).[619] While these two documents focused on the concrete contribution of space to overcoming the crisis, two further important documents laid down the broader strategic framework for space activities.

The 5th Space Council, which took place on 26 September 2008, identified the contribution of space to the Lisbon strategy as one of the new priorities within the ESP. It stated that "space, as a high tech R&D domain and through the economic exploitation of its results, can contribute to reaching the Lisbon goals so as to fulfil the economic, educational, social and environmental ambitions of the EU [...] to achieve the objectives for growth and employment by providing new business opportunities and innovative solutions for various services".[620] The Council further highlighted that space applications in the fields of navigation, telecommunication and EO constitute substantial market opportunities, especially for Small and Medium Enterprises (SMEs), and finally it called for the inclusion of space in the Lead Market Initiative (LMI). The Resolutions adopted at the ESA Ministerial Council on 26 November 2008 were along the same lines. They highlighted the need for the European space sector to be competitive in global markets and they emphasised the central role of SMEs in this regard.[621] However, despite these important decisions, the need for a real paradigm change is still pressing.

1.4.3. The need for a real paradigm change

Two observations can be made regarding the strategic nature of space in Europe. First, the long-term benefits of space can only be fully exploited if its societal implications and economic benefits are fully and effectively taken into account within European macroeconomic policies. Second, there is still a discrepancy between the political intentions expressed in official documents and the absence of concrete measures to support them. This is why the often mentioned paradigm shift – from a technology-push towards a demand-pull approach – is yet to be translated into concrete actions. Three examples illustrate this point: the contribution of space to the EERP, to the European Plan for Innovation and to the LMI.

The inclusion of space in the EERP was supposed to occur through the initiative "factories of the future", one of the three Public-Private Partnerships (PPP) included in the EERP and intended to support the manufacturing industry in the development of new and sustainable technologies. So far, two calls have been issued within FP7 under this title, in July 2009 and July 2010. However, space was not part of them, as the funds were devoted to nanoscience, nanotechnologies, materials and new production, and Information and Communication Technologies (ICT).[622] The inclusion of space in the European Plan for Innovation has also not yet materialised. The Plan is still in its policies design phase and it will not be presented before the fall of 2010. It included a vast public consultation on European innovation policies between September and December 2009, involving, among others, Member States' Ministries, EU Associations, Private Companies, NGOs, Research Centres and Universities. Strikingly, no significant actor in the space sector contributed to this process.[623] Finally, as for the LMI initiative, space was not initially selected as one of the 6 markets included in the programme. A mid-term progress report, which was published in September 2009, carefully analysed the first phase of the initiative and called for a revision of the criteria for selecting new lead market candidates.[624] This development on the one hand suggests that the inclusion of space is not likely to occur immediately after the mid-term report, as it was initially envisaged, but on the other hand it also means that there is a window of opportunity for including space as an area to be exploited.

The strategic nature of space places it at the crossroads of several overlapping political issue-areas. To strengthen this strategic aspect of space, a strong European long-term financial and political commitment to it is necessary, which makes sense from both a structural/international and internal/European point of view. In the international perspective first, Europe has to be competitive in commercial markets, maintain strategic autonomy in key areas and remain a credible partner for international cooperation. At an internal level on the other side, Europe has to anchor space in the long run to the socio-economic landscape, focusing on space applications. There are two key conditions that seem necessary in order to reach this goal: strong political will and consistent public support. As for the first point, there are encouraging signs: the strategic nature of space is better appreciated by decision-makers, the EU has ambitious plans to increase its space budget by a factor of 3 for the period 2014–2021[625] and both ESA and the EU place increasing focus on applications. As for the second point, major efforts still need to be made, as it was evidenced by a recent Eurobarometer survey that showed a mixed picture regarding the support of European citizens for space activities.[626]

In conclusion, looking back at the past two years the space sector has resisted the crisis quite well. Increased public commitment was one of the key reasons for this outcome, both by supporting the commercial space sector and by continuously backing up the institutional space sector, both politically and financially. These facts seem to indicate that the space sector has passed the test of the crisis and that its strategic nature is gradually being recognised by decision-makers. However, the efforts made during the crisis to keep the space sector alive should not be considered as a one-shot initiative. The crisis should instead be considered as an opportunity to fully implement the paradigm change that emerged gradually two decades ago: that is to move from a technology-push approach towards a demand-driven perspective in order to fully exploit the long-term benefits of space applications.

---

575 The Paris-based Euroconsult and the U.S.-based Tauri group for example, regularly publish economic forecast reports and in-depth studies on future economic developments in space.

576 International Monetary Fund. "World Economic Outlook. Financial Stress, Downturns and Recovery". October 2008: xvff.

577 Ibid: 2.

578 International Monetary Fund. "World Economic Outlook Update". 8 July 2010: 2.

579 International Monetary Fund. "World Economic Outlook Update". 8 July 2009: 1ff.

580 International Monetary Fund. "World Economic Outlook Update". 8 July 2010: 2.

581 Those were the central banks from the U.S., Great Britain, Canada, Sweden, Switzerland and the European Central Bank (ECB).

582 International Monetary Fund. "World Economic Outlook Update". 8 July 2009: 1.

583 For all the following figures: International Monetary Fund. "World Economic Outlook Update". 8 July 2010: 2.

584 Steinbruner, John D. Preface of Pasco, Xavier. "A European Approach to Space Security". Cambridge: American Academy of Arts and Science, 2009: viii.

585 OECD. Space 2030 Exploring the Future of Space Applications. Paris: OECD, 2004: 12.

586 Ibid: 35.

587 Pasco, Xavier. "A European Approach to Space Security". Cambridge: American Academy of Arts and Science, 2009: 1.

588 OECD. Space 2030 Exploring the Future of Space Applications. Paris: OECD, 2004: 12.

589 For the specific definitional problems of space economy, see: OECD. The Space Economy at a Glance 2007. Paris: OECD, 2007: 16ff.

590 OECD. The Space Economy at a Glance 2007. Paris: OECD, 2007: 17.
591 OECD. Space 2030 Tackling Society's Challenges. Paris: OECD, 2005: 11.
592 Ibid: 112.
593 Space Foundation. The Space Report 2010. Colorado Springs: Space Foundation, 2010: 30.
594 SIA/Futron. "State of the Satellite Industry Report". June 2010. 22 Sept. 2010. http://www.sia.org/news_events/pressreleases/2010StateofSatelliteIndustryReport(Final).pdf.
595 It should be noted that manufacturing revenues for a given year correspond to the value of the satellites launched during that year, not to the value of satellites ordered during the year.
596 SIA/Futron. "State of the Satellite Industry Report". June 2010. 22 Sept. 2010. http://www.sia.org/news_events/pressreleases/2010StateofSatelliteIndustryReport(Final).pdf.
597 Federal Aviation Administration. Commercial Space Transportation: 2009 Year in Review. Washington DC: FAA, Jan. 2010. 22 Sept. 2010. http://www.faa.gov/about/office_org/headquarters_offices/ast/media/year_in_review_2009.pdf.
598 SIA/Futron. State of the Satellite Industry Report. June 2010. 22 Sept. 2010. http://www.sia.org/news_events/pressreleases/2010StateofSatelliteIndustryReport(Final).pdf.
599 Ibid.
600 Ibid.
601 Rathgeber, Wolfgang. "Space Policies Issues and Trends 2008/2009". 18 May 2009. ESPI Report 18. 29 Sept. 2010. http://www.espi.or.at/images/stories/dokumente/studies/espi%20report%2018.pdf: 65f.
602 Ibid: 18.
603 Pagkratis, Spyros. "Space Policies Issues and Trends 2009/2010". June 2010. ESPI Report 23. 29 Sept. 2010. http://www.espi.or.at/images/stories/dokumente/studies/espi%20report%2023_1.pdf: 24.
604 SIA/Futron. "State of the Satellite Industry Report". June 2010. 22 Sept. 2010. http://www.sia.org/news_events/pressreleases/2010StateofSatelliteIndustryReport(Final).pdf.
605 De Selding, Peter B. "National Export-Credit Agencies Stepping up Satellite Financing". Space News 14 Sept. 2009: 10.
606 De Selding, Peter B. "Panel Urges Britain To Boost Space Spending, Support Exports". Space News 15 Feb. 2010: 15.
607 This means that a GDP rise will not automatically lead to an increase in public space spending, the same logic being valid for a GDP decrease. However, it is obvious that a high GDP growth rate constitutes a positive framework condition for ambitious space programmes.
608 Peeters, Walter. "Forecasting the Consequences of the 'Crash of 2008' on Space Activities". Yearbook on Space Policy 2008/2009: Setting New Trends. Eds. Kai-Uwe Schrogl, Wolfgang Rathgeber, Blandina Baranes and Christophe Venet. Vienna: SpringerWienNewYork, 2010: 167.
609 The previous figures are based on Euroconsult data.
610 Pagkratis, Spyros. "Space Policies Issues and Trends 2009/2010". June 2010. ESPI Report 23. 29 Sept. 2010. http://www.espi.or.at/images/stories/dokumente/studies/espi%20report%2023_1.pdf: 33.
611 De Selding, Peter B. "Italian Space Agency Expects Budget to Remain Flat for 2010". Space News 18 Jan. 2010: 6.
612 Pagkratis, Spyros. "Space Policies Issues and Trends 2009/2010". June 2010. ESPI Report 23. 29 Sept. 2010. http://www.espi.or.at/images/stories/dokumente/studies/espi%20report%2023_1.pdf: 52.
613 Figure based on Euroconsult data.
614 Commission of the European Communities. Communication. The Community and Space: A Coherent Approach. COM (88) 417 final of 26 July 1988. Commission of the European Communities. Communication. The European Community and Space: Challenges, Opportunities and New Actions. COM (92) 360 final of 23 Sept. 1992. Commission of the European Communities. Communication. The European Union and Space: Fostering Applications, Markets and Industrial Competitiveness. COM (96) 617 final of 4 Dec. 1996.

[615] Commission of the European Communities. Communication. The European Community and Space: Challenges, Opportunities and New Actions. COM (92) 360 final of 23 Sept. 1992.
[616] The Lisbon strategy is an ambitious agenda for reform launched by the European Council in 2000. Its overarching goal is to make Europe the most competitive and dynamic knowledge-based economy in the world. It was replaced by the "Strategy 2020" at the beginning of 2010.
[617] Council of the European Union. Brussels European Council. 11 and 12 December 2008. Presidency Conclusions. Doc. 17271/1/08 of 13 February 2009. Brussels: European Union.
[618] Council of the European Union. Council Resolution. The Contribution of Space to Innovation and Competitiveness in the Context of the European Economic Recovery Plan and Further Steps. Doc. 10500/09 of 29 May 2009. Brussels: European Union.
[619] The LMI was launched with a Communication from the European Commission from 9 January 2008. It aims at entering fast growing worldwide markets with a competitive advantage. Six promising emerging markets were identified in the first instance, excluding space.
[620] Council of the European Union. Council Resolution Taking Forward the European Space Policy. Doc. 13569/08 of 29 September 2008. Brussels: European Union.
[621] ESA. Resolution on the Role of Space Delivering Europe's Global Objective. Doc. ESA/C-M/CCVI/Res.1 (final) of 26 November 2008. Paris: ESA.
[622] Call title "Factories of the Future"- 2010. FP7-2010-NMP-ICT-FoF. Published on 30 July 2009. Call title "Factories of the Future"- 2011. FP7-2011-NMP-ICT-FoF. Published on 20 July 2010.
[623] For an overview of the consultation process and its results, see the dedicated page on the DG Enterprise and Industry website: http://ec.europa.eu/enterprise/policies/innovation/future-policy/consultation/results_en.htm.
[624] Commission of the European Communities. Commission Staff Working Document. Lead Market Initiative for Europe. Mid-term Progress Report. SEC (2009) 1198 final of 9 September 2009. Brussels: European Union.
[625] Pagkratis, Spyros. "Space Policies Issues and Trends 2009/2010". June 2010. ESPI Report 23. 29 Sept. 2010. http://www.espi.or.at/images/stories/dokumente/studies/espi%20report%2023_1.pdf: 34.
[626] Commission of the European Communities. Flash Eurobarometer #272. Space Activities of the European Union. Analytical Report. October 2009. Brussels: European Union.

2. The legal personality of the European Union and its effects on the development of space activities in Europe

Lesley Jane Smith

2.1. Introduction

2.1.1. General

With the advent of the Treaty of Lisbon, the legal personality of the former European Community has been transferred to the European Union. This is a logical transition, given that the awkward three Pillar divide across the Community and Union, introduced by the Treaty of Maastricht, has now been eliminated. With the Lisbon Treaty, the relations between the Treaty on European Union (TEU) and the Treaty on the Functioning of the European Union (TFEU) have been finally put on a par.[627] At the same time, the Lisbon Treaty introduces specific, but nevertheless limited powers for the Union in relation to its space activities.[628] Moreover, the Union's space competences are to be exercised in parallel to those existing at national level.[629] This limitation is significant, given that the European Union's two major space projects to date – Galileo and GMES – have been initiated using powers that existed prior to the introduction of the space competence in the Treaty of Lisbon. The Galileo project was based on the provisions of Art. 154 EC (Art 170 TFEU), while the GMES programme was introduced under the rules governing funding for research and development.[630]

This chapter reviews the transfer of legal personality from the Community to the Union, and offers some reflections on the form and extent of the Union's new space competencies since the Treaty of Lisbon entered into force.[631] It discusses whether the Union's new space powers stand to secure its space activities on a longer term basis, both from a programmatic and an institutional perspective.

2.1.2. European space organisations and activities

Over the past fifty years, Europe has developed significant technical and scientific achievements in space under the leadership of the various independent European space organisations that were instated at inter-governmental level from the sixties onwards. Some of these organisations continue to operate within Europe at intergovernmental level,[632] others have since been privatised.[633] Independent of their legal status, the capabilities that have emerged from these organisations leave Europe with a strong legacy of space experience and a well-developed industrial sector. These institutions and stakeholders manage and control key space assets that count towards Europe's civilian and strategic space effort.[634] With telecoms and broadcasting as the major areas of commercial use of space, the European commercial satellite community is also well represented.[635] Europe's intergovernmental European Space Agency (ESA), while independent of the EU, has been operating for several years as the backbone or the *maître d'ouvrage* to the Community or Union in managing the Galileo GNSS project. This has taken place within the context of a special cooperation agreement concluded with the European Space Agency in 2003.[636] ESA is an intergovernmental organisation that grew out of a recognised need among European States to coordinate and cater for technical expertise across the scientific and programmatic uses of space. These institutions and organisations were therefore well established and successful, even before the concept of a Community or Union space competence was first ventured.[637]

2.1.3. Tackling EU space activities

A discussion of the Union's legal personality and its space competences belies the question as to how Europe is currently tackling what should now become a structured approach to its space activities.[638] The European Union's first efforts towards common space programmes and a space policy have been mapped out in key position papers over the past decade, now culminating in the provisions of Art. 189 and Art. 4 Treaty of Lisbon (TFEU).[639] The provisions of Art 189(1) TFEU are as follows:

(1) To promote scientific and technical progress, industrial competitiveness and the implementation of its policies, the Union shall draw up a European space policy. To this end, it may promote joint initiatives, support research and technological development and coordinate the efforts needed for the exploration and exploitation of space.

The subsequent subparagraph Art 189(2) provides for the Union to promulgate the necessary measures in support of these activities, while Art. 189(3) provides a treaty basis for institutionalising the current cooperation with ESA.[640] These powers are at the same time circumscribed. Art 189(2) excludes the Union from undertaking legal harmonisation in the field of space, an aspect that could have

Fig. 2: *The Lisbon Summit (source: EU Council).*

helped align national space laws that are currently being promulgated within Europe and beyond.[641] Art 4(3) in turn restricts the Union's competence in the areas of technological development and space, by providing for Member States to retain their own competence in these areas.

2.1.4. The National and EU paradigm

While the EU's initial entry into space activities neatly averted issues of national sovereignty, the Lisbon Treaty provisions on a space competence suggest that there is at least consensus about the degree of convergence required among the Member States to meet the call for a coordinated space effort.[642] The EU, as an international organisation and supranational community of states, faces the challenge of catering for the paradigm, where national and Union competences co-exist. Both from a political and from a legal perspective, the EU and its Member States are required under Art. 5 of the TEU to position the activity at the level most suited – be this national or Union – when undertaking common efforts within an EU coordinated space programme. This requires an assessment as to efficiency and balance, based on the imperatives of subsidiarity and proportionality, legal tools that provide procedural and substantive control on the level of activities in question.[643]

The insertion of two main legal provisions into the Lisbon Treaties governing the Union's shared space competence, Art. 189 and Art. 4, is therefore a logical continuation from what has been a period of concerted efforts towards consolidating a valuable space sector for Europe at an EU level. While the EU did not belong to the original European international organisations involved in space, it now has the competence to coordinate space activities along with those of its Member States, as well as formulating its own space policy.[644]

2.2. Legal personality in international law

2.2.1. General

Legal personality is an attribute of power or competence, also expressed as capacity, for states and international organisations to act. In the context of international law, it ensures their recognition as full legal subjects. Legal personality is also conferred on international organisations, the limits to their powers being set by their founding constitutions.[645] States, in contrast, have inherent national sovereignty over their territory and people, and are omnipotent, subject to the requirement that they conduct their international affairs within the confines of international law.[646] One of the most common forms of exercising legal personality encountered is a state's treaty making powers.

The steady expansion of the Union's constitutional (and geographical) borders, however, begs the question of how a community that is not a fully-fledged federation may conduct or coordinate its space affairs at an international level, while its Member States hold their concurrent (sovereign) competences. International space law is a field with a comprehensive body of UN Convention law, and accompanying Declarations, Resolutions and other international soft law regulation.[647] However, the extent to which the five UN space treaties bind the Union is an issue separate from whether or not the Union should formally accede to them. A State must consent to a treaty before it can be bound by its terms.[648] The EU itself is bound by the principles of international law, and in this respect, already bound by the provisions of the space treaties, in so far as they represent general principles of international law.[649] The Registration Convention specifically provides for international organisations to make declarations accepting its provisions, and ESA has done so in relation to the first test satellites for Galileo, Giove A and B.[650] ESA has formally notified acceptance of its international treaty obligations under Article XXII of the Liability Convention,[651] as well as under Article VII of the Registration Convention (REG), through a declaration on rights and liabilities. By virtue of its *sui generis* status as part of the international legal order, the European Union is bound by these principles as a matter of customary international law.[652] Whether accession to the UN space treaties is essential for the Union's own activities over and above ratification by its Member States begs the question of its legal personality in international law: firstly, the majority of the EU Member States are already signatories to these treaties;[653] secondly, the progressive alignment of ESA Member States with those of the EU reduces the inconsistency between ESA and EU membership as regards those legal instruments that are binding on both the EU and its Member States.. Galileo belongs to the EU, which exposes it as owner and as a supranational organisation to international responsibility and liability under international space law for damage that may occur. There are precedents showing how international governmental organisations, notably the European Space Agency, have regulated their common liability for space activities in the past. The Declaration by Certain European Governments relating to the Guiana rocket launch site used by Vega, Ariane and Soyuz was based on the premise that France acted as launching state, with a ceiling and apportionment of concurrent liability being accepted by the European Space Agency towards France.[654] Further open issues such as developing a liability regime for the Galileo Project to regulate issues of third party liability are currently under consideration, thereby completing a system of comprehensive liability for damage from Galileo space activities.[655]

2.2.2. Legal personality and the EU

With one simple sentence "*The Union shall have legal capacity*", Art. 47 of the TEU confers legal capacity and with this personality on the Union. Legal personality enables the Union, as it did the Community, to engage in international affairs, to enter into treaties and agreements at international level in the interest of the Union, in so far as these fall within its spheres of competence.[656] In short, it confers recognition of the Union at an international level, and allows it to take action where prescribed by its governing treaties.[657] Most importantly, the Union can speak on behalf of its Member States with one voice in those areas where common goals and policies are pursued. Such characteristics as these form part of the constitutional construct that has developed with the Communities', and now the Union's, powers over time.[658]

These characteristics contribute to the inalienable *acquis communautaire,* the foundation of community laws and legal relations that apply across the Union at any one time.

2.3. Personality, capacity and competence distinguished

2.3.1. Constitutionality

The Union can only accede to treaties where it has both the capacity and the competence to do so.[659] This "dual" requirement is a lever on the exercise of powers devolved on the Union which is exercised by the European Court of Justice when the Community accedes to international treaties and agreements.[660] Judicial review by the Court of Justice is of constitutional nature, involving an assessment of where the Union's external and internal competences lie. Legal personality and capacity are pre-requisites to the assessment. The question is at all times "whether the intensity of the arrangement, whatever its denomination, is such as to involve a cession of national powers in favour of Community competence in the field of application of the rules concerned".[661] Not only must the Union have the capacity to enter into the area of activity; its competence must derive directly from its governing treaties and the area of activity affected. Art. 1 of the TEU, with its reference to "the Union", on which Member States confer competences to attain objectives they have "in common", is a reminder that space now falls into this category of common goals. Member States refer to this in terms of parallel competences, an unwritten "constitutional" arrangement.

2.3.2. Jurisdiction of the Court of Justice

Most aspects of space activities have an inherently international component and fall into the area of external competences, or foreign affairs. The task of assessing whether the Union has the competence in its external relations to accede to international treaties has traditionally fallen to the European Court of Justice. Over the years, the EU has successfully concluded various forms of agreements ranging from foreign trade to more complex stabilisation pacts.[662] It has also concluded EU membership of international economic organisations, notably the WTO, where this has been seen in the Community's interest.[663] This particular group of agreements to which Member State and the EU belong together are referred to as mixed agreements. They signify the areas where the Community and now the Union occupied joint powers to act. Technically, therefore, it would be open to the Union to undertake such agreements relating to space activities, where the activity in question is seen to belong within the competence of the EU.[664]

This level of shared competence or at least agreement on objectives, has enabled the Union to develop relations with the European Space Agency already prior to

Fig. 3: *The European Court of Justice (source: Wolfgang von Brauchitsch/Bloomberg News).*

the advent of the Lisbon Treaty. The EU-ESA cooperation took the form of a simple cooperation agreement between two intergovernmental organisations, without any formal accession by one international organisation to the other. Much has been written about this agreement in European and international law, including proposals as to which organisation might legally accede to the other.[665] Art. 189(3) may do away with discussions about inter-institutional cooperation but it offers no permanent combined organisational structure. Art. 189(3) empowers the Union to arrange its *modus operandi* with ESA, without the need for any further treaty amendment. However, the Position Paper of the ESA Member States in Preparation of the VII Space Council in 2010 is significant in that it clearly shows that Member States are not willing to see encroachments by the Union on ESA's remit.

2.4. Legal personality, the European legal order

2.4.1. Ambit of the new legal order

As indicated, the European Union and European law are squarely placed within the existing international legal order.[666] Hailed as a "new legal order", limited and for the benefit of those states which join, the Community (now Union), in contrast to public international law, boasts the notable distinction of granting enforceable rights, not only on the state parties to the Treaties, but also on their nationals.[667] Such directly effective rights circumscribe an integrated legal community, which, although not a fully federal state, is crafting a union of diverse states towards 'a new stage in the process of creating an ever closer union among the peoples of Europe.[668] Art 42 of the TEU however, goes beyond the Preamble to the TEU by prescribing a common security and defence policy as part of the Common Foreign and Security Policy (CFSP), formerly the 2nd Pillar.

In the context of foreign affairs, Art 24(3) of the TEU further prescribes that "Member States shall support the Union's external and security policy actively and unreservedly in a spirit of loyalty and mutual solidarity and shall comply with the Union's action in this area." These two provisions impose a clear obligation on Member States to actively support the Union, be this in space-based or other form of strategic cooperation.[669] This provides some leeway for including security-related measures within the EU's space policy.

2.4.2. Constraints on legal personality

International law imposes constraints on the European Union, to the extent that it must refrain from conduct that conflicts with existing international treaty obligations.[670] This primacy dictate's purpose is to ensure that its citizens and institutions are not preventede from enjoying the benefits to be derived from these provisions.

The Union's capacity to act is therefore subject to a twofold limitation: firstly as dictated by international law and secondly by the limitations imposed on the Union by its Member States, as formulated within the consecutive treaties, protocols and declarations.[671] Taken together, they form the borders of its constitutional powers.[672] This too is relevant to space activities. As indicated, under Art. 4(3) TFEU, the Union's powers to act are limited to acting in parallel to its Member States. Art. 4(3) TFEU cannot preclude or pre-empt action at a national level in relation to space. This needs a *modus vivendi* to identify the workable borderlines of national and European sovereignty.

2.4.3. Competence creep and sovereignty

With each new treaty, the activities at a European Union level have expanded beyond the original scope of its preceding stage of integration to move further forward. The constitutional bulwark between the various versions of the Treaties has facilitated subtle extensions of competence over time.[673] This has inevitably led to an expansion of the sectors governed by Union law, and notably those areas which are *communitarised*. As the failure of the European Constitution showed, there are not only legal, but also inherent political constraints against an over-centralisation of activities at the supranational level. While the Lisbon

Treaty has not brought in any major substantive changes to the scope of the Union's powers, it concedes greater power and attention to its external and international affairs, seated within the Union. The appointment of a High Representative for Foreign Affairs and Security Policy and of a President of the European Union is witness to the importance of these fields and the concern to create an integrated and coordinated external policy. While the formal inclusion of space competence is perhaps new, the provisions limiting the Union's competence to a shared or parallel competence in space necessarily encroach on a field that is inherently linked to national sovereignty, namely foreign or external affairs.[674] This step has called for the Member States to guard their own preserve, and the Treaty provisions are accordingly cautious.

The ultimate control over the legality of the Union's activities for actions under communitarian issues has traditionally been held by the judicial machinery.[675] Actions at all levels by the Union, with the exception of actions falling within the Union's Common Foreign and security Policy CFSP, are subject to judicial review.[676] While this division has enabled the monitoring of the integration process, including potential or real encroachments or compromises on national sovereignty,[677] it encounters its limits when addressing the scope of the Union's activities in outer space. These lie in the areas of activity that involve dual use and not only civilian use of outer space. A system for demarcating sovereign space powers between the Member States and the Union can only operate within the categories of "external and security" duties expressed in Art. 24(3) of the TEU. Many of the civilian aspects of space activities involve clear issues of dual use, making the cut-off between them difficult to define.[678]

2.4.4. Legal personality, European governance and integration

The lack of a clear demarcation between the national and Union space competence has to be seen as a response to what has been an incomplete process of constructing a system of governance for the European space community.[679] This is currently occupied by the Space Council. The continued division of parallel space competences between national and European levels may even pose a stumbling block to creating clear structures for a future space agenda.[680] Space activities are a classic sphere of political hegemony, clearly reflecting national ambitions in space. Aligning the membership of the European Union with that of its expert agency, the European Space Agency, is a comparatively minor step within what now appears a greater agenda, if the EU is to proceed and succeed with its further space programme.

2.5. Demarcation and conferral

2.5.1. Principles

Legal personality is accordingly only the first part of the equation as to whether powers exist and may be exercised at a Union level. The limits within which it can legitimately operate are prescribed by the Treaty.[681] According to the principle of conferral, or of the attribution of power enunciated in Art. 5(1) of the TEU:

> "The limits of the Union's competences are governed by the principle of conferral. The use of Union competences is governed by the principle of subsidiarity and proportionality."

Art. 5(1) emphasises the dynamic nature of the European Union power spectrum against the balancing tools of subsidiarity and proportionality contained in Art 5(3) of the TEU.[682]

2.5.2. Conferral defined

The demarcation between Member States and the Union is therefore central to the operations of the EU; the activity in question must take place at the most appropriate level.[683] The TFEU lists the areas in which the EU has either exclusive (Art. 2(1), Art. 3 of the TFEU), shared (Art. 2(2) of the TFEU), or supportive coordinating competence (Art. 2(5), Art. 6 of the TFEU).[684]

Two forms of collaboration in governance have developed within the Lisbon Strategy: the Open Method of Coordination (OMC) for economic and monetary union, alongside the new supportive and coordinating competence under Art. 6 of the TFEU. The former has developed as a form of governance-sharing for the EU: it allows "bridges to be built where there are black holes of non-decision".[685] It offers a flexible form of governance in areas which – were it not for OMC – might expand the limits of conferred powers too far. "It offers a broad legal base to transfer political will into EC policy and normative standards."[686]

It would appear that the Union's space competence, while clearly shared or parallel, falls to be handled with a similar supportive approach. Art. 6, with its inclusion of industrial and civil society concerns, would imply so. The EU polity is not only exposed to a competence creep, but to a concern to safeguard the achievements of existing inter-governmental organisations. This is why the Draft Position Paper of the ESA Member States indicates in no uncertain terms that there should be no encroachment by the EU on the work of the ESA. The differing procurement rules of the EU and ESA alone prohibit any overlap.

> "The globalisation of economic activity has increased the opportunities of states and policy makers, as well as other stakeholders in the political policy processes, to learn from the experience of policy intervention elsewhere."[687]

Ultimately the demarcation issues will turn on the civil and military interface for space activities. ESA, as an organisation solely competent to act for peaceful purposes, can facilitate this exercise in demarcation for future space activities.

2.6. Structure for regulating space activities in Europe

2.6.1. Challenges

The provisions of Art. 4 and Art. 189 of the TFEU must be read in the light of what the EU Commission President Barroso described in his "Ambitions of Europe in Space" speech as the "fundamental challenges" required to invigorate competitiveness and economic growth for the EU: by contributing to innovation and employment, by combating climate change, and by addressing major issues of transport and security, the European Union can secure a voice at a global level among the leading space powers.[688] In doing so, it not only represents all Member States; it can also rely on the well-established space infrastructures which have paved the way to securing technological and scientific progress in space for Europe. This "late awakening" on the part of the EU to providing a complete space agenda may even be fortuitous: Member States may not have been ready to support such a step at an earlier stage. Although these are shared powers, they are flanked by other security-related issues, such as defence and space situational awareness.[689]

2.6.2. Civilian and military aspects of space

The strategic aspects of space make it an area of activity with strong geo-political overtones. The concern to develop a European Space Situational Awareness (SSA) has given rise to action at all three levels of the European Space Agency (ESA), the European Defence Agency (EDA) and the EU Commission (EC). Latterly, SSA has been included in the EU Space Work Programme of July 2009, with a view to creating independent space capabilities for the EU.[690] SSA has aspects in common with defence capabilities under the Common Foreign and Security

Policy (CFSP). The provisions regulating CFSP were previously located under the since abolished 2nd Pillar on inter-governmental cooperation. Although the Union Treaty continues to retain a "small" IGC pillar in its Art. 24, these provisions now mandate the Member States' commitment towards common action at an EU level.[691] In this respect, the de-pillarisation of CSFP has opened the way for space to become an EU competence, notably in defence and security issues. This in turn has opened the way for expanding further common aspects, such as common military procurement at an EU level.[692]

2.6.3. Common foreign and security policy and agencification

The European Union's space agenda therefore includes not only civilian, but also military capabilities.[693] The post-internal market *agencification* within the EU, while predominantly of internal market origin, has since seen the creation of the European Defence Agency and the European Satellite Centre in Torrejon, Spain, falling within the Council's remit.[694] Conceivably, such developments might not have found consensus at an earlier stage. Now, with Lisbon, the former 2nd Pillar and its field of Common Foreign and Security Policy (CFSP) has been given further legitimacy by the Member States as a common objective through Art. 24 of the TEU.

This provision confers competence on the EU in all areas of foreign policy. It includes the power to "conduct, define and implement a common foreign and security policy, based on the mutual political solidarity among Member States, . . . and the identification of an ever-increasing degree of convergence on Member State's actions".[695] The inclusion of space within these parameters is a task for political consensus and action: it pays tribute to the boundaries of constitutionalism at EU level. Art. 24 reflects the significance of the EU's presence in the international arena. Nevertheless, as the classic field where national hegemony and ambitions are prevalent, space activities remain subject to legitimacy and legal capacity. Art. 24 is clearly circumscribed by the consensus of the Member States under Art. 4(3) of the TFEU.

2.7. Conclusion

The European stakeholders in space have established an impressive track record in crafting European space activities. The immediate conclusions from the Joint EU and ESA Space Council held on 25 November 2010 were to continue supporting the development of the Union's space policy, by moving forward with its space flagships, Galileo and GMES.[696] The focus is now on several priority actions, from ensuring funding, promoting a healthy commercial space sector, to protecting satellites and radio frequencies.

The dynamics of European integration have varied over time, and despite the many attempts to analyse and classify the accompanying process of the developing polity, they depict the political and legal phenomena associated with complex forms of integration of states and the economic dimensions involved in creating a single market with its own currency. Such steps involve inherently political processes of transition and are no longer of economic or legal nature alone.

Nevertheless, a response to the effect of the new legal personality of the European Union on its space activities may well be found in their very economics: financing the European space effort almost inevitably involves some form of public funding. While the Member States are unwilling to dispense with their own sovereign and economic interests in space, a concerted effort at the EU level appears attractive, not only from a budgetary perspective but also in the interest of the various stakeholders ranging from institutions and agencies to satellite operators and the industry.

The European Union has undertaken important steps and made the necessary investment to warrant maintaining the level of expertise it has produced in its space activities. It has formulated various elements towards a definitive space policy. It must now ensure that its major civilian space projects move forward, so that it can respond to the future challenges that society faces and respond with the benefits that space offers. The inclusion of space activities within the Union's competencies under the Lisbon Treaty was a timely measure to sustain Europe's vital contribution to this sector.

[627] Art. 4(1) TEU: "The treaties shall have the same legal value."

[628] The Union's new space competences, regulated under Art. 189 and Art 4, co-exist with the other competences that originally served as a legal basis for the initial key space projects Galileo and GMES. Art 4(3) prescribes its limits.

[629] Art. 4(3) TFEU: "In the areas of research, technological developments and space, the Union shall have competence to carry out activities, in particular to define and implement programmes; however, the exercise of that competence shall not result in member States being prevented from exercising theirs."

[630] These activities were based on the provisions governing Trans-European networks under Art. 154 EC (transport, telecommunication and energy infrastructure). GMES was originally organised within the 6th, thereafter the 7th Framework Programme, Decision 1982/2006/EP and Council of 18 December 2006 concerning 7th Framework Programme of the European Community. Implementation of GMES is now governed by Regulation 911/2010 of European Parliament and Council of 22 September 2010, OJ L 276/1 of 20.10.2010.

[631] The Treaty of Lisbon entered into force on the 1 December 2009, one month after the final ratification by the last Member State, *in casu* Czech Republic.

[632] On the complex structures of the various satellite organisations in Europe (EUTMETSAT, ESOC), including their privatisation (EUTELSAT), see Francis Lyall, Larsen, P., Space Law, A Treatise, 2009, 356–364; for a history of the specialist international agencies and intergovernmental organisations, see P.I.Ph. Diederiks-Verschoor (ed. Kopal), Introduction to Space Law, 2008, ch. 1.
[633] Some of these, notably EUTELSAT, have since been privatised, see Francis Lyall, id.
[634] Among Europe's space assets figure not only the Giove A and B satellites, but the future Galileo satellite fleet. The prime location of the French territorial launching base, the Guiana Space Centre (GSC), Kourou, Guyana, is a key factor in maintaining an independent European space infrastructure.
[635] For further information on the European Satellite Operators Association, see http://www.esoa.net/v2/.
[636] See Framework Agreement Between the European Community and the European Space Agency (hereafter Framework Agreement), Brussels, done 25 November 2003, entered into force 28 May 2004; OJ L 261/64 (2004).
[637] Krige, J & Russo, A, The story of ESRO and ELDO 1958–1973. A history of the European Space Agency, 1958–1987, Vol I, retrieved from www.esa.int/esapub/sp/sp1235/sp1235vlweb.pdf.
[638] Since the conclusion of the Framework Agreement between ESA and EU, an annual Space Council has been instated as from 2004 allowing representatives of the Member States, the EU and ESA to deliberate together; for a complete overview of its agenda, see Council Resolution on Taking Forward the European Space Policy, 26/27 September 2008, approved by the Council of Ministers of the European Space Agency. Further, Nicolas Peter, Space Power and Europe, in the Need for a Conceptual Framework, 59th International Astronautical Congress, (IAC Glasgow) 2008.
[639] The concept of a European Space Policy (ESP) was first promoted in a European Parliament Resolution of 17 September 1981 on Europe's Space Policy (OJ C 260/102, of 12 December 1981), but not followed through until the development of the Galileo project. See: European Commission, Galileo – Involving Europe in a New Generation of Satellite Navigation Services, of 10 February 1999, COM(1999) 54 final. Proposals for inter-institutional cooperation were subsequently formulated by the latter, see European Commission, Towards a coherent approach for Space, of 7 June 1999, SEC (1999) 789. Recent publications and legislation is available relating to ESP and Galileo, See Commission Communication of 26 April 2007 on European Space Policy, COM(2007) final; see further Regulation of the European Parliament and of the Council on the further implementation of the European satellite navigation programmes (EGNOS and Galileo), No. 683/2008/EC, of 9 July 2008; OJ L 196/1 (2008).
[640] Art 189(3) TFEU: "The Union shall establish any appropriate relations with the European Space Agency."
[641] Alignment of national space laws is a goal in itself, in that it achieves consistency with the treaties at international level. Some coordination is possible at UN level via the Legal Subcommittee of the UN Committee on the Peaceful Use of Outer Space, UNCOPOUS. For a comprehensive overview of the activities undertaken by the UN Office of Outer Space, see http://www.oosa.unvienna.org/.
[642] Gerda Horneck, Coriadini, A, Haerendel, G: Towards a European Vision for Space Exploration, Recommendations of the European Space Advisory Group, in: Space Policy (2010). For a critical assessment of the dilatory process of moving towards a definitive European space policy, see K. Madders, Thiebaut W., Carpe Diem: Europe must make a genuine space policy now, in: Space Policy 23 (2007) 7.12.
[643] Art. 5(3)TEU; for further details, see below, p. 14 (**cross-reference**).
[644] This was launched with the Report of the Three Wise Men, (Lothar Späth, et al), on which, see L.J. Smith/Hörl, K.U., Constructing the European Space Policy, in: P. Olla (ed): Commerce in Space: Infrastructure, Technologies and Applications, Univ. Michigan Press (2007), chap. 9.
[645] The ESA Convention is available under http://www.esa.int/convention/ It entered into force on 30 October 1980 but operated de facto from 31 May 1975.
[646] Brownlie, Ian, Principles of International Law, 7. ed., 2008, Oxford, chap. 2, chap. 14.
[647] For details of the signatories and ratifications of the UN Treaties, see http://www.oosa.unvienna.org/oosa/en/SpaceLaw/gares/index.html.

[648] Brownlie, n. 20, above, 13–14; 611, unless these have already achieved the status of customary international law. This is the case, at least for certain provisions of the OST, see Francis Lyall, Larsen, P., n. 6 above, 70–80.
[649] The initial findings of the Court of Justice on the relationship between community law as a *sui generis* part of the international legal order are to be found in Case 26/62, Van Gend en Loos v. Nederlandse Administratie Belastingen [1963] ECR 1. The ERTA judgment, case 22/70 Commission v Council (ERTA) discussed the extent to which the European Community (at that stage) had an implicit treaty making power in those fields of competence ascribed to it. Member states are not allowed to enter into international commitments which could prejudice either the standing or status of the Community's obligations.
[650] ESA has registered the Giove frequency filings with the ITU. Of the IGOs that have accepted the rights and obligations under the space treaties, ESA and EUTMETSAT have made declarations under ARRA, LIAB and REG; EUTELSAT has accepted LIAB.
[651] Convention on International Liability for Damage Caused by Space Objects, 1972.
[652] The recognition of international law obligations has been accepted with regard to the European Union's overriding liability for the Galileo GNSS system, inspired by the Liability Convention, in Article 17 of Council Regulation (EC) on the establishment of structures for the management of European satellite radio-navigation programmes, of 12 July 2004, No. 1321/2004, OJ L 246/1 of 20.7.2004, as amended by Regulation on the further implementation of the European satellite navigation programmes (EGNOS and Galileo), No. 683/2008/EC, of 9 July 2008; OJ L 196/1 (2008).
[653] As of January 2010, of the EU Member States, 22 have ratified the 1967 Outer Space Treaty (OST) and 23 have ratified the 1972 Convention on Liability for Damage from Outer Space (LIAB), while 17 have ratified the 1986 Registration Convention (REG). 3 have ratified the Moon Agreement (MOON). The EU's relationship to the United Nations and other specialised international agencies is regulated in Art. 220 TFEU.
[654] A clear precedent can be found in the Declaration by Certain European Governments on the Launchers Exploration Phase of Ariane, Vega, and Soyuz, from the Guiana Space Centre, done at Paris, March 2007, reprinted as HMSO Command Paper Miscellaneous No. 10 (2009) Cm 7700.
[655] For the background to the various liability issues involved, see Policy Aspects of Third Party Liability in Satellite Navigation ESPI Report 19, Alfredo Roma, Schrogl, K.U, Sánchez Aranzamendi, M.(eds), July 2009; further L.J.Smith, Facing up to Third Party Liability for Space Activities, Some Reflections, in *Proceedings of the International Institute of Space Law 2009* (2010)Where is Paradise? The EU's Navigation System Galileo – Some Comments on Inherent Risks (*or* Paradise Lost), in: IISL/AIAA Proceedings of the 50th Colloquium on the Law of Outer Space (2007), 346–358. A Commission Working Group on GNSS Liability was set up subsequent to presentation of a Draft EU Regulation for GNSS liability presented by Italy in 2007 and has already reported in: European Commission, Working Document, 'Global Satellite Systems (GNSS) Extra Contractual Liability', 24 June 2009, EGPC-09-07-06-02.
[656] See judgments of the ECJ in: case 22/70 Commission v Council (European Road Transport Agreement), ECR [1971] 263; Opinion of ECJ, 1/91 Draft Agreement between EC and EFTA, [1991] ECR; Conferring legal personality on the Union was discussed during the deliberations on the European Convention, but it was felt premature at that stage to compromise sovereignty over foreign affairs, see Amato Report, CONV 305/02 of 1 October 2002, cited in: Philippe de Schoutheete, Andoura, S, The Legal Personality of the European Union, Studia Diplomatica, Vol LV, 2007, n. 1.
[657] Art 3(6) TEU.
[658] A notable trend towards measuring the "constitutionality" of integration is commented on by Erika Szyszcak, Experimental Governance and Open Method of Coordination, European Law Journal (2006), 486–502.
[659] Where Community and now Union rules are adopted to achieve objectives of the Treaty, then Member States may not assume obligations outside the framework of the Treaties that could affect these obligations or alter their scope, see ECJ Opinion 1/91, EFTA, above n.29; Opinion 2/91, ILO [1991] ECR 1061.

[660] Judgments of the ECJ are binding on the parties and to be followed by virtue of Art 260 TFEU.
[661] P.J.G.Kapteyn, Verloren van Themaat, P., (ed Gormley), Introduction to the Law of the European Communities, 3rd. English edition, 1260.
[662] E.g. There has been an Association Agreement between the EU and Turkey since 1963; Turkey is a candidate country to the EU, with a customs union in operation since 31. December 1995. A Stability Pact was concluded between the EU and the Balkans in 1999, and the Stabilisation and Association Process Dialogue is underway with Kosovo.
[663] Opinion 1/94 related to the legality of the Community acceding to the WTO, see further Meinhard Hilf, The ECJ's Opinon 1/94 on the WTO – No Surprise, but Wise? [1995] 6 EJIL, 1–15.
[664] The pre-Lisbon judgments all make clear reference to the differing competences between the Community and the Union, only the former being competent to accede, see n. 35, above.
[665] F.G. von der Dunk, Towards one captain of the European spaceship – why the European Union should join ESA, 19 *Space Policy* (2003), 83–6; L.J. Smith & K.U. Hörl, Constructing the European Space Policy, Past, Present and Future, in P. Olla (Ed.), *Commerce in Space, Infrastructure, Technologies andApplications* (2008), 187–208; T. Hoeber, ESA + EU: Ideology or pragmatic task sharing, 25 *Space Policy* (2009) 206–8; S. Hobe *et al.*, Entwicklung der Europäischen Weltraumagentur als 'implementing agency' der Europäischen Union: Rechtsrahmen und Anpassungserfordernisse, in *Kölner Schriften zum internationalen und europäischem Recht*, Band 17 (2009), 282–339.
[666] See cases at n. 23, above.
[667] Case 21–24/72 International Fruit Company Produktschap N.V. v Produktschap voer Groenten en Fruit, [1972] ECR 1219. This has led to recognition of directly enforceable rights at Union level, by virtue of which legal or natural persons may be entitled, under specific conditions, to enforce rights acquired under European law against their national administrations or governments. Generally, there must have been some failure to transpose the particular provision of European law as provided. Case 26/62 Van Gend en Loos v Adminstratie der Belastingen, above, n 23.
[668] Art. 1 sentence 2, TEU. The ECJ's jurisdiction to grant directly enforceable rights through varied case law relating to the requirements for vertical and even horizontal actions based on directly effective legal provisions, through to government liability, has filled many text books.
[669] See Preamble to European Parliament resolution of 10 July 2008 on space and security (2008/2030 (INI)) at B: "Where the various political and security challenges which the Union is increasingly facing make an autonomous European Space Policy a strategic necessity... ".
[670] Case 41–44/70 International Fruit Company v Commission [1971] ECR 411; case numbers 89, 104, 114, 116, 117, 125–129/85 Ahlström Osakeyhtiö et al. v Commission [1988] ECR 19; T-115/94 Opel Austria v Council [1997] ECR II 39.
[671] The entirety of these provisions and documents combine to form the consolidated version of the Treaties. See Declaration 24 concerning the legal personality of the European Union.
[672] For a discussion as to how the constitutionality of the Union has been primarily developed through deliberative interpretation of the treaties, see Antoine Vauchez, The transnational politics of judicialization; Van Gend en Loos and the making of the EU polity, European Law Journal 16 (2010) 1–28.
[673] S. Weatherill, Competence Creep and Competence Control, in P. Eeckhout & T. Tridimas (Eds.), 23 *Yearbook of European Law* (2004), 6–7: "EU action may 'creep outward' but it does not wholly foreclose State choice in the relevant area. Competence is shared (...) it is the Member states, within the EU framework, that have been the primary actors in the centralizing process of 'creeping competence'".
[674] Jan Wouters, Space in the Treaty of Lisbon, Yearbook of Space Policy, 2009, 116–123; see further, Draft Joint Position of Member States to ESA on issues regarded as critical for the successful preparation of Space Council VII (final), 2010.
[675] Antoine Vauchez, n. 47, above.
[676] Art. 275 The Court of Justice of the European Union shall not have jurisdiction with respect to the provisions relating to the common foreign and security policy, nor with respect to acts adopted on the basis of those provisions. For a critique of case law relating to the competence demarcation in external affairs, see: Christophe Hillion, Wessel, R, Competence Distribution in EU External relations

after ECOWAS: Clarification or Continued Fuzziness? Common Market Law Review 46 (2009) 551–586.

[677] Many space applications have dual use capabilities that technically address military and not exclusively civilian uses.

[678] This is a particular problem in transatlantic space practice as a result of the ITAR rules.

[679] A. Gaubert, A. Lebau, Reforming European Space Governance, Space Policy 25 (2009) 37–44, 42.

[680] id.

[681] Case 41–44/70 International Fruit Company v Commission, n. 45 above.

[682] Art 5(3) 'under the principle of subsidiarity, in areas which do not fall within its exclusive competence, the Union shall act only if and in so far as the objectives of the proposed action cannot be sufficiently achieved by the member States, either at central level or regional and local level, but can rather, by reason of the scale or effects of the proposed action, be better achieved at Union level. This has led the EU to strive for better regulation, including better consultation, see European Commission, Third Progress Report on the Strategy for Simplifying the Regulatory Environment, COM (2009) 19.

[683] For a discussion as to the methodology of the ECJ and its jurisdiction to rule on the limits of the competence issue, see the UK House of Commons Select Committee on European Scrutiny, 29th Report, session 2006–7, available online http://www.publications.parliament.uk/pa/cm200607/cmselect/cmeuleg/41-xxix/4102.htm, on Opinion 2/94 on the EU's Accession to Human Rights Convention. 'In the course of the operation of the common market' imposed little or no constraint on the use of the Article (308) and was not intended to do so. Since then, the Community has extended its objectives to include much that is not primarily concerned with the operation of the economic community. But the **purpose** [our emphasis] of Article 308 remains unchanged: to provide a necessary power, when none is available elsewhere in the Treaty, to attain any Community objective" The ECJ's usual approach to the interpretation of the Treaty is to give effect to what it understands to be the Treaty's purpose (purposive interpretation).

[684] The powers are outlined in Art. 3 (exclusive competence); Art 4 (shared competence) and Art 6 (coordinated or supplementary competence).

[685] Erika Szyszczak, n 31, above.

[686] id. S 490.

[687] Id. 490.

[688] President José M.D. Barroso, "The Ambitions of Europe in Space", Transcript, Speech 09/476, delivered during the Conference on European Space Policy, Brussels, 15 October 2009.

[689] The EU has an early warning facility – Joint Situation Centre (SITCEN) – based in Brussels.

[690] For details of the Space Situational Awareness, see http://www.esa.int/esaMI/SSA/SEMYTICKP6G_0.html.

[691] Art 24(1) TEU: 'The Union's competences in matters relating to foreign and security policy shall cover all areas of foreign policy and all questions relating to the Union's security, including the progressive framing of a common defence policy that might lead to a common defence.

[692] Rules coordinating a large part of procurement for military equipment and services have been in force in the European Union since 2004. The latest reform of EU defence procurement legislation is Directive of the European Parliament and of the Council on the coordination of procedures for the award of certain works contracts, supply contracts and service contracts by contracting authorities or entities in the fields of defence and security, and amending Directives 2004/17/EC and 2004/18/EC, No. 2009/81/EC, of 13 July 2009; OJ L 216/76 (2009). The scope of the Directive is outlined in its Article 4. The Directive entered into force on 21 August 2009. See further, Anglo-French defence cooperation: Entente or bust, Economist, 14 October 2010.

[693] Galileo was initially announced as a uniquely civilian project. While ESA's activities are restricted to peaceful purposes, the EU's own military capabilities are linked to its role within the WEU and the associated EDA.

[694] The European Defence Army, EDA, European Satellite Centre ESC are attached to Council as part of the executive arm.

[695] Art 24(3) TEU; cf. Art 24(1) TEU: "The Union's competence in matters of common foreign and security policy shall cover al areas of foreign policy and all questions relating to the Union's security, including the progressive framing of a common defence policy that might lead to a common defence". See also: Draft Road Map, ESDP and Space 2007, SEC European Space Programme.
[696] See Note, Council of the European Union, Background, Competitiveness Council, Brussels 25 and 26 November 2010, 16253/10 of 17 November 2010.

3. Institutional development of satellite navigation in Europe

An interview of Heike Wieland to ESPI Resident Fellow Spyros Pagkratis

Q: I would like to begin with my first question: what is in your opinion the importance of Galileo for the EU?

A: For the EU, I would say that the main importance lies in the fact that it is the first European space programme that is financed and managed by the European Union together with ESA. It is very important that we have for the first time a real and concrete space programme on an EU level, which is also an effort of the member states of the European Union to do something together in the space area that until now was pretty much dominated either by national efforts, or through ESA that is an international organisation set up for specifically supporting national programmes.

Q: Your answer takes me to the second question, which is what are the principal stakeholders and what are the roles in implementing it?

A: This is a very interesting question, because it depends on where you look to. Maybe I will start with the inner circle: in a strict sense the stakeholders for the moment are the European Union, represented by the European Commission and ESA. However, if you look at it on a broader scale, a lot of people or other stakeholders are involved, like the member states of the European Union, the members of ESA, the European Council, the European Parliament, small and medium enterprises (SMEs), and of course all EU citizens. Returning to your sub-question, it also brings in mind the industry involved Galileo contracts with ESA, which also have a vested interest in contracts coming out of Galileo, both independently and collectively as the space industry sector.

Q: How is this coordination between all parties working and what is their role in the implementation of the programme?

A: Let's just say that if we examine the coordination of the stakeholders in a broader sense, we might find that there is no coordination at all. Admittedly, what we are talking about here is a huge deal of political interests and political tensions. Concrete management at the moment is done by the European Commission, based on its mandate given to it by the EU member states, by the European Council and the European Parliament. The programme also involves coordinating the various interests involved, including the role of third countries such as the US. I have to say that among the different stakeholders here is no real political coordination at the moment.

Q: This takes us to my third question, which is how would you qualify the role of the member states. Member states are represented in different levels, through the Council, through ESA, through heir own involvement in the programme. Would you say that they assume multiple roles in the programme, and how does this work in practice?

A: It is a very interesting question to qualify the role of member states. It is hard for me to answer right now, but I might describe even more the role of the member states, because if you look into member states now from the EU side, it is very important for them to have this Galileo programme, the first real operational space programme in the European Union. This is the realisation of a very important and quite long deal, which was born in the EU Council many-many years ago and has produced already many political documents and decisions. The member states we are talking about were pushing towards the European vision, the European Union vision. From The EU side, member states are known to keep the budget limited, not to spend more money or extend the EU budget and so on, in order to keep control on the programme's budget. On the other hand, we have a quite divergent situation within ESA member States, because ESA member states are also interested in the EU vision, making Galileo a joint effort now. However, it is clear that both sides and the ESA member states have slightly diverging interests. This is because ESA is supporting the national industry through ESA contracts, through the system of ESA that is also called geo-return, according to which the incentive for ESA member states to make Galileo happen is even greater, because it may constitute the means to support their space industry. This situation brings us to a quite vicious circle, because on one hand we have an ESA logic, which is not a vicious logic, but it's just how it is made. On the other hand, we have the EU member states, partly the same as ESA member states, which are know for looking for a system of open competition.

Q: This takes our discussion to my next question, on what is the current status of the European Commission-ESA cooperation, and how could respective competitive and geographical return models be compromised?

A: You know, if you talk about compromise between the competitive market logic and the geographic return logic, I don't think there is a lot of space for it. Nevertheless, you can still try to get the best to do the job.

Q: So how would they work together? I am not asking about qualifying or comparing them, but I am interested in how would this working arrangement between the EC and ESA could develop in the future, based on the fact that they operate in a different logic, as you described.

A: For the moment there is no formal cooperation, there is no official co-existence between the two procurement models. We have a delegation from the European Union, represented by the European Commission, to ESA and this delegation has foreseen that geo-return is not applied. This is also something which underpins a bit the EU-ESA cooperation -I don't know if we can call this cooperation, or rather an EU-ESA agreement. This agreement is much older than the delegation agreement. According to the agreement's provisions and I think we have a quite interesting formulation there -I believe under its article 5- as far as EU is concerned EU's rules apply and as far as ESA is concerned ESA's rules apply. That means that as far as procurement is done by ESA, the geo-return principle applies. However, this is not the case with the delegation agreement. But I'm not so sure if this is going to be the case in the future, because what we have already now is a situation in which there is on one hand a free competition system according to EU rules, but behind the scenes, and now we are talking about political interests and stakeholders who are interested and involved, behind the scenes we have a situation in which member states are pushing to avoid the competition process and implement the principle of geo-return to which they are used to from ESA. I am not saying that the system of geo-return is bad, because through this system you create a kind of protected area for European industries as well. The system has its drawbacks, but it is successful in giving the right responses to the right people on the job. This is not always the case if you have a free competitive system as in the EU, because in the EU system is not really made for such a protected market as it is in space area, which would also be if we are talking about defence for example.

Q: I believe you have also answered now my next question, regarding the programme's governance structure. So, how would you see it evolving in the future?

A: I would say that none of the different governance structures that have been tried in the past has ever really worked well, certainly because of political tensions, but also because of confusing programme management with political management. Furthermore, it seems that no one ever cared about a very simple principle of life, which is selecting the best man for the job. So I think what we should do in the governance, we should have a structure in which we have somebody who is dealing with the project, fully responsible, fully accountable, with a certain political oversight. At the same time we need to limit political oversight to very basic decisions, relating for example to the programme's budget and review, in order to frame it. All other responsibility should be placed in the hands of someone required to deliver. Unfortunately, according to the provisions of the delegation agreement, the European Commission has reserved for itself a lot of rights, controls and so on and so on, putting itself in the shoes of leading the project in every aspect, which is something that was not intended from the beginning. Therefore, my view on this issue is quite clear: for this programme we need someone who is able to do, to deliver, who is accountable for the delivery and who will be mainly supervised on the execution of the major decisions, milestones, cornerstones and basic elements of the programme.

Q: Thank you, this brings me to my next question. We are now entering Galileo's operational deployment phase and this will create some additional budgetary requirements. Do you think that this new phase in the European GNSS development will in fact complicate the relations between the stakeholders, or rather will simplify things?

A: No, that will certainly make things more complicated. My knowledge of recent developments is a little bit limited. However, as far as I know the operational model is not yet fixed. A lot of discussions are still ongoing, creating again a lot of political tensions on who is doing which part of the operation, and who might be looking for what return, if any. In other terms, who will be able to pass to his industry a little bit of the "cake". On top of that, the legal and project structure of the operational phase are also not very clear.

Q: Do you think that these procurement necessities that we will have in the immediate future related to the Galileo deployment phase and the increased budget, especially from the part of the EU, would create some kind of exit from this maze, would that simplify things in any way? The fact we will be handling the final deployment and the EU would have bigger responsibilities in running the programme?

A: That won't simplify anything because at the moment I don't see a trace of any definition how that should look like, who is responsible and who should get what part of the cake.

Q: In your opinion, what would be a suitable working arrangement?

A: For the deployment, or for the operation?

Q: For the deployment.

A: For the deployment, as I told you, for me there is one body in the European environment who is able to do the job, as long as the structure is appropriately empowered to do so, and also take a certain accountability for that and I think it's really that we should not mix again political and project management issues, we should not do it and not even with the technical issues. I don't know if we can really do it with the existing structures, they may be suitable for the initial operation phase, but on the long run they are not very suited for the operation, because they are not made for it, it's just not their mission.

Q: I would move now directly to a question about the commercialisation of the GNSS services. First of all, should the PRS be treated as a commercial, or a strategic asset, which means should its security aspects outweigh commercialisation objectives in any way?

A: I think it is difficult to answer. I believe that PRS is anyway a strategic asset, because it is the same kind of core system as GPS is for the US. There are PRS aspects that could be also commercialised and useful, potentially creating at least some revenues, or at least some kind of limited investment return, but still in the sense of PRS, not in the sense of something broader. PRS should, according to my opinion, really remain what it is, and it is certainly one of the core elements of the system itself, because if you look at GPS, what you get from the GPS on your navigation equipment, or mobile phone etc is in deed the open signal, but what is behind this system is in fact made for the American "PRS".

Q: So, in your opinion, the existence of PRS creates commercial possibilities, but these possibilities should not be the "raison d'être" of this system at all. It would be just a kind of side advantage.

A: Yes I agree on that, I am just saying that you can use PRS on a very small extend of the commercial market, but it should definitely not be the "raison d' être" of the system.

Q: Do you think then that it could be an instrument of foreign policy for the EU, and how would it affect EU's international relations? And I'm talking specifically about the new US national space policy and some advantages on cooperation in the GNSS area it could produce, and again it would be a question of with whom to cooperate in distributing PRS?

A: Yes, to be honest I have never treated that aspect, at least not in relation to the US, so my answer would be quite indicative. The only thing I can tell you on that, is that there is an agreement between the US and the EU, a cooperation agreement in the GNSS and their full interoperability. I don't know how a commercialisation effort of PRS in the future would influence this agreement, given that there is only a very restricted market for PRS, as I already told you. Consequently, I think that there could be some affect on EU-US relations, and I would be contradicting what I said before id I thought otherwise, but I did not really study or discuss this issue in detail before, so I can just give you my personal gut feeling on the subject.

Q: Then we can continue about GMES: for our readers' information, how would you compare the two flagship programmes, Galileo and GMES, in terms of their development and realisation?

A: Again, this is a topic on which I haven't really worked so far. Let's say that the only topic on which I can compare them already is that GMES is also a joint effort in space between ESA and the EU, and that although the two programmes have different structures, they do share similar problems. Both Galileo and GMES are suffering a lot, on the EU side, from a huge amount of political influence, and both are suffering from the fact that when they were originally set up people hoped to have a kind of very interesting business commercial model for the EU, which was nevertheless not thoroughly studied in advance. Therefore, at least until now their commercial aspects have not been overly successful; in fact Galileo's haven't even started yet.

Q: What you are saying, is that in the case of GMES we somehow repeated some of the mistakes, or let's just say some of the complications that we had with GNSS?

A: Absolutely!

Q: So we are not learning from our mistakes . . .

A: I think we are in the middle of a difficult learning process at the moment. I am very much in favour of a lessons learned policy on the EU and also on the ESA side, but I think that the lesson learned so far is that difficulties lay not so much on the technical field, but rather on the political field; and this has led to repeating the same mistakes over and over again.

Q: Thank you for your direct answer. How would you see cooperation with Russia in GNSS, and I am referring to the possible interaction between Galileo and the Russian system Glonass? What kind of cooperation do we currently have with Russia on GNSS and how would you qualify Glonass as a competitor GNSS system that is developing and deploying quite fast at the moment?

A: Again, this is a topic on which I am not so well prepared. I think Europe and Russia do not yet have an agreement, as far as I know. There is a certain attempt to arrive at a mutual agreement on signal resilience and interoperability issues, but I don't know how far they are with that.

Q: Let us move on then to a more specific question: will Galileo be able to compete commercially on a global scale, because according to the timetable we have now in front of us, the Russian constellation is already complete and the Chinese will be also nearing completion by the time the European GNSS will be fully deployed. So with regard to the initial planning that we had some years ago, it now seems that there would be at least three other commercially competitive systems facing Galileo. Do you think that this will affect its commercial position? Should we begin reconsidering its prospects on purely commercial grounds?

A: I am not sure. Galileo was supposed to be a commercial system, offering a number of services available to commercial users, such as the safety of life application. However, I do not think that the EU will be able in the end to field a purely commercial system running. I believe that what we will have would be a system indirectly creating huge benefits for the EU citizen. I would qualify these as social-economic benefits, rather that exclusively commercial. Therefore, I do not also really see the relevance of worrying about the competition. Competition is something that might occur at a certain moment, but I don't think that at any given moment our mobile phones would receive only GPS or Glonass signal either. But now we are talking about a commercial system, while we haven't already talked about the use of the system in general.

Q: So basically, in any case Galileo's prospects on purely commercial grounds are not very good at the moment.

A: No not at the moment and they have not been too good in the past either, and I would like to be perfectly clear on that. Galileo was set up as a commercial services' system, which in my sense would be the kind of service that would be able to create revenues: that is for me the meaning of commercial. There have been some studies in the past with the underline logic of a PPP model, which unfortunately failed. They failed for many reasons and some of them were already discussed, such as the politics behind it and the great number of stakeholders involved, with a lot of different interests at stake moving towards different directions at the same time. Last but not least, it was never really thought through that Galileo could not actually be a real commercial system. There would and should be, and we had made studies in the GSA on that, huge social-economic benefits from its use. I believe it is on these benefits that we should really concentrate on and look into them in detail, instead of all this continued discussion about its commercial prospects.

Q: So you think there should be a change of paradigm?

A: No, I don't think there should be a change of paradigm, but rather a change of attitude and direction in the programme. Let's say that I hope the EU is not talking so much anymore about the system's commercialisation.

Q: So, in this case wouldn't there also be grounds for improving cooperation with the US. The departure from Galileo's purely commercial approach you just described would also imply that the final system could be more open to coordination with the GPS satellites, for example.

A: If you not are talking about the PRS area, which in my opinion is anyhow not really commercial, I think the answer is yes it should be opened; and here I am not only talking as a European, but also as a world citizen. I really do see this kind of technologies as a strong baseline for international cooperation.

Q: How would this probable cooperation affect the GNSS industrial policy in Europe? If there would be some kind of joint development or use?

A: On this issue I don't really see the threat that other people see in it. I think that it is very important for Europe to clarify what is Galileo about. In my opinion, Galileo is about European space, European space research and European space industry; and this also includes space industry of not only a large scale, but also of a medium and small scale. That means that space technologies and European know-how in this field in general, would be in a position to provide concrete benefits to the people living in Europe. However, giving priority to European citizens does not mean that we can't exchange experience, or cooperate with the US. I do not believe that international cooperation in this area would entail any kind of negative consequences for the European industries. On the contrary, I believe that it will create much more synergies and eventually lead to a stronger support for the European space industry, simply because we are not sitting on an island anymore.

Q: On the other hand, Galileo was also conceived from the beginning as an expression of European independence. How much independence do you think we should have, or seek, on an operational as well as industrial level?

A: Again, and I am talking really as a citizen, for me it was never a question of independence, it was never a question of competition, and this view is shared by many of my colleagues working in the European GNSS programme. For me, it is a question of technology development in Europe; a question of not even ownership, but really about know-how, about technology and the ability to have a certain type of industry in Europe as well, and not only in the US or Russia. Striving to acquire and maintain this kind of knowledge and technical know how in Europe does not necessarily mean that the focus should be on being independent. On the contrary, the focus should be on having these industries in Europe, of being able to produce such space based systems for our own benefit, as well as for the benefit of strengthening our cooperation with other countries.

Q: But on the other hand, having this know-how in Europe is in itself a kind of independence.

A: Yes, sure, but there is not only focus on independence, because I think if you have an industrial policy, for example here in Bavaria a lot of investment is made in order to support the local space industry here, to concentrate it in the region and to produce benefits for it. We do have an industrial region close to Munich, which is benefiting quite a lot from this very dedicated support of the Bavarian government. Nevertheless, that does not mean that Bavaria is the only region in Germany, Europe, or the World to have this particular know-how in space technologies. However, there are specific local benefits for the region, by concentrating this technology, this know-how, here, without necessarily focusing on getting independent, or creating a monopoly.

Q: But this is an expression of the industrial complications that existed. Like you said before, there were a number of actors or stakeholders that really tried to get as much as investment return as possible, either directly or indirectly. Isn't this kind of the same thing?

A: Yes, but again it is not really the issue of independence that is in the focus. It is really about creating and supporting this know-how in our countries and in Europe in general, as well as about being able to set up and operate this system. On the other hand, we are not the only ones, developing such technologies, nor are we all alone in our journey, without considering other countries, like the US. So, for me the focus is not on the issue of independence, not at all.

Q: So I guess the question would be which countries should have this know how.

A: That is something to be discussed. I gave the example of Bavaria, but Bavaria understands itself very much as part of the EU. I am talking a lot about Europe, because something that we also believe, and we know this because it is the reason for the existence of ESA, no European country is strong enough to have its own space industry independently from other countries in Europe, that is absolutely not the case, and this is why we need a common European space effort.

Q: But there is a joint interest in independence on an EU level vis-à-vis the rest of the world in certain technologies, or should I say not independence, but at least on acquiring and maintaining a certain number of critical technologies.

A: Right...

Q: Which would mean that this would be a kind of intellectual property issue, rather than of influence on an industrial level?

A: Yes, absolutely. We have engineers in Europe, we have a space industry, and we don't have to go outside Europe to look for equipment, in the case for example of a signal receiver or a certain type of clock. We don't have to go outside Europe, simply because we have this kind of expertise in our common European house. Consequently, we can create a lot of benefits for our in-house research area, meaning the EU research area. Again, this approach does not necessarily have to focus on the issue of independence.

Q: I think the point is of maintaining capabilities, and I am talking about development capabilities, without focusing so much on operational independence. So you believe that we should at least acquire a minimum of technological know-how and industrial capabilities, and preserve them?

A: Yes, absolutely.

Q: If I am not mistaken, Galileo was from the beginning a kind of vehicle for such a policy, is that right?

A: Yes, exactly.

Q: If this was the case, what changed and we got focused so much on the issue of commercialising Galileo's services?

A: The commercialisation issue, and I would be talking again about the programme's focus, shifted the programme's direction to a very different level, because if you talk about technology, if you talk about technology ownership, if you talk about socio-economic benefits, then you also have another justification for expenditures, and you have a very different view on certain problems, as well as on how they should be managed and financed. If you talk about something that is commercial, you talk about different financing structures, you talk about competition and you talk a lot about some of the issues I mentioned previously, which are not really applicable to the current Galileo system set up.

Q: Why is that, in your opinion?

A: Because Galileo, as it is set up right now and also as reality shows, is not something that is really for commercial use. It was supposed to offer commercial services, produce revenues and so on. However, that is not the logic of such system, because it is not mature enough yet. We might talk about it again in 20 or 30 years, but for the moment we are talking about technologies in Europe and nothing else.

Q: Thank you, for the last question I would like to talk about the legal aspects of PRS commercialisation. As far as I understand -correct me if I am wrong- PRS commercialisation could be considered as a kind of compromise between having some kind of revenues on the one hand and providing for a lot of free access services on the other. Would you share this estimate, and how do you think we could possibly commercialise the services that we ourselves are practically offering for free at the same time?

A: As you know very well, you can only sell something that has a value for someone who is able to pay a certain price for it.

Q: Is this clear for PRS?

A: No. This is why I was talking before about a very limited market, because you might use the benefit of having an encrypted signal, which could be a good value for money. However, this kind of product would not, in my opinion, be attractive to the "normal" user, but rather to the public users, such as military services, police services, boarder surveillance etc. These would be services at a very restricted level, and for such users it could be quite interesting to have a signal that is not open to everyone, or that it could not be jammed by everybody, incorporating high security and accuracy standards. Again, we are talking here about a form of use that is, according to my opinion, rather limited and restricted, also in respect to the market sector it would target. On top of that, we would be entering in the topic that we discussed before, regarding Europe's relations with third countries, and especially with the US.

Q: I see. In this respect, what could be the role of the ITU in regulating GNSS use, and would you foresee any legal implications emanating from PRS signal commercialisation, because of its accuracy?

A: ITU is not regulating the use of Galileo. As far as I understand, the ITU is responsible for the registration of the frequencies used by the different signals used by the satellites. The ITU could have a real role, but here I am far beyond my competence, in mediating in the case of problems regarding frequencies, but I don't know if one can talk of regulating Galileo, at least I am not aware of any discussion in this respect.

Q: Ok. As far as the question of liability is concerned, who do you think should have the authority, but also the responsibility, for GNSS in Europe?

A: That will be a different matter. For me it was always quite clear: the liability is addressed first of all to the owner, because it is the owner who is operating potentially dangerous services, including the system itself. This is an underlining principle, especially used in aviation. So there could be a potential liability of the owner first, and secondly you may potentially have a liability of the operator, because he is also running the system. Then, kit might also depend on how you have contractualisationship, for example, between users and the operator and/or someone else like the EU. Then, you might add another layer of complications: if the EU makes certain promises concerning the quality of the signal it is providing, there may also be a liability not only from the fact of the ownership itself, but also from raising expectations from the signal provided.

Q: So in your opinion, all these matters need to be resolved prior the use of Galileo, especially regarding the PRS signal?

A: If you talk about the commercialisation of PRS, then you have certainly to think about it first. I also think that the EU would be well advised to look deeper into that for a moment, which is something I think they really try to avoid. This is because it is a very complicated and burdensome issue, but you have to look into it because you can not deny ownership and so on, but you should at least measure what you are going to promise.

Q: Thank you very much, is there anything else that you would like to add, especially as far as the governance issues that we discussed in the beginning are concerned?

A: As far as governance issues are concerned, I think that we really need to think about lessons learned, about what went wrong in the past and what we should avoid in the future. I think we need to do this prior to taking any further programme implementation steps. We raised some points already today during our discussion. In my opinion, everything begins from establishing clear competence mandates, dealing with clear decision programme procedures and clear accountability for them. I believe it's all about having a clear road for responsibility: who does what and who has which role and accountability. As long as you don't get this one straight, I don't think you will ever have a stable system, because what you will get would be a bits' and pieces thing. This bits' and pieces approach leads to making decisions that are quite short-term minded, simply because they are mostly of technical nature. Furthermore, acting in this mind frame only allows you to tackle problems one by one, as they come. However, we have found too often that a problem coming up was basically not something that just fell from the sky, but something that was actually foreseeable, albeit ignored on a political level. This is not the way to manage such a programme. Of course, I can only talk about what happened until February 2011, when I left the programme.

Q: What do you think should be the solution, what should be the administrative instance that should take care of this?

A: This is a question that is difficult to answer. I do not believe I would be able to provide a definite answer to a question that a lot of highly competent people have been considering for a number of years now. Nevertheless, as a first step I think we should at least straighten out who has the programme's leadership, who has its political guidance, and who is actually responsible for implementing it. I believe it would be best to have a maximum of two entities and not more: the one political and the other technical. As I explained before, the technical manager of the programme should enjoy a certain degree of independence, but he should also shoulder the project's accountability and responsibility. In any case however, I think we should avoid mixing the management of the political, technical and operational levels. As far as international cooperation on GNSS is concerned, something that I would also like to keep in mind is that, when we enter the system's operational phase, we should start thinking getting private industry involved. Of course, Galileo's model is for the time being a contract model. This reality corresponds to the fact that so far Galileo, with or without the PRS commercial uses, is not in fact a commercial system. Consequently, both the programme's contracting and operating models should also bear this in mind.

Q: Should the system's operational exploitation follow the industrial model we have had so far, or should this change too?

A: The industrial model is at the moment highly impacted by the political model. I am not very familiar with the programme's industrial aspects, either as a whole, or as far as specific industry stakeholders are concerned. Therefore, I can not really make any distinctions between different companies that are building satellites, this is absolutely not what I would know or interfere with. However, I do think that if you open a reasonable competition procedure, bearing in mind that as I said Galileo's operations will not be really commercial, then I believe we should be able to get the best players on board, without meddling around with artificial shares.

Q: Thank you very much for your time and for our very interesting discussion today!

A: Thank you!

Fig. 4: *Artist's impression of a Galileo Satellite (source: ESA).*

4. The new 2010 U.S. space policy

Michael Sheehan

4.1. Introduction

The Obama Administration released the new US Space Policy on 28 June 2010. This was slightly unusual because Presidents normally review space policy during their second term, not the first, because of the low political salience of space policy, though Presidents Reagan and George H W Bush did so in their first term, during a period when space policy was highly controversial. The document is divided into three sections, on principles, goals and guidelines, as the Bush document was. These sections are important because they indicate where the priorities of US space policy lie. As Garnett has noted, "in retrospect at least, policy is revealed by a *series* of decisions, and in prospect it is revealed by general statements of purpose."[697] This is why the space policy document is important, not only in terms of the principles and goals outlined, which reflect core values repeated in virtually every presidential space policy since the start of the space age, but also because the guidelines suggest the areas where the government is determined to act. The long-term aspirations outlined in a policy document need to be distinguished from the objectives that the government is actually going to seek to achieve during its term of office. The question therefore is not so much what aspirations are outlined in the policy, but rather what is the administration actually intending to do?

The Obama policy covers the broad sweep of all aspects of US space policy and some observers have described it as the first substantial updating of the 1996 Clinton policy.[698] This is misleading however, since the Bush administration chose to cover the same ground in two policy statements, one civilian and one military oriented, as well as documents dealing with specific policies, such as the GPS satellite system, rather than a single document as the Clinton and Obama administrations did. It is also worth noting that a great deal of the 2010 document repeats almost verbatim, the contents of the much criticised 2006 policy.

While the announcement of a space policy by a new administration encourages the idea that it represents significant new initiatives, as would be expected with economic policy for example, the reality is that the document shows striking similarities to the G W Bush administration space policy, which in turn followed the Clinton policy closely. This has been a feature of US space policy historically, the policies of new administrations build on and modify those of their predecessors, rather than dramatically altering them. US space policy has in fact been marked by a consistency of principles and policy goals since its inception in the late 1950's. These core ideas are freedom of access to space and free passage through it for all nations, an emphasis on the peaceful use of space while reserving the right to use space for purposes of national self-defence, and seeing spacecraft as sovereign national vehicles, but denying the existence of sovereignty in space itself or on heavenly bodies. In order to accomplish these objectives, the US has divided responsibility between three complementary, but distinct programmes, conducted by NASA (civil), the Department of Defence, (military) and the intelligence community.

4.2. Key Features of the Obama Space Policy

The Obama document highlights a number of key themes that distinguish it from the approach taken by the previous administration, for example a new

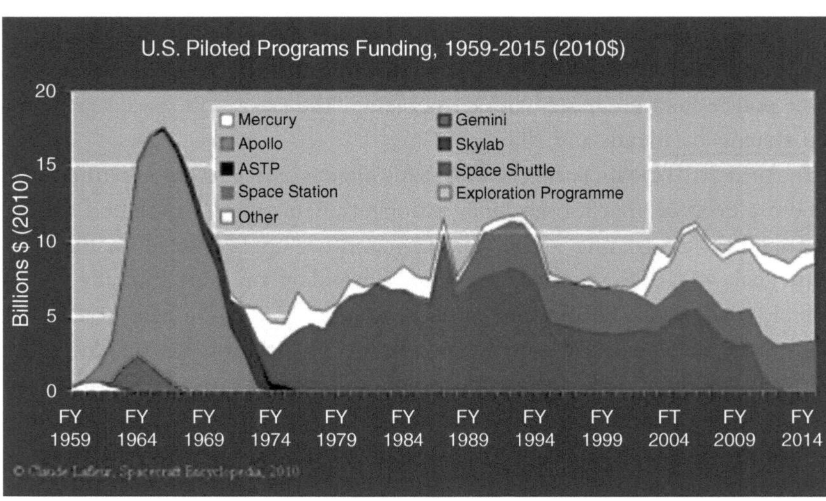

Fig. 5: *The amount spent by the United States on piloted spaceflight from 1959 to 2015.*

emphasis on international cooperation. The changes in the document also reflect the evolution of the environment in which space policy is now being constructed, for example the growing emphasis on the role of the private sector. The new policy also lays stress on a new concept "the sustainability of the space environment", though it does not make clear exactly what activities are thought to help or hinder achieving this objective.

The new policy reflects the evolving space regime in the way that it seeks to partner with commercial organisations for the transport of crew and supplies to the ISS and to begin manned missions to "new destinations" by 2025. The 1996 space policy referred only to manned missions in low-Earth orbit and launcher technology development proposals reflected this, with a focus on reusable shuttle follow-on technologies.

It puts a strong emphasis on international cooperation. There may be a budgetary logic for this, in an era where funding is likely to be constrained, but it reflects also the Obama administrations preference for multilateralism, unlike the previous administration, and a recognition of the increasingly sophisticated space programmes of a number of other countries and organisations. US government departments and agencies are specifically encouraged to identify potential areas of cooperation and the emphasis on cooperation is a theme found throughout the document. Certainly the language is far less militant than the 2006 Bush document and the Obama administration emphasises that "in fact one of our central goals is to promote peaceful cooperation and collaboration in space". However it should be noted that the 2006 policy also spoke of the pursuit of international cooperation to further the exploration and peaceful use of space. Where the emphasis differs is that the earlier policy saw cooperation more in terms of encouraging other states to follow the US lead. It spoke of "diplomatic and public diplomacy effortsto build an understanding of and support for US national space policies and programs" and encouraging "the use of US space capabilities and systems by friends and allies".

In April 2010 Obama cancelled NASA's planned programme to return to the Moon, a decision that former lunar astronauts Armstrong, Lovell and Cernan described as "devastating". The administration has opted instead for the long-term goals of sending crewed missions beyond lunar orbit, initially for an asteroid rendezvous, (2025), and subsequently to a landing on Mars, (2030). In a very different geopolitical context however, there is little evidence that the White House is willing to make the sustained political efforts to win the funding from Congress that would be required to sustain such a venture, unlike President Kennedy's 1961 lunar commitment. The robotic exploration of the Solar System is to continue, with some missions having the additional goal of scouting "locations for future human missions".

This is a major change from the 2004 Bush administration *Vision for Space Exploration* policy document .[699] The 2004 document called for the development of a new manned exploration vehicle, the *Crew Exploration Vehicle*, to provide the first long-range manned spacecraft since the Apollo. The vehicle would be used to return American astronauts to the Moon between 2015 and 2020, and this would be used as a stepping stone for the manned exploration of the Solar System. The crew exploration vehicle was to be tested no later than 2010 and to be operational by 2014. Lunar rovers, based on the Mars *Spirit* design would explore the lunar surface beginning in 2008, and the manned lunar presence would become increasingly long-stay to develop techniques and technologies, and exploit lunar resources to allow subsequent exploration beyond the Moon, beginning with Mars.[700] NASA was directed to review all existing plans and direct them towards the new goals. NASA declared that it would "make use of destinations like the Moon and near-Earth asteroids to test and demonstrate new exploration capabilities".[701] The asteroid mission would be subsequent to manned lunar landings. By shifting the objective away from the Moon, the Obama policy not only puts back a vigorous manned exploration programme by a decade, it also raises major challenges for the Japanese programme, which had adopted a lunar focus in order to allow for effective co-operation with NASA.

4.3. The gap

NASA plans to close down the space shuttle programme in the first half of 2011 after the final mission and with the completion of the International Space Station. However the planned successor manned systems were not due to enter service until 2015 at the earliest. This five year period when the United States would lack a manned spacecraft of its own is commonly called "the gap". It is not the first time the United States has been in this position. There was a six-year gap between the final Apollo mission in 1975 and the first flight of the space shuttle in 1981. Nevertheless, this is a serious concern for the US, which will be forced to rely on other countries for the transportation of its astronauts to the space station during this period. The former NASA Administrator Michael Griffin called the gap "unseemly in the extreme" and it marks a clear retreat from traditional commitment, reflected in the 2005 NASA Authorisation Act to "possess the capability for human access to space on a continuous basis".[702] In April 2010 however President Obama announced the cancellation of the launcher element of the new Orion spacecraft, but the new policy asserts that that the operational life of the International Space Station is to be extended from 2016 to at least 2020, and "likely beyond". Because of the gap, crewed missions to the ISS will have to be accomplished using the Russian Soyuz spacecraft. The policy declares that commercial companies are to be encouraged to take over as providers of crew transfer vehicles to and from the ISS. But the Orion crew capsule would continue to be developed as a "lifeboat" for the ISS.

In many ways the abandonment of the manned lunar missions means that the administration is largely committed to completing the Clinton space policy. In civil space, the Clinton policy was built around the construction of the ISS, operation of the shuttle fleet and robotic exploration of the solar system, particular the landing of robotic explorers on Mars. The long-term programme to identify planets around other stars was also flagged. NASA was tasked to develop 'smaller, more capable spacecraft' to perform these missions. Acquiring spacecraft from the private sector was encouraged, but with significant caveats. However, there was no commitment to developing a manned deep-space exploration programme, unlike the 2006 and 2010 documents. The call for a manned asteroid rendezvous mission does represent a significant difference, but since the required vehicle will not be ready before 2025, the commitment will be hostage to the policies of successor Administrations, and the objective may be modified.

4.4. Next generation spacecraft

There is an emphasis on the potential of nuclear power systems, which was not featured in the 2004 NASA *Vision for Space* plan, although it was quite prominent in the 2006 Space Policy document. In the 2006 document the purpose for the nuclear systems was not made clear, so it was reasonable to assume that it related more to specialised military microsatellites, rather than large propulsion systems for manned missions.[703] However the Bush document also referred to nuclear power systems for non-government spacecraft, where the operator would be responsible for safe operation. The 1996 Clinton space policy also gave the Department of Energy a requirement to "maintain the necessary capability to support space missions which may require the use of space nuclear power systems". These would not be used in Earth orbit without the specific approval of the President, a requirement repeated in the 2006 document. However, while the Clinton and Bush documents refer to both government and commercial spacecraft in this regard, the 2010 policy only describes government systems. One section on nuclear space systems in the Clinton document was classified and did not appear in the published version. The 2010 policy commits NASA to develop "next-generation" launch systems, including new US rocket engine technologies, and declares that the US will "develop and use space nuclear power systems where such systems safely enable or significantly enhance space exploration or operational capabilities".[704] In relation to this the Secretary of Energy and Secretary of Transportation are to cooperate in the licensing of activities involving spacecraft with nuclear power systems.

This should perhaps be understood in conjunction with the Presidents April 2010 space policy statement. That cancelled the launcher element of the *Constellation* spacecraft, but at the same time spoke of a US commitment to manned missions to the asteroids and Mars using "new spacecraft designed for long journeys to allow us to begin the first ever crewed missions beyond the Moon into deep space".[705] The President declared that the US must begin development of a new heavy-lift rocket no later than 2015, but that it must be based on "new propulsion technologies". The new Space Policy document clearly suggests that these propulsion technologies may be nuclear, which would be a dramatic innovation, with significant safety issues. It is noticeable also that this statement drops the Bush/NASA objective of using manned lunar missions to prepare for deep space exploration and moves directly to the asteroid mission that the 2004 Vision for Space Exploration document lined with the lunar programme. In a speech outlining the policy, President Obama declared both that the new deep-space spacecraft would be ready by 2025 and that "we'll start by sending astronauts to an asteroid for the first time in history".[706]

4.5. US space policy governance

The policy document gives additional guidance in terms of three identified sectors, commercial, civil and national security. This section of the document clearly envisages a significant shrinking of NASA's historic role. The US government is to "purchase and use commercial space capabilities and services" from the commercial sector "to the maximum practical extent", and government agencies and departments are to "refrain from conducting United States Government space activities that preclude, discourage, or compete with US commercial space activities unless required by national security or public safety".[707] These statements simply repeat with little change, the wording of the 2006 space policy, and indeed the 1996 policy, which similarly stated that the US government "shall not conduct activities with commercial applications that preclude or deter commercial space activities except for reasons of national security or public safety". The 2010 policy also encourages the transfer of routine operational space activities to the commercial sector, and making US space technology and infrastructure available

Tab. 1: *Costs of US piloted programs by Claude Lafleur, Monday, March 8, 2010 (source: U.S. Office of Management and Budget).*

Year	NASA budget		
	Nominal	Fed Budget (%)	Constant 2007 Dollars
1996	13,881	0.89	16,457
2000	13,428	0.75	14,926
2004	15,152	0.66	15,559
2006	15,125	0.57	16,085
2007	15,861	0.58	15,861
2008	17,318	0.60	17,138
2009	17,782		
2010	18,724		
2011	19,000		
2012 (est.)	19,450		
2013 (est.)	19,960		
2014 (est.)	20,600		
2015 (est.)	20,990		

to the commercial sector. Of significance for the European Union and China is the policy's emphasis on making the US space industry more competitive in international markets, particularly in relation to satellite manufacturing and services and space launch applications, though this too echoes the 2006 document.

The Obama administration has indicated that it plans to revive the National Aeronautics and Space Council, (NASC). This body was created under the National Aeronautics and Space Act in 1958 and was a highly effective body in its early years. However, it was abolished in 1973 and then recreated by President H W Bush in 1989 as the National Space Council. The incoming Clinton administration abolished it once more in 1993. The Obama administration argues that a revived NASC is needed to "report to the President and oversee and coordinate civilian, commercial and national security space activities".[708] It might be argued that a new body will simply create a forum for political and bureaucratic struggle between NASA and the White House. However, given the administrations division of space policy into discrete sectors, all seen as important, a high powered advisory council may facilitate development of coherent space policy.

4.6. The national security dimension

The Obama document continues the emphasis on the importance of space for US security seen in earlier administrations, and reiterates a commitment to defeat any efforts by adversaries to attack US or allied space systems. As with the 2006 policy, the Obama policy also places strong emphasis on "protecting US global access to, and operation in, the radiofrequency spectrum". This may reflect both Iraqi attempts to jam US military satellites during the 2003 Gulf War and the lengthy dispute with the EU over the *Galileo* satellite constellation. Despite a softening of tone, the new policy in fact echoes many of the concerns of the Bush administration.

In particular the 2010 policy echoes the Bush document language in asserting the need to invest in capabilities to "deter, defend and if necessary, defeat efforts to interfere with or attack US or allied space systems". The new policy emphasises that the US will continue to pursue measures to enhance the survivability of its satellites. This embraces not only continuity with earlier administrations in stressing efforts to enhance the protection of key satellites and associated infrastructure, but also, in the reference to "relationships", a new recognition of the role diplomacy can play. However, a novel element is the statement that it will also "develop and exercise capabilities and plans for operating in and through a degraded, disrupted or denied space environment for the purposes of maintaining mission-essential functions". This seems to suggest that the administration is accepting the inevitability of anti-satellite warfare in any future large-scale conflict. In this regard it is significant that the new policy also notes that 'options for mission assurance may include rapid restoration of space assets and drawing on allied, foreign and commercial assets where necessary. In practice the development of a rapid replacement capability would make US satellites less attractive targets in wartime, reducing the pressures for space weaponisation, pressures that adversaries would inevitably feel given the US capacity, reiterated in the new policy, for "the space support, force enhancement, space control and force application missions".[709] This also demonstrates continuity with the 1996 as well as the 2006 policy.

The Clinton administration had also argued that the US would seek to develop space control capabilities to ensure its freedom of action in space, but that it would do so only when such actions were "consistent with treaty obligations". Like the Bush and Obama documents, the Clinton space policy asserted that the US will acquire the capability for "deterring, warning and if necessary, defending against enemy attack" and "countering if necessary, space systems and services used for hostile purposes". The Obama policy places these issues within an allied as well as a national context, declaring that the U.S. will employ measures to "... defend our space systems and contribute to the defense of allied space systems".

The language used in the 2006 G W Bush Space Policy document alarmed many observers, because it appeared more open to the idea of eventual space weaponisation, though the administration denied this. [710] The policy reaffirmed the traditional US position being "committed to the exploration and use of outer space by all nations for peaceful purposes and for the benefit of all humanity". However, the document overwhelmingly emphasised the security dimension, though there was also encouragement for private enterprise, a theme repeated in the Obama policy. Unlike Obama however, space diplomacy was seen more in terms of persuading other states to follow the US lead, rather than embracing genuine multilateralism. In particular, there was a firm opposition to any arms control initiative that might restrict US military space options. However, the strong language again concealed the continuities with the policies of preceding administrations. The 1996 Clinton space policy did however see a role for arms control, and was open to agreements provided that they were "equitable, adequately verifiable and enhance the security of the United States and our allies". This commitment was dropped from the 2006 policy, but has re-emerged with almost identical wording in the 2010 Obama policy.

Shortly after taking office the Obama administration seemed to commit itself to space arms control. The White House web-site declared that the administration would seek to negotiate a ban on weapons that "interfere with military and commercial satellites". This commitment was later quietly dropped.[711] The new space policy like the Clinton policy is agnostic on space arms control, saying it will pursue "confidence building measures" in space and "consider proposals and concepts for arms control measures if they are equitable, effectively verifiable and enhance the national security of the United States and its allies".[712] It is unlikely that the administration will pursue a comprehensive arms control agreement or one specifically on anti-satellite technologies. More probable is an effort to develop "rules of the road" or confidence building measures in relation to space usage.

Like the Bush administration policy, the 1996 document rejected territorial claims in space and asserted that the US considered "the space systems of any nation to be national property with the right of passage through and operations in space without interference" and that "purposeful interference with space systems shall be viewed as an infringement of sovereign rights". [713] This passage attracted criticism in the 2006 document even though it simply repeated the earlier Clinton administration policy, which had been unremarked. It is once again repeated verbatim in the 2010 Obama policy.

Whereas the Clinton document described America's overall goals as being "to enhance knowledge of the Earth, the solar system and the universe through human and robotic exploration", and to "strengthen and maintain the national security of the United States", the Bush policy emphasised the need to "strengthen the nations space leadership and ensure that space capabilities are available in time to further US national security, homeland security, and foreign policy objectives" and to "enable unhindered US operations in and through space to defend our interests there". Even more controversially, the 2006 policy asserted a claim to the right to deny access to space to anyone "hostile to US interests". However, both the Clinton and Bush policies instructed the Department of Defence to pursue capabilities for "force enhancement, space control and force application" missions.[714] This terminology is also repeated verbatim in the Obama policy.

In relation to the national security intelligence gathering role, the document strengthens the previous administrations statement on monitoring foreign space programmes. The 2006 document spoke simply of providing "a robust foreign space intelligence collection and analysis capability". The 2010 document outlines a requirement for the Director of National Intelligence to have a specific focus on the space related activities of other states. The DNI shall, "provide robust, timely and effective collection, processing, analysis and dissemination of information on foreign space and supporting information system activities" and "develop and enhance innovative analytic tools and techniques to use and share information from traditional and non-traditional sources for understanding foreign space-related activities".[715] It should be noted that some sections of the policy are classified and are not in the published version.

Although it is addressed in the civil rather than the national security section, the long sections on environmental earth observation and weather, and land remote sensing can also be seen as falling into the broader definition of security. NASA will remain the lead organisation for satellite development in relation to environmental observation. NASA will also lead, alongside the US Geological Survey, research and monitoring of natural and human-induced changes to the earth's land surface and inland waters. This part of the document is not dissimilar to the 2006 policy, which had the same broad themes. A novel feature of the new policy however is the emphasis placed upon the "long-term sustainability" of the space environment itself. This new emphasis appears in the goals section of the policy, but is not linked to specific policy initiatives or objectives other than in relation to orbital debris and collision prevention measures, so that its implications are not fully spelt out.

The Obama administration seems more relaxed about multilateralism in addressing security related issues than its predecessor, seeking for example to cooperate with other nations in terms of space surveillance for debris monitoring. This had been prefigured in Obama space policy position papers published during the 2008 election campaign, which referred to developing an international approach to minimising space debris', and "enhancing capabilities for space situational awareness". These documents also spoke of opposing space weaponisation and developing with other nations "rules of the road" for space to ensure all nations have a common understanding of acceptable behaviour'.[716] In relation to this, the new Obama space policy introduces a key concept of "stability" in space, which it deems to be in the vital national interests of the United States. While this concept is introduced to US space policy for the first time in this document, it is not defined.

4.7. Conclusions

Reaction to the Announcement of the 2010 Obama space policy has tended to mix disappointment at the abandonment of the manned lunar return objective, with relief that the policy marks a sharp break with the previous Bush policy and is seen as being either a valuable updating of the US position or a return to the balanced space policies of the Clinton era. A detailed analysis of the Obama document in comparison to the earlier policies shows that these judgements are misplaced. There is a degree of novelty, for example in the abandonment of the manned lunar programme, but for the most part the Obama policy largely repeats the Bush document, including in areas that attracted great criticism in 2006, but have apparently passed without public concern in 2010. While the lunar mission has been dropped, the asteroid and Mars goals were already part of NASA's long-term plans. However, the abandonment of the lunar return objective is a setback for NASA's need for a clear focus for its manned programme, and to that extent represents a major shift in US policy. As in 2004, NASA will now have to translate these aspirations into a set of goals and programmes and hope that they will not suffer the fate of their predecessors and see them abandoned by the next administration.

In the longer term, the new policy has serious implications for NASA, which faces dramatic changes. Near-Earth space utilisation activities will largely transfer to the private sector, though NASA will remain key to deep space *exploration.* Like earlier space policy revisions, the Obama administrations will ultimately be assessed not in terms of the new rhetoric or philosophical guidelines outlined for the space programme, but rather in the light of the actual budgetary commitments which would bring the vision to life. The administrations of Presidents

H W Bush and G W Bush were both strong on lofty rhetoric, but weak in terms of committing resources, so that the visions were never realised. The danger in the shift to the asteroid/Mars goal is that a similar failure to materialise will be the final result.

697 J Garnett, "Defence Policy Making", J Baylis, K Booth, J Garnett and P Williams, *Contemporary Strategy*, 2nd edn, (New York, Holmes and Meier, 1987), p. 2.
698 Presidential Decision Directive/NSC-49/NSTC-8, *National Space Policy,* 14 December 1996.
699 *Vision for Space Exploration*, (The White House, 14 January, 2004).
700 *Vision for Space Exploration*, (The White House, 14 January, 2004).
701 *The Vision for Space Exploration*, (NASA, Washington DC, February, 2004), p. 3.
702 US Public Law 109–155, section 501.
703 Joan Johnson-Freese, "The New US Space Policy: A Turn Towards Militancy?", *Issues in Science and Technology*, (Winter, 2007).
704 *National Space Policy of the United States of America*, (Washington, DC, 28 June 2010).
705 Jonathan Amos, "Obama sets Mars goal for America", *BBC News*, 15 April 2010.
706 Jonathan Amos, "Obama sets Mars goal for America", *BBC News*, 15 April 2010.
707 *National Space Policy of the United States of America*, (Washington, DC, 28 June 2010).
708 Office of Science and Technology Policy, *Issues: Technology*, http://www.ostp.gov/cs/issues/technology.
709 *National Space Policy of the United States of America*, (Washington, DC, 28 June 2010).
710 "Bush Sets Defense as a Space Priority", *Washington Post*, 18 October, 2006.
711 Victoria Samson, "Making a Mark in Space: An Analysis of Obama's Options for a New US Space Policy", *Arms Control Today*, October, 2009. http://www.armscontrol.org/act/2009_10Samson.
712 "*Fact Sheet: The National Space Policy*", The White House, 28 June, 2010.
713 Presidential Decision Directive/NSC-49/NSTC-8, *National Space Policy,* 14 December 1996.
714 "US Nixes Arms Control in New Space Policy", *Arms Control Today*, November, 2006. http://www.armscontrol.org/act/2006_11/ACSpace.
715 *National Space Policy of the United States of America*, (Washington, DC, 28 June 2010).
716 Obama for America, *Advancing the Frontiers of Space Exploration*, nd, www.fladems.com/page/Obama_space.pdf (17 August, 2008).

5. The potential for transatlantic cooperation in the International Space Station programme and space exploration

Ian Pryke

5.1. Introduction

Europe has cooperated with the United States and other Partners in the International Space Station (ISS) programme since its inception. In the area of exploration Europe has played a key role, both at the European level through ESA and at the national level through a number of national space agencies in the development of "The Global Exploration Strategy (GES): The Framework for Coordination"[717] released in May 2007. ESA has also worked with NASA on a comparative assessment of lunar architecture concepts.

As the U.S. and Europe both debate the future direction of their civil space programmes, continued cooperation in the ISS and in exploration merits serious consideration. It is not the purpose to this paper to define specific co-operation options, but to examine the environment in which those options will be developed.

5.2. International cooperation in the broader space policy context

International cooperation has played a major role in the implementation of programmes by national and regional space organisations around the world.

The U.S. and Europe have had significant cooperation in space science, meteorology, Earth observation, human spaceflight and more recently planning for long term space exploration. Historically the U.S has played the lead role in much of this cooperation, but as European capabilities have matured the relationship has evolved and many areas has reached the level of partnership among equals.

Although both the U.S. and Europe look to other parts of the world when seeking cooperative partners, the relationships between their respective space agencies remain strong. It therefore seems obvious that transatlantic civil space cooperation will continue in the future.

In order to address potential cooperation relating to the ISS and space exploration it is useful to look more broadly at relevant space policies on each side of the Atlantic. While it is not the purpose of this paper to undertake a detailed analysis of these policies, their respective views on international cooperation and where ISS and exploration fit within the broader context merits consideration.

5.2.1. The Obama Administration's Civil Space Policy[718]

The first indication of an Obama Administration's approach to civil space policy was the release by his campaign, in August 2008 of "Advancing the Frontiers of Space Exploration",[719] which stated that "*As President, Barack Obama will establish a robust and balanced civilian space program*" and "*will reach out to include international partners*".

Specific sections dealt with "*Space Science and Exploration*" and "*Promoting International Cooperation and Keeping Space secure*". The former, when referring to Shuttle retirement and the attendant gap in U.S. human space flight capability, proposed to "*enlist international partners to provide International Space Station (ISS) cargo re-supply and eventually alternate means of sending crews to the ISS*". With respect to enhancing ISS cooperation, "*America must take the next step and use the ISS as a strategic tool in diplomatic relations with non-traditional partners*" and "*will consider options to extend ISS operations beyond 2016*". The second section stated that "*Space exploration must be a global effort. Barack Obama will use space as a strategic tool of U.S. diplomacy to strengthen relations with allies, reduce future conflicts, and engage members of the developing world.*" Concerning "*Collaborating on Exploration*" it recognized the U.S. needs "*to fully involve international partners in future exploration plans to help reduce costs and to continue close ties with (its) ISS partners*". It also referenced GES activities noting an intention to "*continue but intensify this effort*".

In May 2009, following President Obama's January inauguration, the Administration directed NASA to establish a committee to "*conduct an independent review of ongoing U.S. human spaceflight plans and programs, as well as alternatives, to ensure the nation is pursuing the best trajectory for the future of human spaceflight*".[720] The Committee's final report[721] released in October 2009 endorsed extension of ISS operations, and the importance of international cooperation in future space exploration. However, it questioned the viability and direction of NASA's exploration activities, the Constellation programme in particular, and explored various options for redirecting exploration planning.

NASA's FY2011 budget request was released 1 February, 2010. Of particular relevance were the announcements in the request[722,723] to:

- *"Extend operations of the ISS past its previously planned retirement date of 2016."* An accompanying Joint Statement[724] by the NASA Administrator and the Director of the Office of Science and Technology Policy referred to *"NASA (working together with its international partners) to extend the operation of the ISS, likely to 2020 or beyond.*
- Cancel the Constellation Programme replacing it with an exploration programme directed, in the near term, to research and development of enabling technologies such that *"future human and robotic exploration missions (would be) both highly capable and affordable"*.
- Undertake "a *steady stream of precursor robotic exploration missions to scout locations and demonstrate technologies to increase the safety and capability of future human missions and provide scientific dividends"*.

Portions of the NASA budget request elicited a strong negative reaction from both houses of the U.S. Congress, particularly as regards plans to cancel the Constellation program and rely first on the Russians for access to the ISS for an extended period and then on an as yet to be developed U.S. commercial human transportation capability.

In April, 2010 President Obama, addressing[725] the Conference on "The American Space Program for the 21 st Century" at KSC, talked of extending ISS life *"likely more than five years"*. Regarding Constellation, he indicated that rather than outright cancellation NASA was being directed to develop a "rescue vehicle" (ISS lifeboat) *"build(ing) on the good work already done on the Orion crew capsule"*, and to work towards the eventual development of a heavy lift vehicle. He also talked of the next human mission beyond low Earth orbit being to an asteroid, quoting 2025 as a target date. At no time in his speech, when addressing exploration, did he mention international cooperation. This was not seen as any change in policy, but as function of venue and audience.

The June, 2010 release of the Administration's National Space Policy[726] superseded the Bush Administration's 2006 policy. One of its major goals is *"Expanding international cooperation on mutually beneficial space activities to: broaden and extend the benefits of space; further the peaceful use of space; and enhance collection and partnership in sharing of space-derived information"*. It calls upon U.S. Government departments and agencies to:

- *"Promote appropriate cost- and risk-sharing among participating nations in international partnerships"*
- "Augment U.S. capabilities by leveraging existing and planned space capabilities of allies and space partners"
- "Identify potential areas for international cooperation that may include, but are not limited to: space science; space exploration, including human spaceflight activities; space nuclear power to support space science and exploration; space transportation;...."

The policy reiterates the intention to *"continue operation of the ISS in cooperation with... international partners, likely to 2020 and beyond"*, and talks of a bold new approach to space exploration and the beginning of human missions to new destinations beyond the moon by 2025. Reference is also made to sending humans to orbit Mars by the mid-2030s. Overall the new policy is viewed by the majority of space policy analysts as being more favorable towards international cooperation as compared to that of the previous Administration.

In June 2010 NASA's Deputy Administrator, while visiting Europe, stated that *"NASA has a long history of international cooperation. We intend to broaden and deepen those relationships as we seek to implement the president's new U.S. space exploration enterprise"*.

At the time of drafting this article the U.S. Administration and Congress is still in discussions on the new civil space policy and its programmatic implementation. One important point of agreement has been the desire to see the operational life of the ISS extended well beyond 2016, with 2020 often mentioned. NASA and the other "cooperating agencies" are currently working on certifying ISS hardware through 2028.

On the issue of future of space exploration however, there have been marked differences of opinion on the path to be followed, both between the Administration and the Congress and between the House of Representatives and the Senate. The House favored a continuation of certain elements of the Constellation program and the development of a heavy lift vehicle on a faster timescale than that foreseen by the Administration. The Senate took a position somewhat between the Administration's and that of the House. After much debate Congress passed the NASA Authorization Act of 2010 based on the Senate language, which was signed into law[727] by the President on 11 October 2010.

The effect of recent elections, in which control of the House switched from the Democrats to the Republicans, on NASA in general and exploration in particular remains to be seen, particularly in the short term as concerns NASA appropriations for the current fiscal year (FY2011).

5.2.2. European Civil Space Policy

Development of a European Space Policy by the European Union and the European Space Agency began in the late '90s, both organizations recognizing the need for closer cooperation on space matters. In late 2003 the EU and ESA Councils adopted the "Framework Agreement between the European Community and the European Space Agency".[728] It entered into force in May 2004 calling for "*coherent and progressive development of an overall European Space Policy*".

When the Treaty Establishing a Constitution for Europe[729] was signed in October 2004 it contained references to "*A European space programme*", directing the EU to "*establish any appropriate relations with the European Space Agency*". (Note: All space related language in this treaty was included verbatim when the Treaty of Lisbon[730] was adopted at the end of 2007).

That November, the "European Space Council" met for the first time, providing a forum for ministers of the EU and ESA member states to discuss development of an overall European space programme. Two meetings followed in 2005 and at the fourth meeting in May 2007 a Resolution[731] was adopted, as a joint European Commission/ESA document, recognizing that "*Europe is among the leading space-faring actors in the world and remains committed to maintaining its position both via strengthened intra-European and international cooperation*". This document, together with the April 2007[732] Communication from the European Commission, constituted the first comprehensive policy framework covering European space activities.

The Communication recognised that "*Europe needs an effective space policy to enable it to exert global leadership in selected policy areas in accordance with European interests and values*" and that the EU, ESA and their member states needed to develop "*a joint international relations strategy in space*". As regards the ISS and exploration the communication stated that "*Europe needs to achive optimum utilisation of the ISS; prepare for a visible, affordable and robust exploration programme, involving the development and demonstration of innovative technologies and capabilities for the robotic exploration of Mars, to search for evidence of life and understand the planet's habitability*".

The Resolution itself "*emphasises the political and scientific importance of the ISS and of exploration ... reaffirms the continued strong and unified European commitment regarding its ISS contributions*" and notes "*that the continuity of the ISS partnership is an asset for future exploration endeavours*". Emphasis is also placed on "*the importance of proactive ESA participation in the preparation of future international exploration programmes, with the objective of ensuring a significant targeted and coordinated European role in this endeavour*".

The 5th meeting of the Council in September 2008 approved a resolution on "Taking Forward the European Space Policy".[733] Solar system exploration was recognised as one of the *"priority domains for implementing international cooperation"* and as a *"political and global endeavour"*. Consequently *"Europe should undertake its action within a worldwide programme"* necessitating the development of a *"common (European) vision and long-term planning for exploration, ensuring key positions for Europe ... based on its domains of excellence"*.

When the Council met for the 6th time in June 2009 its main focus was on the contribution of space to innovation and competitiveness in the context of the European economic recovery programme. Space exploration was noted in the Council's resolution as having the potential to provide a major impact on innovation and referenced a forthcoming *"High level political conference on space exploration"* as *"a first step towards the elaboration in due time of a fully-fledged political vision on "Europe and Exploration" encompassing a long-term strategy/ roadmap and an international cooperation scheme"*.

This 1 st EU-ESA International Conference on Human Space Exploration took place in October, 2009, Ministers concluding that the EU and ESA, in cooperation, should:

- *Continue work on the development of common (exploration) objectives;*
- *Improve communications with international partners;*
- *Elaborate a roadmap, a set of robotic and human scenarios and a set of priorities for a visible and significant role of EU/ESA in an international exploration initiative;*
- *Explore an implementation mechanism (inc funding schemes);*
- *Report progress at a follow-on conference in 2010;*

The follow-on conference took place in October 2010. Concluding that *"space exploration is a driver for innovation, technological development and scientific knowledge which can bring about tangible benefits for citizens"*,[734] delegations agreed on the need for action in four main areas, inviting the EU, ESA and their member states to take appropriate action, concerning:

- *Technologies as an enabler for space exploration*
- *Space Transportation for exploration*
- *Exploitation of the ISS as a platform for exploration* (Inc. supporting extension to at least 2020, making utilisation accessible to all ESA and EU Member States, and studying with other Partners the potential opening of ISS utilisation to additional non-European participants.)
- *International high-level cooperation*

It is therefore obvious that throughout the development and early years of the implementation of a European Space Policy, support for continued involvement in the ISS programme has remained strong, as has Europe's desire to identify an appropriate role for itself in any large scale international space exploration endeavour.

5.3. Future prospects for transatlantic cooperation

It is not intended to provide a detailed history of activities in the U.S. and Europe relating to the programmes in question. However, an appreciation of how the different programme areas have evolved is necessary in order to understand the context within which future prospects can be viewed.

5.3.1. The International Space Station

The ISS has been an international programme since its inception. President Reagan, in his State of the Union Address[735] in January 1984, directed NASA "*to develop a permanently manned space station and to do it within a decade*" instructing the agency to "*invite other countries to participate so we can strengthen peace, build prosperity, and expand freedom for all who share our goals*". This invitation was accepted by certain European nations, Canada and Japan. At the time the USSR, the only other entity with a human space flight capability and its own space station (MIR), was not considered a potential partner, as it did not meet the "*friends and allies*" criteria. The programme proceeded under the name "Space Station Freedom".

Following the demise of the Soviet Union, Russia was invited, in 1993, to join the partnership. This required a renegotiation of the multilateral Intergovernmental Agreement (IGA) signed by all participating states, and the network of bilateral Memoranda of Understanding (MoU) between NASA and the cooperating agencies of the other Partners. The programme was renamed the International Space Station. In the case of Europe, ten ESA member states were involved, the Agency taking on the role of Europe's cooperating agency. Italy participates in the ISS through ESA. However, its space agency (ASI) also entered into a bilateral MoU with NASA to provide three Multi-Purpose Logistics Modules to the programme.

Assembly of the ISS began in November 1998, the first laboratory (U.S. – Destiny) being added in February 2001. Following the loss of the Shuttle

Fig. 6: *The ISS (source: NASA).*

Columbia in February 2003 there was a hiatus in assembly that lasted until July 2005. U.S. Core Complete was declared with the delivery of the U.S. Node 2 in October 2007, and six person crew capability was achieved in March 2009. Europe's Columbus Laboratory was installed in February 2008.

Current ISS operations rely on the U.S. Shuttle and the Russian Soyuz for crew transportation, and on the U.S. Shuttle, Russian Progress, European Automated Transfer Vehicle (ATV) and Japanese H-II Transfer Vehicle (HTV) for logistics resupply. The ISS partnership has now been in operation for over a decade and the Station has been permanently occupied since November 2000. When the Bush administration announced its Vision for Space Exploration[736] in January 2004 NASA was directed to:

- *"Complete assembly of the ISS, including the U.S. components that support U.S. space exploration goals and those provided by foreign partners, planned for the end of the decade"*
- *"Conduct ISS activities in a manner consistent with U.S. obligations contained in agreements between the Unites States and other partners in the ISS."*

The same document discussed the resources needed to pursue stated exploration goals; NASA funding after FY2009 projected to only keep pace with inflation. The majority of exploration funding was to be found by terminating the Shuttle program once ISS assembly was complete and ceasing ISS operations at the end of FY2016. FY2016 was therefore carried as the official termination date, although the NASA Administrator when the Vision was announced and his successor both stated that they did not expect ISS operations to be terminated at that date, only a few years after assembly would have been completed, thus achieving full research potential.

The Obama Administration's NASA budget submit for FY2011 and the subsequently published National Space Policy talk of continuing ISS operations to 2020 and possibly beyond. This has been welcomed by Europe (and the other ISS partners). ESA's Director General, is on record, including his 17 June, 2009 address to the Augustine Committee,[737] as supporting the idea of such an extension. While not referring to a specific year he has expressed the opinion that "*we use the ISS as a laboratory... as long as the benefits are worth the costs*". ESA is still working with its Member States on approving the necessary funding to meet Europe's share of ISS common operations costs beyond 2015 and on the possibility of obtaining increased funding for station utilization in coming years.

The NASA Authorization Act of 2010 includes language supporting "*full and complete utilization of the ISS through at least 2020*" Budget appropriators in both houses have also voiced their support, although they have yet to achieve consensus on NASA's FY2011 Appropriations bill. Assuming that the ISS will therefore be in operation until at least 2020 the potential for transatlantic cooperation can be addressed, from two different viewpoints.

5.3.1.1. Potential transatlantic cooperation within the context of the existing ISS Partnership

The Partnership has been successfully assembling and operating the ISS for well over a decade. In a joint statement[738], following its February 2010 meeting, the ISS Multilateral Coordination Board (MCB)[739] "*confirmed that there are no identified technical constraints to continuing ISS operations beyond 2015*", indicating a preparedness "*to begin implementation of such a decision when it is taken*".

These statements were reiterated at the ISS Heads of Agency meeting in March, 2010[740] where a "*strong mutual interest*" was expressed "*in continuing operations and utilisation for as long as the benefits of ISS exploitation are demonstrated*". Recognizing that a NASA FY2011 budget, consistent with the Administration's request, "*would allow the United States to support the continuation of ISS operations and utilisation activities to at least 2020*" agency heads "*emphasised their common intent to undertake the necessary procedures within their respective governments to reach consensus... on the continuation of the ISS to the next decade*".

Future on-orbit research opportunities offered by a completed ISS with a crew compliment of six were acknowledged, along with its use as a test bed which would "*allow the partnership to experiment with more integrated international operations and research, paving the way for enhanced collaboration on future international missions*".

At a subsequent September 2010 MCB meeting it was announced[741] that "*the government of Japan has approved continuing space station operations beyond 2016*". Reference was also made to "*the approval of the government of the Russian Federation for continuation to 2020*" and to the fact that "*ESA and CSA are working with their respective governments to reach consensus about the continuation of the station*".

It is therefore clear that any potential transatlantic cooperation must be reviewed within a broader full partnership context. Such prospects could include:

- Further utilisation, by other Partners and their respective research communities of facilities already placed aboard the station by one Partner. This could include opportunities for non-US involvement in the US National Laboratory project, e.g. education.[742]
- Bartering of a Partner's utilisation rights in excess of that Partner's requirements.
- Further contributions to the overall logistics resupply of the Station (inc. industry-to-industry teaming, which is already taking place).
- Development and eventual implementation of a plan for ISS end-of life.

The development of the International Docking System Standard by the ISS Partners, released in October 2010, will have implications for future cooperation (inc. possibly with non-partners) and can also be expected to have relevance to future exploration activities.

Consideration, at the government or implementing agency level, of such prospects could take place on a bilateral (e.g. U.S./European) or multilateral basis.

5.3.1.2. Potential transatlantic cooperation involving nations that are not ISS programme Partners

There have been a number of suggestions regarding bringing new Partners into the station programme, e.g the Chinese with their human spaceflight capability. Such an action would imply:

- Renegotiation of the current Partnership agreements which do not contain mechanisms for automatically adding new Partners.
- Re-computation of the apportionment of ISS resources (Power, pressurized volume, crew time, etc) each Partner receives in return for their hardware and other contributions to the station and its operation.

This should not be undertaken lightly. The negotiations that brought Russia into the programme were of comparable length and complexity to the negotiations that established the original partnership!

A much more likely scenario is involving "non-Partner participants" in station utilization. Such involvement is covered under IGA Article 9 (Utilization) Section 3:

> *"Each Partner may use and select users for its allocations for any purpose consistent with the object of this Agreement and provisions set forth in the MOUs and implementing arrangements, except that:*
> *(a) any proposed use of a user element by a non-Partner or private entity under the jurisdiction of a non-Partner shall require prior notification to and timely consensus among all Partners through their Cooperating Agencies; and*
> *(b) the Partner providing the element shall determine whether a contemplated use of that element is for peaceful purposes,. . .."*

A specific European related issue that will have to be addressed is the proposal, from the Second International Conference on Space Exploration, that *"ISS utilisation is made accessible to all ESA and EU Member States to optimize and broaden European scientific, technological and operational returns"*.

Another potential cooperation opportunity that could be explored involves linkage between the ISS and future exploration activities. Certain European (and other non-US Partner's) exploration contributions might be "book-kept" against Europe's contribution to future ISS operations costs.

In the future, should the U.S. and Europe decide that it is in their mutual interest to involve non-Partner nations in the utilisation of the ISS they would still need to seek agreement from the other Partners. Considering such involvement will involve ascertaining how such candidates propose to utilize the Station, which Partner(s) would contribute the necessary resources, and what contributions the candidates would make to benefit the Partner(s) in question and to the programme as a whole.

5.3.2. Space exploration

Both the US and Europe have been working on the development of exploration plans throughout the previous decade, international cooperation playing an important role in their thinking.

5.3.2.1. Implementing an "exploration vision" in the U.S.

President Bush's "Vision" of 2004 set four ambitious goals:

- "*Implement a sustained and affordable human and robotic program to explore the solar system and beyond*":
- "*Extend human presence across the solar system, starting with a human return to the Moon by the year 2020, in preparation for human exploration of Mars and other destinations*":
- "*Develop the innovative technologies, knowledge and infrastructures both to explore and to support decisions about the destinations for human exploration*": and
- "*Promote international and commercial participation in exploration to further U.S. scientific, security and economic interests.*"

and stated that in its implementation the U.S. should "*pursue opportunities for international participation to support U.S space exploration goals*".

Subsequently, the Report of the President's Commission on Implementation of United States Space Exploration Policy[743] noted that "*how our international partners will participate in the vision will depend on the specifics of the architecture that will be established by the United States and the value potential partners bring to the elements of the mission*" and recommended "*that NASA pursue international partnerships based upon an architecture that would encourage global investment in support of the vision*".

NASA established the Exploration Systems Mission Directorate to work on defining such an architecture including a Crew Exploration Vehicle (CEV), the next generation system (post Shuttle) for U.S. human space transportation, to be brought on-line no later than 2014. The CEV was not intended to provide access to the ISS, but to be used for exploration missions beyond low Earth orbit. Various architecture options were developed, but none were deemed feasible within expected budget envelopes.

In November 2005 NASA issued the final Report of its Exploration Systems Architecture Study,[744] which had been initiated in May 2005. The CEV was now seen as a means of accessing the ISS with the goal of accelerating its development and bringing it into operation in 2011. The study also examined the cost and benefits of developing a Shuttle-derived Heavy Lift Launch Vehicle for use in lunar and Mars exploration. Numerous design reference missions were studied, from ISS crew and cargo transport, to lunar (sortie and outpost) missions and Mars exploration. These efforts resulted in an overall Architecture Roadmap encompassing development of a CEV along with an appropriate launcher, robotic precursor missions to the Moon, development of a heavy lift launch vehicle, an Earth Departure Stage and a Lunar Lander, and plans for a Lunar Outpost.

Around the same time NASA established the Constellation programme centered on an initial capability comprising an:

- Orion Crew Capsule
- ARES 1 crew launch vehicle

A lunar capability was also planned including an:

- ARES V heavy lift cargo launch vehicle
- Earth Departure Stage
- Altair Lunar Lander

NASA made it clear that they intended to develop a lunar transportation architecture alone, but would welcome other nations proposing contributions to an overall lunar return/outpost capability.

The overall plan was endorsed by the U.S. Congress with its passage of the NASA FY2005 Authorization Act[745] and reaffirmed in the FY2008 Authorization Act. However, as the Constellation programme evolved it had to contend with annual budget appropriations which fell well short of initial estimates. Over time, this resulted in the curtailment of study efforts related to future human Mars exploration. Work on ARES V and the Altair lander has also been deferred.

The current situation in the U.S. as regards the future of space exploration remains unclear. The NASA Authorization Act of 2010 now provides clarification as to the direction NASA is expected to follow in its exploration activities, terminating the Constellation programme and its lunar centric orientation, while directing NASA to work on:

- A Multipurpose crew vehicle that *"shall achieve operational capability no later than December 31, 2016"*.
- A Space Launch System *"capable of accessing, at a minimum, the full range of destinations envisioned in the NASA Authorization Act of 2008, and including cis-lunar space, Lagrangian points, the Moon, near-Earth objects, and Mars and its moons, as well as being capable of providing, when used in conjunction with the multipurpose crew vehicle.. a continuing backup capability for supplying and supporting ISS cargo requirements or crew delivery requirements not otherwise met by available commercial or partner supplied vehicles"*.
- Exploration technology development and robotic precursor missions.

However, NASA's appropriations bill for FY2011 has yet to be finalised, and NASA finds itself operating on a "continuing resolution" (currently running until March 4 2011) requiring it to maintain spending on Constellation program activities.

5.3.2.2. Implementing a European exploration strategy

When the Bush Administration announced its "Vision", ESA and the European Community were already coordinating on space matters. Europe, through ESA, had been evolving its Aurora Programme since 2001, its primary objective being *"to create, and then implement, a European long-term plan for the robotic and human exploration of the solar system, with Mars, the Moon and the asteroids as the most likely targets"*.[746] Within Europe particular emphasis was given to Mars exploration. In agreeing to work with the U.S. and other nations on an exploration strategy built around the U.S. "Vision" Europe had to reorient its thinking to give more prominence to the Moon, while working on developing its own long-term strategy for space exploration.

While working with NASA and other non-European space agencies on GES related activities, and working internally to define its overall exploration strategy, Europe has carried out numerous studies relating to different potential components of an overall exploration architecture. These include lunar and Mars robotic missions, both orbiters and landers, a Crew Space Transportation System Study undertaken with Roskosmos on a Soyuz based spacecraft for journeys beyond LEO, and work on variations of the ATV, to provide a return capability and possibly evolve to a crew carrying capability. ESA and European industry are also working on studies of an autonomous lunar lander capability that could eventually be used for cargo and logistics delivery.

5.3.2.3. The US and Europe working together

Europe, along with a number of other nations, initiated discussions on space exploration with the U.S. in the months following the 2004 announcement of the "Vision". These included the development of "The Global Exploration Strategy" between 2005 and May 2007. The strategy identified five general themes in which space exploration was considered to provide benefits to society, discussed different potential location based exploration scenarios, the Moon and Mars in particular, and proposed *"the future establishment of a formal, though non-binding and voluntary, coordination mechanism among interested space agencies to aid in the implementation of the strategy*". This mechanism has since been implemented as the "International Space Exploration Coordination Group" (ISECG). European members are ESA, ASI, CNES, DLR and BNSC/UKSA. A fundamental principle of GES activities, recognised during discussions, was that "*while general agreement exists on broad exploration themes, individual space agencies are required to pursue their unique scientific, technological and social objectives at a scale and pace dictated by national priorities. Thus successful cooperation can only occur with thorough discussion of shared interests and capabilities*".

In this spirit, in January of 2008, NASA and ESA initiated a joint activity to evaluate if their respective lunar architecture concepts could support each other's exploration plans. They issued a joint report[747] in July that year addressing three scenarios concerning potential ESA contributions to a lunar exploration programme:

- *Scenario 1: ESA Provision of Stand-Alone Capabilities*:
  - Automated Lunar Cargo Landing System
  - Communication and Navigation Systems
- *Scenario 2: ESA Development of Crew Transportation Architecture Elements*:
  - Human Crew transportation to LEO via a human rated Ariane 5 and a crew transportation vehicle
  - Orbital Infrastructures
- *Scenario 3: ESA Development of Dedicated Lunar Surface Exploration Elements*
  - Surface Habitation Elements, or
  - Surface Rover

Given NASA's plans to develop an independent lunar transportation system, ESA's potential contribution of lunar surface elements which NASA, due to funding limitations, could not contemplate starting to develop before 2011, suggested a particularly interesting area of study with respect to future cooperation.

In related areas, June 2009 saw the initiation of the Mars Exploration Joint Initiative (robotic), and September 2009 saw the signing of a Memorandum of Understanding on cooperation in civil space transportation (inc. human spaceflight).

5.3.2.4. The future

The President's signature of the NASA FY 2011 Authorization Act means that a new U.S. exploration policy, markedly different from that of the Bush "Vision", is now "the law of the land". The scheduling and timing of the implementation of the new policy, however, will be dependent on funds appropriated, and the NASA Appropriations for FY 2011 bill is still being discussed in the Congress. The outcome of these deliberations, and their implications as regards the future path of U.S. space exploration, will have major ramifications on potential US-European cooperation in this area.

Given the number of space agencies currently engaged in ISECG planning, any US-European cooperation will have to take into account this broader international interest and potential involvement.

A U.S. human exploration programme, focused in the near term on a mission to an asteroid, as opposed to a lunar return with its attendant need for the development of an associated surface infrastructure, raises the question of what role potential international partners could play. As concerns the possibility of partnering in the development of the required transportation capability, this would run counter to the approach adopted for the implementation of the previous Administration's "Vision". Should the U.S. maintain this approach, international cooperation opportunities could be constrained in the near future. The Second International Conference on Space Exploration, however, saw "*international cooperation as a sound and cost effective way to ensure more resilient space architecture to and beyond LEO*" and called for "*further reflection on an international common space exploration transportation policy*".

Regarding the development of future exploration enabling technologies and robotic precursor missions, NASA is currently developing a series of Technology Roadmaps (including ones related to robotic and human exploration) which are being reviewed by the National Research Council. Meanwhile, Europe is working to establish its own long-term road maps and associated programmes for technology, which will form the bases for subsequent discussion with the U.S. and other potential partners. Technology transfer issues are likely to surface when potential cooperation is discussed. The Obama Administration is implementing plans for reforming the U.S. export control system. How this will effect bi-lateral and multi-lateral discussions on future cooperation in exploration has yet to be determined.

Despite the rhetoric on both sides of the Atlantic on the importance of international cooperation in exploration, specifics on exactly how such cooperation could take place still need to be clarified. The establishment of an international high level exploration forum to promote coordinated strategic guidance and international cooperation, as proposed by Europe, could play an important role in this area.

5.4. Conclusion

The potential for transatlantic cooperation in the ISS program and in space exploration has to be assessed within the context of current and foreseen civil space policies and the relevant programmatic content and plans of the parties concerned.

In the case of the ISS this can be based on the virtual certainty that the Partnership will reach agreement on an extension of operations out to at least the year 2020. However, any such future cooperation between the US and Europe, including decisions on bringing new participants into the programme, will have to addressed within the broader framework of the Partnership as a whole.

In the case of exploration; while both parties see international cooperation playing an important role in any large scale endeavour, prospects again need to be reviewed in a broader international context, the scope of which has yet to be clearly defined. There is also the added uncertainty as to the paths the parties will eventually decide to take in implementing their own exploration planning. Clear guidance from the highest political levels on both sides of the Atlantic will be essential.

It has also to the borne in mind that President Obama's current term has passed its mid-way point. Should he not be elected for a second term the potential exists for a further radical reorientation in U.S. space exploration plans.

---

[717] The Global Exploration Strategy: The Framework for Coordination, May 2007.

[718] A more detailed assessment can be found starting on page X): "The New 2010 U.S. Space Policy" by Professor Michael Sheehan, University of Swansea.

[719] OBAMA 08: Advancing the Frontiers of Space Exploration.

[720] Seeking a Human Spaceflight Program Worthy of a Great Nation – Report of the Review of U.S. Human. Space Flight Plans Committee, October 2009, Appendix D, Item 3.

[721] See footnote 4 above.

[722] NASA Fiscal Year 2011 Budget Estimates: Overview.

[723] Office of Management and Budget, FY 2011 NASA Fact Sheet.

[724] Launching a New Era of Space Exploration: Joint Statement from NASA Administrator Bolden and John. P. Holdren, Director, Office of Science and Technology Policy, February 1, 2010.

[725] Remarks by the President on Space Exploration in the 21st Century, KSC, April 15, 2010.

[726] National Space Policy of the United States, June 28, 2010.

[727] NASA Authorization Act of 2010, P.L 111–267.

[728] Framework Agreement between the European Community and the European Space Agency; Official. Journal of the European Union, 6/8/04, L261.264.

[729] Treaty Establishing a Constitution for Europe, Title III, Chapter III Section 9 (Research and. Technological Development and Space), Article III-248, 29 October 2004.
[730] Treaty of Lisbon amending the Treaty on European Union and the Treaty establishing the European Community, signed at Lisbon, 13 December 2007, Title XIX, Article 189; Official Journal of the European Union C 306 Volume 50, 17 December 2007.
[731] 4th Space Council Resolution on Space Policy, 22 May 2007.
[732] European Space Policy; Communication from the Commission to the Council and European Parliament, Brussels, 26.4.2007; COM(2007) 212 Final.
[733] Taking Forward European Space Policy; Council of the European Union, Brussels, 29 September 2008.
[734] Conclusions of the Second International Conference on Space Exploration by the Belgian Presidency of the EU, the European Commission, the Chair of the ESA Council at ministerial level and the European Space Agency on 21 October 2010.
[735] President Ronald Reagan, Address Before a Joint Session of the Congress on the State of the Union. January 25, 1984.
[736] A Renewed Spirit of Discovery: The President's Vision for U.S. Space exploration, President George W. Bush, January 14, 2004.
[737] Augustine Committee: Review of U.S. Space Flight Plans Committee, Statement by Jean-Jaques. Dordain, Director General of the European Space Agency, 17 June 2009.
[738] MCB Joint Statement Representing Common Views on the Future of the ISS, 3 February, 2010.
[739] The Multilateral Coordination Board (MCB) is the highest level management body established under the ISS Agreements to ensure coordination of activities of the partners related to the operation and utilization of the station.
[740] Statement: International Space Agency Heads of Agency Meeting, 11 March, 2010.
[741] NASA RELEASE 10-228 of 22 September, 2010.
[742] ISS National Laboratory Education Concept Development Report, December 2006, Page 4.
[743] A Journey to Inspire, Innovate and Discover, June 2004.
[744] NASA TM-2005-214062.
[745] NASA FY2005 Authorization Act, P.L. 109–155, Title 1, Section 101.
[746] ESA Fact Sheet: "Aurora's Origins", last updated 9 January 2006.
[747] The NASA-ESA Comparative Architecture Assessment, 9 July 2008.

6. Trends in shaping space policies around the world

Deganit Paikowsky and Isaac Ben Israel

6.1. Introduction

This paper aims to note the primary trends in space policy in the period from June 2009 to June 2010, a busy year for space activities worldwide. Many nations entered the process of reevaluating their space programmes and future policies; governments' space spending reached about a third (86.17 billion dollars) of the global space market (261.61 billion dollars), reflecting an aggregate growth rate of 16%, which demonstrates the value attributed to space activity.[748]

The United States continues to be the main actor and the most advanced space faring nation, with an estimated budget of 64.42 billion dollars, which accounts for 25% of the global market. It is followed and challenged by Russia, Europe (mainly as the ESA), China and India. The space club continues to grow. In recent years medium-sized and small states are interested in catching up with the traditional space faring nations by demonstrating similar capabilities in order to enjoy the added strategic, political and social values related to space activities. The period covered by this article reflects the continuation of this trend.

The rapidly growing space market makes it logical to assume that states that need space applications for daily use would turn to the procurement of technology and services. Nevertheless, evidence shows that many nations share the objective of developing a national expertise in space, demonstrating at least some indigenous space hardware production capabilities in order to join the "space club". Nevertheless, the space club is an informal club. As more and more nations expand their space activities and capabilities, there will be a need to organize and coordinate their activities. This process may demand the formalization of a space club.

6.1.1. Space as a symbol and a means of power

States that aspire to position themselves as more powerful and influential within the international community (or to preserve their status), use space programmes to demonstrate their power and convince the world, as well as their own citizens, of their high status among other nations. The ability to develop and launch a satellite into space testifies to a high level of technological capability. This is even more significant as progress is geared towards peace, and military force cannot be used as much as in the past. Because it is becoming increasingly difficult to create deterrence by traditional means, states must find alternative means to increase their deterrent capability – instead of making a show of their military strength, they must rely more and more on demonstrating other capabilities.[749] Presenting technological capabilities of a peaceful nature but with clear dual use potential, like these related to space, increases a states' status, power and deterrent capabilities.

This trend is evident across a wide range of nations, regardless of their size. It is manifested either in the form of upgrading existing programmes, or in the growing number of national space agencies and the increase in the overall international government space budgets. In the years 2006–2008 the total international government space budgets excluding the U.S. accounted for 6% of the global space market (12.46–16.44 billion dollars). In the year 2009, international government space budgets excluding the U.S. accounted for 8% (21.75 billion dollars) of the global market.[750]

The growing number of space agencies worldwide also illustrates the continuous interest nation-states show in having a national capacity to develop, produce, and operate space systems. In the early 1980s there were less than 20 national space agencies. Almost thirty years later, in the year 2009, there were almost 60 operating agencies. Britain and Australia, which are discussed below, traditionally refused to organize their space activities via a formal agency. But in 2009–2010 both of them reached the conclusion that only by forming a national space agency they would be able to fully exploit the potential of their capabilities and expertise in space technologies. The following statement by Lord Mandelson, Britain's Secretary of State for Business, Innovations and Skills, explains the British new approach: "As a focal point for this activity, we're launching our new Space Agency. This will have the muscle it needs to coordinate space policy and boost our international standing. It will bring together all UK civil space activities under one single management and give this sector the support it needs to grow."[751]

Another aspect of this trend is the upgrading of existing programmes. In the past few years, Russia has reemerged as a space faring nation, boosting its space budget and restoring its space activities' potential and capability. Europe increasingly acts as a unified actor in the field of space, forming European space policies on various objectives and concerns. India continues to expand its space programme in two directions. First, by its ambitions to excel in space exploration embarking on human space-flight missions; second, by expanding its space activities into the realm of national security and military activity.

South Korea has been investing great efforts to upgrade its capability. Increasing funds for Korea's space programme is part of these efforts. Korea aims at bringing its space industry to a new level, as Seoul seeks to end its reliance on other states and keep pace with global developments.[752] In South Africa as well, independence and self-reliance play a role in the renewal of the space programme. Kazakhstan is another example of a nation that aspires to upgrade its space program. The Kazakh government expects to exploit the Soviet/Russian space infrastructure left in the country in order to develop a robust space programme and industry. This is part of the government's overall strategy to position Kazakhstan high in the international community, especially by developing satellite communication capabilities.[753] Having a national capacity to explore and use space is of great significance to Iran too, as part of its overall struggle with the "imperialist powers".

Examining the space programmes and policies of many space faring nations allows for several conclusions regarding the major trends that emerge from their activities: (a) there is greater emphasis on international cooperation, (b) more nations expand their space activities to include national security missions, (c) the growing space market motivates nations to improve their industrial scale, capabilities and competitiveness by decreasing costs, improving and expanding the use of space applications and adopting efforts to miniaturise space technologies and products.

6.1.2. Greater international cooperation

The perception of space as a global commons, along with the fact that global economy and security are increasingly reliant on space, motivate the international community to find ways to cooperate and share global utilities from space. Hence, in the last year there is greater evidence of bilateral and multilateral ventures in space, as well as of more actions taken towards a greater coordination in space activities on a global scale. One example of this trend is the initiative to reach a U.N. space policy in order to better respond to the evolving challenges of the international space arena.[754]

In the period covered in this paper there were many cooperation agreements signed between space agencies. To name but a few: the Kazakh and Japanese space agencies signed cooperation accords in January 2010; Ukraine and China signed cooperation accords through 2015; Brazil and Belgium in October 2009; the U.S. and India expanded civil space cooperation between them; Russia and India consider a joint Moon mission; and India also signed an agreement with South Korea in January 2010.

The future of global activity in space would be even more dependent on international cooperation. The growing reliance of daily activities on space assets increases their importance and concomitantly their value. Cooperation is needed for the development of measures to assure their intact operation. The high costs involved in developing advanced space technologies for space research and exploration makes cooperation between nations a rational strategy for achieving worldwide human aspirations to go where no human has gone before. Many nations have become conscious of this fact worldwide. Jean-Jacque Dordain, Director General of the ESA, explained this perception very well in a speech in March 2009: "My dream is that the young generation perceives international cooperation not only as a tool, but as a culture, because the future requires a global view and a culture of international cooperation. The future is global, not individual, and it is certainly the most important message coming from space so far that the future of planet Earth and its inhabitants has to be addressed from a global standpoint."[755]

6.1.3. Expansion of national security space missions

The increasing reliance on space-based systems for day-to-day activities on Earth, along with the growing number of reported satellite jamming events, lead nations to search ways of ensuring their access to space and their freedom of action in it. As a result, the number of space security programmes worldwide is increasing. In 2009 there was a 12% increase in the overall governmental military space budget (32 billion dollars).[756]

Concomitantly, there is a growing debate over the legitimacy and regulations regarding military space activities, especially in relation with "counter space operations" that prevent adversaries from interfering with the use of national space assets, as well as with the mitigation of the space debris problem. An important example is the case of Iran's satellite jamming activity against BBC broadcasts, which was widely criticised and condemned by leading European Union countries and the UN-ITU.[757] However, as it was noted by the Space Security Index of 2010: "despite efforts to construct a robust regulatory framework for space activities, the international community has been unable to reach consensus on an overreaching and legally binding space security treaty that reflects the current challenges facing an ever more complex domain".[758] On a national level, more nations, such as Australia, U.S., Japan, China, Russia, and India, took the issue of space security into account in their policies, programmes and statements. For example, after the successful Chinese test of an ASAT system in January 2007, India also declared its aspiration to explore the option of developing ASAT.[759] This important step was part of India's overall effort to expand its space activity into the realm of national security and military activity.[760]

Furthermore, a growing number of nations now seek to develop space situational awareness (SSA) capabilities. Improved international SSA capabilities can have a positive effect on the sustainability of outer space, because it would increase transparency. If shared, this information could also upgrade confidence in the international community, because it would allow for a better chance predict or prevent harmful interference with space assets. Nevertheless, it could also be used for negating the use of satellites. If so, it could have dangerous implications for the space environment.

Lastly, the high costs of developing and maintaining space-based systems in addition to technological advancements in space activities make the dual-use model more effective and affordable. The benefits of using the dual-use model are also recognised as a useful way of increasing the political cost of the harmful interference with space assets.

6.1.4. Commercialisation and industrial scale

In the last few years, world space activities are becoming increasingly commercialised. In 2009, commercial satellite infrastructure and commercial satellite services activities accounted for two thirds of the global space market.[761] As a result, many nations adjust their space policy towards achieving the development of an innovative infrastructure, as well as a wider more competitive industrial basis.

Furthermore, there is a growing trend for public-private partnerships and dual use ventures, which is expected to intensify in the coming years. The primary growing drivers are security missions, environmental monitoring (including climate change research), and energy supply. National agencies are looking to optimize the return of their investments by developing indigenous capabilities and emphasising the need for local industrial capabilities.

Another important issue is that nations are more concerned with the cost and time schedule management of their projects. Hence they are looking to make space activities more affordable by focusing on developing small scale – light weight satellites and miniaturising related technologies.

Although the number of nations that are active in space is growing as noted above, the following section will focus on the processes that took place in the United States, U.K., Australia, Israel and Singapore, as they represent the most notable changes during the period under examination.

6.2. American space policy and future space activity

The fact that the National space budget of the United States of America constituted 25% of the total space market in 2009 makes the U.S. the focal point for observing trends in space policies. Understandably, many nations look up to the U.S. when considering their own space activities and policies, by carefully observing U.S. objectives, goals and actions. The implications and effect of the new U.S. space programme made public by the Obama Administration in June 2010 is yet to be scrutinised and evaluated, but it will surely shade a light to the road ahead. In spite of this, it is reasonable to say that world space activities in the last year were greatly affected by events and measures that have taken place in the U.S.

The Obama Administration, which came into office in January 2009, inherited the 2006 Space-Policy established by the Bush Administration in the context of very different economic and political circumstances. In May 2009, only five months after entering office, the Obama Administration announced the creation of a "Review of United States Human Space Flight Plans Committee", also known as the Augustine Commission. On the one hand, the readiness of the new Administration to deal with the setting of a new space policy so early in its tenure is a sharp and positive departure from the longer periods required by both the Bush and Clinton Administrations that waited years before making policy decisions regarding space and NASA in particular. On the other hand, the speed of the Obama Administration's resolve in this matter has put the American space community into turmoil of uncertainty and instability at a very early stage.

The Augustine Commission released its final report and recommendations on 22 October 2009, heating up the debate over space activities in the United States.

Fig. 7: *U.S. President Barack Obama speaking at NASA Kennedy Space Center (source: NASA/Bill Ingalls).*

In the wake of the findings of the Augustine Commission, the Obama Administration unveiled a new direction for NASA in its budget request for 2011. Until then, there were three main programmes that occupied NASA: (a) the space shuttle, (b) the International Space Station (ISS), and (c) the Constellation Programme. The last two, are of international nature as they greatly rely on cooperation with international partners.[762] Hence, all discussions within the U.S. administration over the future of these projects attract a lot of international attention.

The ISS was set to be decommissioned in 2015, but the consensus is that the project should be extended until at least 2020, especially to avoid any loss of credibility vis-à-vis its international partners. President Obama supports this direction and has requested that this extension is reflected in the budget.

The future of the Constellation Programme on the other hand, is less certain. The Augustine Commission concluded that in light of delays and increasing costs in its development, the viability of the Constellation Programme should be re-examined. The committee went on to suggest that a more collaborative and commercially oriented effort with revised goals would be more feasible and cost-effective. The 2011 NASA budget practically calls for the cancellation of the Constellation Programme.

On 28 June 2010, the Obama Administration released a new National Space Policy. The document outlines the Administration's perspective and agenda about the significance of U.S. presence in space for the country's economy and national security. Overall, the goal of the new space policy is to strengthen U.S. leadership in space-related science, technology and industrial bases. In order to achieve this goal U.S. space organisations and agencies are required to follow several guidelines, including to "conduct basic and applied research that increases capabilities and decreases costs, where this research is best supported by the government; encourage an innovative and entrepreneurial commercial space sector; and help ensure the availability of space-related industrial capabilities in support of critical government functions".[763]

Many of the principles, goals and objectives of this document are found in earlier space policies and reflect long-standing U.S. views on the use of outer space activities and the objectives pursued through it. Nevertheless, the Obama Administration policy adds several new terms, such as sustainability, responsible behavior, and stability. It also emphasises the importance of expanding international cooperation with U.S. allies, even when it comes to space security concerns.[764]

In the field of space security, the Obama Space Policy emphasises the need to develop and implement plans, procedures, techniques and capabilities necessary to conduct critical national security space-enabled missions, by rapidly restoring space assets and leveraging allied, foreign and commercial space and non-space capabilities in order to help in accomplishing these missions.[765]

The Obama Administration Space Policy signals the U.S. direction for the coming years, but eventually it will be the concrete U.S. decisions, actions, allocation of funds and positions taken in international forums and cooperation ventures that will reveal the true nature of the U.S. Space Policy and activities in the years to come.[766] When it comes to the guidelines regarding greater international cooperation, the challenges ahead involve improving and enabling cooperation with allies, both on a technical and an operational level.

In conclusion, the uncertainty and instability of the American space programme is currently assessed by other spacefaring nations that await for more clarity on future opportunities and possibilities regarding American space activities.

6.3. United Kingdom

Although during the 1950s and 1960s the U.K. was one of the world's leading nations in space activities, especially in the field of rocketry, and it had successfully developed a satellite launch capability, it decided in the late 1960s to discontinue its launch programme. Consequently, the British launch into space in October 1971 was the first and last one. Traditionally, the U.K.'s main preoccupation was to make space technology more cost-effective. As a result, the country relied heavily on the United States, ESA and commercial companies in order to satisfy its space related operational requirements. Over the years, the U.K. space activity has centred on areas of high commercial potential, such as Earth observation, communications, navigation and space science for environmental and economic development purposes.[767] Currently, the U.K. is only the fourth largest contributor to ESA.

In the last few years, there has been a growing debate in the U.K. over the scope and size of its space activities. Gradually, it was recognized that the U.K.'s space programme should be re-examined. In 2009, a dedicated task-force was appointed in order to map out the future opportunities of the country's space sector in the world space market. The primary objective was to establish the U.K. as one of the world's leading space nations, increase the U.K.'s share of the global space market to 10% and transform the U.K. space sector into a prominent part of the country's economy. The task-force's work and recommendations were summarised in a report entitled "A U.K. Space Innovation and Growth Strategy 2010–2030", which was released in December 2009.[768] The release was followed by a decision to establish a dedicated space agency to direct the country's space policy and activities, stating that "with coordinated action we can create a comparative advantage for the U.K. technology and services. We can secure greater wealth creation, more jobs and enhanced intellectual leadership".[769]

6.4. Australia

Throughout the years, Australia has shown little interest in having a national space capability. The Australian government has not made any major efforts to develop indigenous space technologies. As a result, Australia relies heavily on commercial suppliers to meet its operational demands and takes pride in being a "sophisticated user" of space applications. Nevertheless, 2009 saw a significant change in the Australian approach to space activities.

In 2008, there were several attempts to change the Australian space policy. Several papers on this issue were published[770] and the Senate Economics Committee released a detailed report on their space inquiry entitled: *"Lost in Space? – Setting a New Direction for Australia's Space Science and Industry Sector."* The report called for the establishment of an Australian Space Agency and for the immediate implementation of all necessary steps to coordinate Australia's space activities and reduce its reliance on other countries in the area of space technology. The committee also recommended that a Space Industry Advisory Council should be established, comprising of industry, government, academic and defense officials, chaired by the Minister for Innovation, Industry, Science and Research. This council would guide the development of the national space agency.[771]

In a response released in November 2009, the Australian government noted the recommendations of the committee. Furthermore, it committed in the 2009–2010 budget 48.6 million Australian dollars to establish an Australian Space Science Programme over a period of four years, in order to improve the country's capacity to independently develop and utilise space technology.[772] This development marks a significant change in the Australian approach to space activities.

6.5. Israel

In Israel, the space community underwent a long and comprehensive process of reevaluating its space related objectives and policies in the last few years. This process reached a peak in November 2009, when the President of Israel Shimon Peres and the Prime Minister Benjamin Netanyahu appointed a task-force[773] to examine the Israeli space programme and recommend a framework for a new national space program. The main objective of the task-force was to focus on civilian and scientific applications that would allow Israel to develop a greater industrial scale and competitiveness in the growing global space market. The task-force submitted its report and recommendations in June 2010.[774]

The report outlines Israel's strengths, weaknesses, opportunities and challenges for achieving its goals in space. The task-force document argued that Israel has a great potential to lead in space technologies development in specific areas, but it is gradually losing its competitive edge because of insufficient investments. Therefore, governmental action and subsidies are needed. Sufficient funds backed by government support could upgrade Israel's competitive edge, placing it among the top five space faring nations. For this reason, the task-force recommended to invest in space research and activities 300 million new Israeli shekel annually for a period of five years, in addition to defence related expenditures. The research areas suggested for funding included satellite miniaturisation, communication, fundamental and applied research. It was also stated in the report that Israel should promote international cooperation with other established and emerging space faring nations. The report was adopted by both the Israeli President and Prime Minister. The aim of the government is to include the new programme in the 2011 national budget and therefore achieve some progress towards its declared objectives already in 2011.[775]

6.6. Singapore

In the last few years Singapore is looking for ways to increase its status in Asia. Positioning itself as an emerging space faring nation is one of the strategies taken for achieving this goal. For this reason, Singapore is searching for ways to increase its activities and capabilities in the global civil space market. For example, Singapore designed and developed the X-Sat LEO micro-satellite,[776] which is scheduled to be launched at the end of 2010. This is a technology demonstration project undertaken as collaboration between the Nanyang Technological University and different Singaporean organisations. Its main mission is imaging over Singapore and satellite-based advanced data acquisition and messaging over the Indian and Pacific Ocean. Another example of this process is the annual space show hosted by Singapore in the last few years. The show, which is the largest in Asia, aims to bring together leading aerospace industries and agencies under one roof for better discussion and cooperation. For Singapore, here is a technological and an economic potential. The statement by the senior Minister of State for trade and industry and education, S Iswaran, at the opening of Global Space and Technology Convention-Satellite Technology Asia on 28 January 2010 that Singapore can serve as a catalyst for further growth of the space industry in Asia manifests the importance attributed to space.[777]

To conclude, the global space environment is rapidly growing and constantly changing. The examples provided in this chapter reflect the new trends in space policy identified in the last year. All five nations' policies examined above demonstrate the importance attributed to indigenous capabilities, international cooperation, greater commercialisation and industrial production scale, as well as to the sustainability of space-based systems.

---

[748] The Space Report 2010, Space Foundation, p. 39.

[749] Ben Israel, Isaac. Presentation. Space Research and its Applications. Yuval Ne'eman Workshop for Science, technology and Security, Tel-Aviv University, December 2002. In 2005, Prof. Ben Israel was nominated as Chairman of Israel's Space Agency.

[750] Data was obtained from the Space Foundation 2006–2010 annual space reports.

[751] http://webarchive.nationalarchives.gov.uk/+/http://www.bis.gov.uk/news/Speeches/mandelson-uk-space-agency-launch.

[752] NTIS, World News Connection, ROK Daily: Korea to Invest W316 Bil. in Space Research *Chosun Ilbo WWW*-Text, Thursday, January 17, 2008.

[753] For more information: http://centralasianewswire.com/viewstory.aspx?id=2046, Accessed November 29, 2010.

[754] Yepes, C.A., "Towards a UN Space Policy", ESPI Perspectives, 23, Vienna, European Space Policy Institute, (June 2009).

[755] International Cooperation in Space, Remarks of Jean-Jacques Dordain, Director General of the European Space Agency at the 40th Anniversary of the Universities Space Research Association (USRA), 26 March 2009, Available at: http://www.usra.edu/galleries/default-file/09Symp_Dordain.pdf accessed on 28 November, 2010.

[756] Data was obtained from the Space Foundation annual 2010 Space Report and Euroconsult reports.

[757] Theodoulou, Michael, "Tehran Told to End Satellite Jamming", The National, 21 March 2010. http://www.thenational.ae/apps/pbcs.dll/article?AID=/20100322/FOREIGN/703219849/1002/FOREIGN 13 July 2010.

[758] Space Security Index 2010, p. 2.

[759] Milowicki, G., and Johnson-Freese, J., "Strategic Choices: Examining the United States Military Response to the Chinese Anti-Satellite Test," *Astropolitics*, 2008, Vol. 6, No. 1, p. 5 (1–21).

[760] For a detailed information on the Indian approach to national security space activity, doctrine and motivation see: Nair, K.K., *Space-The Frontiers of Modern Defense*, (New-Delhi: Knowledge World in association with the Center for Air-Power Studies), 2006.

[761] The Space Report 2010, p. 30.

[762] It should be noted that so far, no international contracts were signed regarding the Constellation Programme.

[763] National Space Policy of the United States of America, 28 June 2010, p. 5. http://www.whitehouse.gov/sites/default/files/national_space_policy_6-28-10.pdf 12 July 2010.

[764] Kueter, Jeff, "Evaluating the Obama National Space Policy: Continuity and New Priorities", Marshal Institute Policy Outlook, (July 2010), p. 1. http://www.marshall.org/pdf/materials/900.pdf 12 July 2010.

[765] National Space Policy of the United States of America, 28 June 2010, p. 17. http://www.whitehouse.gov/sites/default/files/national_space_policy_6-28-10.pdf 12 July 2010.
[766] Kueter, J., (July 2010).
[767] Burleson, Daphne, Space Programs Outside the United States, Jefferson, North Carolina: McFarland & Company Publishers, (2005): 306.
[768] A U.K. Space Innovation and Growth Strategy 2010–2030: 4.
[769] Ibid: 9.
[770] Among these papers are: Biddington, Brett, "Skin in The Game: Realizing Australia's National Interests in Space to 2025" The Kokoda Foundation, paper no. 7, (2008),. Holt, Lyle., "Integrating Space Efforts into Australia's Joint Operations," Australian Defense Force Journal: 175, (2008): 51–65.
[771] For more on the issue of future Australian space advancement: Merrett, Nicholas. "New Directions for the Heavens From National Security Statements," Australian Defense-Business Review, (January-February 2009): 43–47.
[772] Australian Government Response to the Inquiry By the Senate Standing Committee on Economics Into the Current State of Australia's Space Science and Industry Sector, November 2009.
[773] The task force was headed by Mr. Menachem Greenblum, Director General Ministry of Science and Technology and Prof. Isaac Ben Israel, Chairman of the Israeli Space Agency.
[774] Paikowsky, Deganit, and Levi, Ram, "Space as a National Project – An Israeli Space Programme for a Sustainable Israeli Space Industry, Presidential Task-Force for Space Activity Final Report", Jerusalem: Israel Ministry of Science and Technology (June 2010), Hebrew.
[775] Coren, Ora, "Reaching for the Stars – A new space race, fueled more by profit than by national pride, has begun, and Israel wants in", *Haaretz daily news paper*, (5 August 2010), http://www.haaretz.com/print-edition/business/reaching-for-the-stars-1.306093. 6 August 2010.
[776] For additional information on the X-Sat mission: http://www.dlr.de/iaa.symp/Portaldata/49/Resources/dokumente/archiv4/IAA-B4-0506P.pdf.
[777] http://www.zdnetasia.com/s-pore-space-industry-nascent-but-progressing-62062359.htm, accessed on November 28, 2010.

7. Space applications after Copenhagen

Simonetta Cheli

7.1. The challenges of a changing world

The vulnerability of society to climate extremes and their consequences, such as rising temperatures, floods, wildfires, etc., has become one of the highest priority issues on the political agenda of world leaders. It is one of the most discussed issues in global economic, social, scientific and political fora. Concern about climate change has now become part of public consciousness and dialogue.

This new dimension and awareness of the global environment and its challenges can be attributed to various factors. One of these is the Fourth Assessment Report of the Intergovernmental Panel on Climate Change (IPCC) where consensus at scientific level emerged on the fact that "warming of the climate system is unequivocal" and, moreover, that "most of the observed increase in global average temperature since the mid 20th century is likely due to the observed increase in anthropogenic greenhouse gas concentration".[778]

The rate at which global climate change is happening is the most pressing environmental challenge we face today. The consequences of global warming are far reaching, potentially affecting natural resources such as water and, consequently global food production and influencing the sea level and the rate and severity of natural hazards. In the last century mankind has driven greenhouse gases concentrations beyond the maximum reached during the last one million years. We have become responsible for 70% of the nitrogen and 95% of the phosphorus cycle on Earth and have reduced tropical forest areas by 50%.

To determine whether these recent human induced changes could ultimately destabilise the Earth ecosystem, the consequences of human activities have to be fully understood and quantified. Two issues are at stake: sustainability and biodiversity.

Human life draws heavily on the availability of natural resources: fresh water, food, clean air, building materials. In the interest of future generations we must secure ways to ensure that the functioning of our life support system and the ability of the ecosystems to deliver vital goods and services is maintained.

On the biodiversity side, we must ensure the future diversity of species on Earth and the richness of life for future generations. Human impact on the ecosystem at a regional level results in widely different patterns, such as deforestation, forest fires, fossil fuel burning, land-use management, use of fresh water, etc. Different local and regional phenomena and different types of regional management have to be considered; the sum of them has a major impact on what can be called the "System Earth".

For all these challenges there is a need at the international level to look for solutions and to monitor the scientific aspects. Space represents an important tool to support such actions.

In 2006, the Stern Review on the Economics of Climate Change, which was prepared for the British Government by the economist Nicolas Stern, was released. It analysed the effects on the world economy of climate change. The conclusions of the Stern Review show that the benefits of strong and early action on climate change considerably outweigh the costs.

It had been suggested that 1% of the annual global gross domestic product (GDP) would have to be invested in order to avoid the worst effects of climate change. In June 2008, Stern increased the estimate to 2% of GDP to account for a faster than expected rate of climate change.[779]

7.2. Copenhagen's accomplishments

The United Nations Climate Change Conference in Copenhagen in 2009 (COP 15) was, perhaps because of excessive expectations, believed to be the moment in history in which humanity would have the opportunity to rise to the challenge and take major decisions related to the climate change debate. Predecessor milestone gatherings were, inter alia, the Bali meeting in 2007 and the 1992 Parties of the UN Framework Convention on Climate Change, where it was agreed by members to launch negotiations in order to strengthen action taken against climate change.

The Copenhagen Climate Summit did not however result in the breakthrough that so many had hoped for. The outcome responded only partially to the high expectations before the Conference. Despite the pessimism of the press and the general feeling that COP 15 fell short of its aspired objectives, the Conference did however provide the world with clear signals that governments would like to see action against global climate change move forward. In addition, some important steps were indeed taken.

In a public Hearing on climate change in Brussels on 14 April 2010, Yvo de Boer, Executive Secretary of the United Nations Framework Convention on Climate Change, said that Copenhagen was an important event "as it raised climate change policy where it belongs: [at] the highest political level ... it advanced negotiations significantly on the infrastructure needed for well functioning global climate change cooperation. Lastly, COP 15 produced the Copenhagen Accord, which is a [statement of] political intent to constrain carbon and to respond to climate change".[780]

The Accord sets a two degrees Celsius temperature limit and includes provisions to review this goal by 2015. It also includes short term financing of 30 billion U.S. dollars with a balanced allocation between adaptation and mitigation planning for developing countries up to 2012.

Much of the credit for this partial success must go to rapidly developing countries like Brazil, China, Indonesia and South Africa. They produced plans to tackle these emissions and to have these plans internationally monitored and verified. For the first time in history there is [now] a voluntary partnership between North and South on climate change cooperation, backed by emission targets and intentions. More than one hundred countries have subscribed to the Copenhagen Accord. Thirty six developing countries have communicated information on their mitigation plans, either in economy wide terms or in specific actions.

Developed countries pledged 30 billion U.S. dollars of climate support to developing economies and said that those funds would possibly lead to 100 billion U.S. dollars by 2020. Although targets and actions by 2020 are insufficient, they represent a clear indication that the world wants to move towards an economic growth path that is more sustainable.

Possibly the best outcome of Copenhagen relates to forestry – up to 20% of global greenhouse gas emissions are linked to forestry. Paying developing countries to conserve rather than to cut down their forests could curb these emissions and generate important benefits to local and national economies. The United Nations Environment Programme (UNEP) and the UN Food and Agriculture Organisation (FAO) are carrying out a UN collaborative programme to reduce emissions from deforestation and forest degradation.

Despite some significant steps forward in terms of emissions, Copenhagen has left a gap between where science says emissions need to be in 2020 (to limit the temperature rise to 2 C or less in 2050) and where they stand today.

The next step is the Convention Meeting in Cancun in December 2010, where what remained incomplete in Copenhagen needs to be completed. Industrialised countries need to make firm commitments to take the lead in establishing legal means to achieve the emission reduction targets. In addition, a fully operational architecture needs to be agreed that makes it possible for developing countries to act on climate change in all key areas: adaptation, mitigation, finance, technology, forests and capacity building.

7.3. The contribution of space to climate

The mapping and understanding of climate change is a complex undertaking. It is essential to provide factual, objective evidence to contribute to the scientific models developed. From that point on, links have to be established to monitoring of the Environmental Conventions, to public debate and to concrete political action. Satellites in this respect deliver data related to environmental monitoring and climate change in a reliable way. Satellite data also helps to find ways to adapt to some of the consequences of climate change.

Satellite data has improved our ability to monitor and understand how atmospheric accumulations of greenhouse gases (GHGs) change over time. The ESA Envisat satellite has produced data on greenhouse gases, CO2 and methane, that detect the evolution of global warming. Data of this type are critical for establishing baselines by which to measure emission reduction programmes. Satellite instruments are also useful for checking CO2 emissions in the atmosphere produced by forest fires. Forest fires are relevant not only because they destroy forests, but also because they are a major cause of global air pollution.

In 1998, the El Nino phenomenon helped to create fires across Borneo that emitted 2.5 billion tonnes of CO2 into the atmosphere, equivalent to Europe's entire carbon emission for that year.

Satellites can also monitor glaciers. Glaciers are the most reliable indicator of climate change due to the major influence they have on water availability. They are thus of great interest to scientists. The ongoing intense political and public debate on how rapidly the Himalayan glaciers are retreating highlights the need to monitor glaciers worldwide.

Measurements from ESA's Envisat satellite have contributed to tackling changes in Greenland's glaciers. Tandem missions (of ESA's ERS-2 and Envisat Satellites) in 2008 and 2009 collected data over the Arctic and Antarctic that showed that polar glaciers are moving faster than previously expected.

Finally, in the field of forest carbon tracking Earth Observation data enables the use of archives of data to analyse the last three decades of forest dynamics. This is important in the context of the Group on Earth Observations (GEO), an international initiative started by several space agencies in order to coordinate the definition, development and validation of robust Earth Observation tools and methodologies to provide periodic evaluations of carbon storage in forests for further operational use.

Measurement of the global deforestation rate via satellite monitoring supports the implementation of the UNFCC/REDD (Reducing Emissions from Deforestation and Forest Degradation in Developing Countries) initiative. In addition to missions that are already operational, such measurements will also be

Fig. 8: *Artist's view of Envisat (source: ESA).*

taken in the future by the Sentinel-2 satellite mission currently under development by ESA in the context of the GMES Programme (Global Monitoring for Environment and Security).

In addition, satellite observations have the advantage of going across political borders. The importance of global satellite observations for understanding climate change has also been recognised at an international level.

In the context of the United Nations Framework Convention on Climate Change (UNFCC) and the International Panel of Climate Change (IPCC), the Global Observing System (GCOS) has defined a set of Essential Climate Variables (ECVs) that will be systematically monitored, in order to quantify the state of our climate in an objective and effective way.

The Global Climate Observing System (GCOS) was established in 1992 to ensure that the high quality observations needed to address climate change related issues are obtained and made available to all users. GCOS defined a set of forty-four ECVs.[781] Twenty-five of these variables can be measured by space sensors and can therefore receive major contributions from satellites. Datasets, including historical earth observation data from archives, are essential for measuring key parameters of climate change such as greenhouse gases concentrations, sea ice extent and thickness, sea surface temperature and ocean salinity. The IPCC mentions that "Scientific evidence for warming of the climate system is unequivocal".[782]

Satellites provide key data to the scientific community to improve understanding of the Earth System, detect trends in climate and environment parameters and help to predict the future climate. At the same time, data from satellites support decision makers in the implementation of relevant environmental policies and in the definition of strategies to adapt and mitigate the effects of climate change.

In 2006, the Committee on Earth Observation Satellites (CEOS) that is the primary international forum for coordination of Earth Observation space based systems provided a coordinated response by space agencies to the data needs expressed through GCOS, identifying more than fifty actions to be performed by space agencies all over the world. In this context ESA made the commitment to contribute to the implementation plan of GCOS.

In this context, in 2008 the European Space Agency initiated "The Climate Change Initiative" (CCI), which aims to systematically generate, preserve and provide access to long term datasets to support the requirements of GCOS in the field of Essential Climate Variables.

The ESA Climate Change Initiative includes recalibration, periodic reprocessing, algorithm development, product generation and validation, and quality assessment of climate records in the context of climate models.

This response by ESA is coordinated with other key partners at the European level such as the European Commission, Eumetsat and the European Centre for Medium range Weather Forecasting (ECMWF). ESA is also contributing to an international effort to coordinate the work of space agencies with respect to climate change in the CEOS context that includes all major space agencies worldwide. This recently set up coordination mechanism on climate change will ensure that activities in this field are carried out coherently, guarantee the best use of the data, and ensure that the best rationalisation of available and planned satellite resources is achieved.

ESA is also cooperating closely with its Member States, using its best resources in the academic field, research institutes and industry to contribute to the programme and is also discussing bilateral collaboration with international partners in this field, such as NASA and NOAA, which have recently initiated a new Climate Services Activity.

ESA recently agreed to retrofit its data policy for Envisat and ERS-2, in order to grant open and free access to data. For the GMES Missions, called Sentinels, a similar data policy was recently agreed by ESA Member States. It should be approved by the European Union by the end of 2010.

A free and open data policy at the European level, similar to the one set up in the United States, has strong relevance for climate change activities. Earth observation data can in fact be used more widely to support environmental climate actions in this context.

On 18 December 2009 in Copenhagen, U.S. President Obama said "The problem actually is not going to be verification in the sense that this international consultation and analysis mechanism will actually tell us a lot of what we need to know and the truth is that we can actually monitor a lot of what takes place through satellite imagery and so forth. So I think we're going to have a pretty good sense of what countries are doing".[783]

The intervention by President Obama confirms the fact that politics and science are not two separate domains, especially in the climate change debate. This U.S. approach was recently confirmed by the New U.S. Space Policy issued on 28 June 2010.[784] This Policy reflects the 21st century's globalisation of space activities and calls for the expansion of international cooperation, including in the field of climate change. The US government intends to promote policies internationally that will facilitate full, open and timely access to government environment data, while at the same time accelerating the development of new Earth observation missions.

Space can contribute to climate monitoring for assessing and predicting climate change: this can be achieved through systematic observations and, as climate is a global phenomenon, the information required should be of a global scale. Furthermore, as was previously mentioned, space is also essential for the monitoring of emissions mitigation and adaptation procedures; this can be achieved through focussed observations of aspects of local environments in order to gather evidence of the implementation of adopted policies, or of the failure to do so.

7.4. The role of the European Union

Climate change has been a flagship policy of the European Union for a long time, and certainly an area where Europe has acquired a leadership role since 2001 when the United States withdrew from the Kyoto protocol. Europe has in fact imposed itself on the international scene as a "champion" of the fight against climate change.

With the adoption of the "climate – energy" deal in December 2008, Europe committed itself to reducing its gas emissions by 20% by 2020. It also promised to increase such efforts to 30% in view of the Copenhagen summit, provided that an agreement could be reached at COP 15, which unfortunately was not the case.

The new EU climate change policy will be consistent with the so called Europe 2020 strategy of making the EU more competitive. For example, the 2020 strategy seeks to "establish a vision of structural and technological changes required to move to a low carbon, resource efficient economy by 2050".[785]

The linking of EU climate change policy together with other EU policies is important, first because climate change must be embedded in the overall EU strategic approach and secondly because climate change has become an element of the EU's industrialisation and competitiveness strategy. In this respect, EU climate policy is also linked to its energy strategy.

In an intervention at the European Parliament on 10 February 2010, the President of the European Commission José Manuel Barroso mentioned that it is "necessary to build a new economic model based on [a] knowledge and innovation based economy, carbon emissions and high level of employment".

The most convincing sign of leadership the EU could show would be to implement concrete and determined actions towards becoming the most "climate friendly" region of the world. The Europe 2020 strategy has put together "greener" economic growth at the heart of the Union's vision for a resource efficient future Europe that will create new jobs and boost energy security.

Since COP 15, the European Union has taken a number of actions. In March 2010, it approved a post Copenhagen communication[786] and in April 2010 a staff working document on innovative financial tools,[787] as well as a communication on the assessment of costs and options for raising the 2020 GHG emission target from a 20% to a 30% reduction rate[788] were [prepared] [released].

The European Spring Council (25-26 March 2010) agreed on a stepwise approach for setting a Roadmap in Bonn in order to take negotiations forward and arrive at concrete decisions in Cancun.

The European Union retains the ambition of a legally binding agreement, but it has re-examined its overall strategy in post Copenhagen negotiations and the 2020 package of measures in energy and climate change. It has determined as its final objective the COP-17 in South Africa, scheduled to take place at the end of 2011.

7.5. Towards Mexico

A major goal for the Cancun Meeting at the end of 2010 will be to acknowledge the improvement of mechanisms and the new carbon market as a means to generate financial flows to developing countries.

Commitments of a legal nature are needed from industrialised countries in order to capture the emission reduction targets and to meet needs in the areas of adaptation, mitigation, finance and capacity building.

Some of the key remaining issues to be resolved are the need to build a robust and transparent emissions and performance accounting framework and to secure its long term funding.

Fig. 9: *The Copenhagen Summit (source: The Guardian).*

The medium-term goal is to achieve a balanced set of concrete action-oriented decisions in Cancun and to continue work on reaching a legally binding agreement in the South Africa summit, in 2011.

[778] IPCC 4th Assessment Report (2007). www.ipcc.ch/publicationsanddata/publications and datareports.htm.

[779] Stern Review on the Economics of Climate Change, 700 pages by Nicolas Stern, 30 October 2006.

[780] Public Hearing on Climate Change, Brussels, 14 April 2010 Address by Yvo de Boer, Executive Secretary, United Nations Framework Convention on Climate Change (pages 1–5).

[781] http://www.wmo.int/pages/prog/GCOS/index/php.

[782] IPCC 4th assessment Report (See footnote I).

[783] The White House, Office of the Press Secretary, Remarks by the President during press availability in Copenhagen Bella Centre, Copenhagen, Denmark, 18 December 2009.

[784] The National Space policy of the United States of America. 28, June 2010, 8 pages, The White House.

[785] See flagship initiative on "Resource Efficient Europe" in European Commission, Europe 2010. A strategy for smart sustainable and inclusive growth. Communication from the Commission COM (2010) 2020 final of 03 March 2010. Brussels: European Union.

[786] European Commission, International Climate policy post Copenhagen; Acting now to reintegrate global actions on climate chance – SEC (2010) 261 of 09 March 2010. Brussels: European Union.

[787] European Commission, Innovative Financing at global level, Commission Staff Working Document – SEC (2010) 409 final of 01 April 2010. Brussels: European Union.

[788] European Commission, Analysis of options to move beyond 20% greenhouse gas emission reductions and assessing the risk of carbon leakage concentration for the commission to the European Parliament, the Council, the European Economic and Social Committee and the Committee of the Regions COM (2010) 265 final of 26 May 2010. Brussels: European Union.

8. Satellite Earth observation and disaster management – lessons and needs after the Indian Ocean tsunami and the Haiti earthquake

Ray Harris

8.1. Introduction

The Indian Ocean tsunami in 2004 and the earthquake in Haiti in 2010 were two of the largest and most significant natural disasters to occur on Earth. The Indian Ocean tsunami of 26 December 2004 was created from an earthquake west of the Indonesian island of Sumatra that registered a magnitude on the Richter scale of between 9.1 and 9.3, the second largest earthquake ever recorded on Earth. The tsunami resulted in the loss of over 200,000 lives. The Haiti earthquake of 12 January 2010 was of a lower magnitude, 7.0 on the Richter scale, but also resulted in the loss of over 200,000 lives because the earthquake epicentre was very close to the population of the Haiti capital Port au Prince. The January 2010 earthquake was the worst earthquake to hit Haiti in 200 years.

The use of satellite remote sensing data after both of these disasters was put in place rapidly and extensively, and satellite images of the disaster areas were provided to relief teams and others within two days of the events. The purpose of this paper is to examine the lessons learned from the use of satellite Earth observation data of the two disasters and to identify future needs based on the experiences gained. The orientation of the paper is to make the examinations in the light of space policy rather than space technology or space science.

There was extensive publication of remote sensing science papers after the Indian Ocean tsunami. Special issues of two of the main scientific journals in remote sensing were published in 2007: both the *International Journal of Remote Sensing* and the *IEEE Transactions on Geoscience and Remote Sensing* published a special issue devoted to the tsunami itself in the case of the former and to disaster monitoring, assessment and prediction more generally in the case of the latter. However, although there has been extensive coverage of the Indian Ocean tsunami and the Haiti earthquake in science publications, there have been few papers that examine the policy dimensions of the use of satellite remote sensing data to help with these two major events and with natural disasters more generally, which is perhaps surprising as it is the policy dimensions of the use of space data in natural disasters that will assist most with preparing for future disasters.

One international policy that is relevant is the set of 15 principles that make up the UN Principles on Remote Sensing. This paper examines the context provided by the UN Principles and then goes on to discuss one of the main vehicles for the use of satellite remote sensing data in the two natural disasters, the International Charter on Space and Major Disasters. The paper then examines the wider question of data from space for science and concludes by identifying, from a space policy perspective, lessons learned and future needs to improve the role of satellite remote sensing in responding to major disasters.

The paper concentrates on the use of Earth observation data from space. Other space technologies that are particularly useful when responding to disasters are satellite communications and satellite navigation. Satellite telephones have allowed users on the ground to communicate with their home base via a geostationary satellite. Very Small Aperture Terminals (VSATs) require an antenna on the ground of about 1 m diameter and can provide internet access, while the Broadband Global Area Network (BGAN) also provides internet access normally via a laptop used in line of sight to an Inmarsat satellite. Satellite navigation is now common place by using the US Global Positioning System (GPS), a capability that will be enhanced in the future when Europe launches its Galileo satellite navigation system.

8.2. UN principles on remote sensing

After approximately 15 years of discussion and negotiation, in December 1986 the members of the United Nations reached agreement on the *Principles relating to remote sensing of the Earth from space.*[789] The Principles have a wide scope, but two principles in particular are concerned with space and disaster management, namely Principle X and Principle XI. To take Principle X first:

Remote sensing shall promote the protection of the Earth's natural environment. To this end, States participating in remote sensing activities that have identified information in their possession that can be used to avert any phenomenon harmful to the Earth's natural environment shall disclose such information to States concerned.

Principle X appeared in draft form in the UN Principles discussion as early as 1974 and by 1977 the text was in its more or less finished form, suggesting that reaching agreement on the value of remote sensing data for environmental protection did not prove difficult.

The core concept in Principle X is that of the good neighbour. A State that has acquired remote sensing data that shows potential harm to another State should provide the third State with that information, for example France should provide SPOT data to Haiti if processing of the SPOT data shows information that can be used to avert any natural environment phenomenon harmful to Haiti. The responsibilities in Principle X are among States, not least because the 15 UN Principles are agreements among UN member States. As such it is the responsibility of States to provide the information to affected States, although of course in practice this State responsibility is typically carried out by designated agencies.

Principle XI is similar to Principle X but has a focus of protecting mankind from natural disasters:

Remote sensing shall promote the protection of mankind from natural disasters. To this end, States participating in remote sensing activities that have identified processed data and analysed information in their possession that may be useful to States affected by natural disasters, or likely to be affected by impending natural disasters, shall transmit such data and information to States concerned as promptly as possible.

Principle XI has three refinements or developments compared to Principle X: the principle is more explicit on the data-owning State having either processed data or analysed information in its possession rather than just the general term of information; the principle is explicit on the future with its reference to impending natural disasters, emphasising environmental prediction; and some element of time is provided although it is a rather weak "as promptly as possible".

The 15 UN Principles have embedded within them other concepts that are relevant to space and disaster management, especially the use of remote sensing to benefit all countries specifically including the Less Economically Developed Countries (Principles II and IV), and the promotion and intensification of international cooperation on remote sensing (Principles V, VI, VIII and XIII).

8.3. International Charter

8.3.1. The Charter

At the UNISPACE III Conference held in Vienna in 1999 the International Charter *Space and Major Disasters* was announced by the European Space Agency

(ESA) and the French Space Agency (CNES), followed by the joining of the Canadian Space Agency (CSA) in 2000 shortly before the Charter became fully operational on 1 November 2000. Since then there has been an increase in the number of members, and Table 2 gives the Charter membership position in May 2010.

Tab. 2: *The members of the International Charter Space and Major Disasters, October 2010. Source: http://www.disasterscharter.org/web/charter/home.*

Member	Participant(s)	Satellite resources
CNES France	Centre national d'etudes spatiales Spotimage NSPO (Taiwan)	SPOT Formosat
CNSA China	China National Space Administration	FY, SJ, ZY satellite series
CONAE Argentina	Comision Nacional de Actividades Espaciales	SAC-C
CSA Canada	Canadian Space Agency	Radarsat
DLR	Deutsches Zentrum für Luft und Raumfahrt	TerraSAR-X TanDEM-X
DMCii Disaster Management Constellation	CNTS Algeria NSRD Nigeria Tubitak-BILTEN BNSC/SSTL BNSC/Qinetiq	ALSAT-1 NigeriaSat BILSAT-1 UK-DMC TopSat
ESA Europe	European Space Agency	ERS, Envisat
ISRO India	Indian Space Research Organisation	IRS
JAXA Japan	Japan Aerospace Exploration Agency	ALOS
NOAA USA	National Oceanic and Atmospheric Administration	POES, GOES
USGS USA	United States Geological Survey DigitalGlobe GeoEye	Landsat Quickbird GeoEye-1

There are two primary objectives of the International Charter.[790]

- Supply during periods of crisis, to States or communities whose population, activities or property are exposed to an imminent risk, or are already victims, of natural or technological disasters, data providing a basis for critical information for the anticipation and management of potential crises.
- Participation, by means of this data and of the information and services resulting from the exploitation of space facilities, in the organisation of emergency assistance or reconstruction and subsequent operations.

Under the aim of the Charter only authorised users are allowed to request and then initially receive Earth observation data. These authorised users are typically the civil protection, rescue, civil defence and security bodies of the participating country that has entered into the formal Charter agreement. The Earth observation data used under the International Charter are not openly distributed to any organisation that might happen to be interested, such as a research group in a university for example.

When a disaster occurs an authorised user calls a single point of contact with a data acquisition request.[791] The desk officer who receives the call works with an emergency on-call officer (or technical team) to identify the potential satellite resources available for the location in question, to plan satellite data acquisition and to task the satellite(s). The participating agencies task their satellite(s) and resolve any conflicts with their own, planned acquisitions. The images are acquired by the satellite(s), interpreted by one or more specialist teams and then the images and derived maps are delivered to the authorised user. One key data policy feature of the International Charter is that the authorised users are provided with the Earth observation data by the participating space data suppliers free of charge, no matter what the charging policy for the same data normally is.[792]

The number of activations of the Charter is approximately 40 per year.[793] In 2008 and 2009 there were 40 activations each year, and in 2010 there were 24 activations by the end of June. The largest category of activations has been in response to floods. For example, in 2009 there were 21 activations in response to flooding events, such as floods in Vietnam in July 2009 and floods in Georgia, USA in September 2009.

8.3.2. Indian Ocean tsunami

As noted in the introduction, the Indian Ocean tsunami of 26 December 2004 resulted in the loss of over 200,000 lives, mostly in Indonesia, Sri Lanka and

India. The coastal regions of India, Sri Lanka, Thailand, Indonesia, Maldives, Malaysia and Myanmar were all severely affected, while Bangladesh, the Seychelles, Somalia, Kenya, and Tanzania also suffered damage and loss of life. The Charter was activated by a request from the Indian Space Research Organisation (ISRO) on 26 December 2004, and the data was project managed by ISRO, the National Remote Sensing Agency of India (NRSA), the UN Office for Outer Space Affairs (UNOOSA) and the French space agency (CNES).

Under the Charter there were three categories of map information produced. First, there were regional maps that showed the extent of the potential damage over the whole Indian Ocean coastal area. By 28 December 2004 (i.e. two days after the event) a regional map of the tsunami-affected areas had been produced by NASA and the USGS for UNOSAT. The map shows land lying below 20 m, and therefore susceptible to damage by the tsunami, as derived from the SRTM30/ETOPO2 data set, plus land cover information derived from the Modis instrument on the US Terra satellite. SRTM30 data are land surface altitude data from the Shuttle Radar Topography Mission (SRTM) re-mapped at a spatial resolution of 30 arc-seconds (approximately 1 km).

Second, there were image maps that showed the effects of the tsunami on specific regions. These were commonly shown as before and after images. Images captured by the Indian IRS-P6 AWiFS instrument of Trinkat Island in the Nicobar Islands group show what was a single island on 21 December 2004 had become three separate islands on 26 December 2004 because of flooding during the tsunami.[794] Images in the visible and near infrared parts of the electromagnetic spectrum are affected by cloud, which was commonly the case on and shortly after 26 December 2004. Radar has the ability to penetrate cloud and image the surface, so several of the before and after image maps use radar data from ESA's Envisat ASAR operating at C-band (around 5 cm wavelength) and with a spatial resolution of 30 m. The Envisat ASAR images taken after the tsunami show clearly many coastal areas submerged by the sea. Radar has the extra advantage for flooding in that the radar backscatter responds to surface roughness. Flooding is typically characterised by a change from a rough land surface of vegetation or buildings to a smooth surface of water, which means that flooding is relatively easy to see on radar images.

Third, there were images that showed the detail of the impact of the tsunami on individual buildings, fields and forest areas. Figure 10 shows two Ikonos images[795] of a region of Aceh, Sumatra, Indonesia, each covering a surface area of 2.59×2.59 km with a pixel size of approximately 2 m. The image on the left was acquired on 13 January 2003, i.e. before the tsunami event, and the image on the right was acquired on 29 December 2004, i.e. 5 days after the tsunami. At this spatial

Fig. 10: *Ikonos images of part of Aceh, Sumatra, Indonesia with a pixel size of approximately 2 m (source: CRISP, http://www.crisp.nus.edu.sg/tsunami/tsunami.html, accessed 22 July 2010).*

resolution it is possible to see in detail the flooding of large areas of agriculture, roads and buildings down to the level of individual fields, buildings and parts of roads.

8.3.3. Haiti earthquake

On 12 January 2010 a major earthquake of magnitude 7.0 on the Richter scale struck 16 km south of Port au Prince, Haiti on the Enriquillo fault line, followed by several aftershocks of magnitude over 5.0. The deaths, casualties and damage affected about 5.4 million people; the number was so large mainly because of the poor state of economic and social development of Haiti. On 13 January 2010 the Charter was activated by a group of organisations: the French Civil Protection Agency, UNOOSA on behalf of the UN Peacekeeping Mission in Haiti, Public Safety of Canada and the US Geological Survey (USGS).

In the case of the Indian Ocean tsunami of 2004 the main effects concerned flooding of low lying coastal areas. Earth observation data that showed the spatial extent of flooding were the preferred data. In the case of the Haiti earthquake the main effects were in direct connection with the earthquake and so the most useful Earth observation data were those that could show the physical effects of the earthquake itself. There were two broad categories of Earth observation data used after the Haiti earthquake: (1) optical wavelength data often with a very high spatial resolution of the order of 1 m and (2) radar data.[796] The optical wavelength data included image data from GeoEye-1 (USA, 0.41 m pixels),

QuickBird (USA, 0.6 m pixels), Kompsat-2 (Korea, 1 m pixels), SPOT-5 (France, 2.5 m pixels), ALOS AVNIR (Japan, 10 m pixels) and Huanjing-1 A/B (China, 30 m pixels). Visual and machine-aided image interpretation of these optical data enabled the rapid production of maps of the affected areas such as the following.

- Gathering areas for the population
- Location of public buildings affected by the earthquake
- Damage assessment for major buildings and infrastructures
- Obstacles on bridges and roads

Some of these maps were produced as early as 14 January 2010, that is 48 hours after the earthquake itself and 24 hours after the Earth observation data were acquired. An important characteristic of the image maps was that they were geo-rectified and were accompanied by a scale and a key: this is vital for users in the field who need information in a form that is easy to use and fits with other map data they possess.

The radar data were contributed from Radarsat (Canada), ERS-2 and Envisat (Europe), TerraSAR-X (Germany), Cosmo-SkyMed (Italy) and ALOS PALSAR (Japan). One interesting use of the radar data was the application of the technique of SAR interferometry to create maps of vertical surface deformation and horizontal surface movement that resulted from the earthquake. SAR interferometry (InSAR) uses the phase differences in the radar wave in the range direction from the radar antenna to the target from two different positions of the SAR antenna and was the basis, for example, of the Shuttle Radar Topography Mission. The vertical resolution of InSAR is of the order of half the wavelength of the radar system, which means that for C-band systems such as Envisat and Radarsat that have a wavelength of around 5 cm the maximum vertical resolution of InSAR is of the order of 2.5 cm, although in practice the vertical resolution is typically not as good as this. The use of SAR interferometry allowed the production of maps of surface height and surface height changes with contours of 12 cm. Maps showing horizontal displacement of up to 2 m were also produced.

8.3.4. Access and accuracy

The amount of Earth observation data made available after the Haiti earthquake was clearly very large. But the images produced from these data raise two policy questions, namely concerns over access and accuracy. First, there is the question of access. As noted earlier, the Earth observation data made available under the Charter are only for the use of the requesting organisation and the members of the team carrying out the related work. This restriction has extended more widely and the GEO Haiti Event Supersite Website illustrates such an exclusion.[797] Regarding the ALOS data from Japan there are four restrictions noted on the GEO site:

1. The [ALOS] data sets are to be utilized only for the requested purposes of the GEO task.
2. The data shall not be re-distributed to another party.
3. All copyright of [ALOS] PALSAR data belongs to JAXA and METI;[798] thus, copyright should be indicated as © METI, JAXA.
4. GEO Secretariat to report to JAXA the name (or affiliation) of each user and how the data was used.

The use of the ALOS data is therefore restricted to a relatively small number of individuals or organisations, those who are carrying out work on a recognised GEO task.

The International Charter was initially designed on a best-efforts basis to use the then existing Earth observation data resources for a specific humanitarian role. The Charter can be regarded as an operational system for those who are authorised users, but not an operational system for all users. Operational remote sensing has been a declared goal for many decades, but even the Charter with a defined humanitarian objective is only operational within narrow limits.

Second, there is the question of accuracy. There are many differences between the image maps produced showing damage in the Port au Prince region, even those image maps purporting to show the same type of information such as building damage. InSAR images are useful and interesting in a scientific context but they are still experimental outputs. They are hard for the layman to understand, not usually presented as image maps and different InSAR images of the same area show different surface deformation effects. They are not normally in a form that non-experts can use in the field.

8.3.5. Other disaster information systems

The International Charter is one of several information systems that use Earth observation data to provide information at times of disasters or emergencies. Table 3 gives examples of some of the other major disaster information systems

Tab. 3: *A summary of selected disaster management information systems that provide Earth observation data.*

Name	Characteristics	Web site
Center for Satellite Based Crisis Information	Rapid provision of Earth observation data products for humanitarian relief activities and for civil security	www.zki.dlr.de
Disaster Management Constellation	A proof of concept constellation, capable of multispectral imaging of any part of the world every day because of the large number of satellites in the constellation. Low cost satellites owned by Algeria, China, Nigeria, Spain, Turkey and the UK	www.dmcii.com
RESPOND	Part of GMES, works with the humanitarian community to improve access to maps, satellite imagery and geographic information	www.respond-int.org
SAFER	A pre-operational version of the GMES Emergency Response Service, 2009–2011	www.emergencyresponse.eu
Thomson Reuters AlertNet	Rapid alert of humanitarian organisations to disasters mainly through the mechanism of journalism	www.alertnet.org
UN-SPIDER	UN gateway to space information for disaster management support	www.oosa.unvienna.org/oosa/unspider/index.html

and a note on their characteristics. The list in Table 3 is not exhaustive, yet it shows the variety of systems already in existence to respond to disasters. These range from the Disaster Management Constellation which is a set of similar, low-cost satellites that provide rapid optical image data with a spatial resolution of around 30 m (see also the paper by Sandau in this volume), through to the AlertNet system of Thomson Reuters that has its foundation in the provision of up to date information about disasters through news channels.

8.4. Data policy trends for science and research

As noted earlier, the International Charter provides data free of charge to the authorised users. This immediately raises the question of data policy. How extensive is the list of authorised users? For how long can the authorised users use the data? Can the authorised users give the data for free to their neighbouring organisations that have a legitimate interest in the data for disaster management? Why is it not possible for research scientists to have access to the disaster area data if they are carrying out research that assists those affected by the disaster?

There do appear to be some trends in data policy that may enable some answers to these questions to develop. The US federal government has for some time had a data policy that all federally produced data (including Earth observation data from space) should be made available to users for the cost of fulfilling a user request (COFUR). COFUR is also termed marginal cost by many. The science and technology ministers of the member countries of the Organisation for Economic Co-operation and Development (OECD) agreed in 2007 that for research data that are gathered using public funds for the purposes of producing publicly accessible knowledge then:[799]

. . . access [to the data] on equal terms for the international research community [should be available] at the lowest possible cost, preferably at no more than the marginal cost of dissemination.

The Programme Board for Earth Observation (PB-EO) of the European Space Agency (ESA) has approved a new data policy for ERS-2, Envisat, Earth Explorer and Sentinel missions. The new data policy ensures that any user has the right to access the data; that licences for the use of the data are free of charge; and that online access is provided with a user registration process whereby users accept a set of generic terms and conditions for use of the data.

In the UK the Ordnance Survey has changed its data policy to provide certain digital map data free of charge.[800] After an open consultation in 2009, government policy changed in 2010 to create a suite of digital products called OS OpenData that are free of charge to use and with no restrictions on re-use. The free data in the OS OpenData package are at map scales of around 1:25,000 plus digital point and boundary data.

The Group on Earth Observation (GEO) has, under the leadership of the Committee on Data for Science and Technology (CODATA), adopted a set of high level data sharing principles to guide the sharing of relevant Earth observation data contributed to GEOSS. The three data sharing principles are:[801]

- There will be full and open exchange of data, metadata and products shared within GEOSS, recognizing relevant international instruments and national policies and legislation;
- All shared data, metadata and products will be made available with minimum time delay and at minimum cost;
- All shared data, metadata and products being free of charge will be encouraged for research and education.

The International Council for Science is developing a World Data System (WDS) to draw together in a more coherent way the former data centres and geophysical services that it stimulated originally in the 1950s.[802] The World Data System Scientific Committee is discussing during 2010 the use of the three data sharing principles adopted by GEOSS for implementation for all the data in the World Data System.

The trend is clear. More and more organisations are producing data policies for all or part of their data to provide the data either free of all charges or at the marginal cost of reproduction and delivery, especially for research and education use. For disasters there is a moral dimension or pressure to provide data free of charge and very rapidly to respond to emergencies. This still leaves the ever-present question of who pays for the data. This is a matter of policy. In the USA the policy is to provide Earth observation data funded by the government and then achieve gearing by encouraging others to use the data and add value either in a commercial, or a science sense. In Europe and other parts of the world the policy is to fund Earth observation by government until it can become a sustainable sector, at which point government can exit and the sector can operate by itself. In the case of disasters, there would seem always the need for Earth observation data to be provided without a concern for paying a fee, which in turn suggests government support or charity support.

8.5. Lessons

In 2002 the UN organised a workshop on the use of space technology for disaster management in Addis Ababa. Bessis et al[803] explored the lessons learned after 20 months of operation of the International Charter. The authors identified nine points for improvement. The points on the need for better access to high resolution data, faster turnaround times, better use of space telecommunications, improved user feedback and the need to avoid conflicts with commercial coverage appear to have been resolved. With encouragement from the USGS and from the European Commission the commercial providers have provided more very high resolution data of disasters free of charge to the user community. For example, GeoEye stated on its web site[804] in 2010:

When a crisis on the scale of the Haitian earthquake occurs, we are all moved to help. GeoEye has done just that, by providing its satellite imagery of the devastation in Port-au-Prince for free to relief agencies, governments and the media.

From experience with the Indian Ocean tsunami and the Haiti earthquake (amongst others) it is clear that there is a very rapid response time to produce image maps using Earth observation data. The image maps are typically produced within one day of the analysts receiving the data and the satellite data acquisition is typically within one day of the disaster event. The image maps produced within the International Charter were geo-rectified and in a form that can be readily used by the disaster relief users, although it must be noted that Earth observation imagery does not replace expert assessment in the field.

Where there is still room for further development from the list of lessons identified by Bessis et al is in the fit of sensors with disasters, the selectivity of data with respect to each disaster and importantly capacity building by end users to make better use of Earth observation data. There has been better working between Earth observation experts and disaster management experts since the start of the International Charter. There has been a growth of data from different types of sensor, but this may well confuse more than enlighten as end users have difficulty understanding InSAR products for example. The large number of different data types contributed by satellite owners to disaster management, combined with open web access to much of the data, has had the benefit of providing several independent perspectives on disasters. The question raised here is whether the end users can cope with the multiplicity of independent views when time to respond is at a premium. The list of different disaster management systems in Table 3 also raises the question of complexity and implicitly the question of international policy coordination.

The development of higher capacity satellite communications such as VSAT and BGAN has provided the opportunity to send image maps of disasters to users on the ground within days or hours of a disaster event. Sometimes these events have a single impact, such as the Chile earthquake in 2010, but on other occasions the disaster develops over a period of time, such as the forest fires in Russia or the flooding in Pakistan in 2010, and so the needs also develop over time.

8.6. Future needs

It is likely that the effects of natural disasters such as floods, droughts, landslides, fires and earthquakes will increase in the future because of the increasing urbanisation of the world's population, the exploitation of marginal land resources and the effects of climate change (see also the paper by Cheli in this volume). These factors may explain why already it is flooding that is the largest activation category in the International Charter.

The systems that respond to disasters listed in Table 3 (and others beside) are still best efforts and are not what can be regarded as operational. The International Charter could develop into a more binding instrument with (say) European Union involvement, or it could migrate to become part of GEO, although it would be wise to follow Roy Gibson's advice at the GEO symposium in November 2009 that GEO needs much stronger financial and political support to succeed.[805] Users expect Earth observation to provide operational support and the GEO structure could be a vehicle for this operational support. The SAFER project is still a pre-operational version of the GMES Emergency Response, so still funded as an evaluation rather than an operational system although with the expectation that the operational system will develop once SAFER concludes successfully.

The spatial resolution of civilian Earth observation systems has been edging towards those of defence systems since the 1960s. Now that we have optical systems and radar systems providing data with a spatial resolution of less than 1 m then perhaps a point of convergence for practical purposes has been reached, at least for responding to disasters? This then raises the question of whether better Earth observation systems can be provided by being explicit about such convergence, accepting that the dual use of civil and military assets can benefit both sectors.[806] In practical terms the acquisition of images from defence Earth observation satellites at times of disasters proves impossible or at best very difficult, yet these satellite resources could provide useful information to respond to disasters.

Geophysical data showing surface deformation resulting from earthquake damage will have greater value when it is presented in a way that can be easily used and integrated with other map data sets.[807] After the Shuttle Radar Topography Mission there have been TerraSAR-X and TanDEM-X. These X-band radar systems will allow the production of a digital elevation map of the globe with a vertical resolution of the order of 2 m and a spatial resolution of the order of 12 m, and their data will allow surface change maps to be created at times of earthquakes and landslides. These data will be at their most valuable when they are geo-registered and presented in a form that is compatible with other geographical data sets.

Government ministers are implicated in Earth observation data policy through their approval of (1) the OECD principles and guidelines for access to research data from public funding and (2) the GEO data sharing principles planned to be approved by ministers in November 2010. This may provide more weight for Earth observation data provided in the case of disasters to be made available free of charge to all users.

8.7. Conclusion

Satellite Earth observation has a unique role in disaster management in that the data can show the spatial extent of a disaster at a time when finding out the extent of the disaster on the ground is difficult or impossible because of the disaster itself. Earthquakes, landslides and floods all damage or destroy roads and bridges so that surface transport becomes impossible. Floods resulting from heavy rain can be characterised by periods of thick cloud that prevent aerial survey planes from flying and collecting images. From the experience of dealing with the Indian Ocean tsunami and the Haiti earthquake the value of geo-rectified satellite Earth observation images and derived information products delivered within a few days of the disaster has been clearly shown. As users gain more experience of Earth observation image maps and build their own capacity to use the maps then the role of space data is likely to increase.

---

[789] United Nations General Assembly A/RES/41/65, 3 December 1986. See N Jasentuliyana 1988 United Nations Principles on Remote Sensing, *Space Policy* 4(4), 281–84 and F von der Dunk 2002 United Nations Principles on Remote Sensing and the user, in Harris R, *Earth Observation Data Policy and Europe*, Lisse: A A Balkema, 29–40.
[790] See the International Charter web site, http://www.disasterscharter.org/charter.
[791] A Ito 2010 The Disaster Charter and highlighting issues of Haiti earthquake, Current legal issues for satellite Earth observation, Vienna: European Space Policy Institute, report 25, 22–27.
[792] A Ito and L F Martinez 2005 Issues in the implementation of the International Charter on Space and Major Disasters, *Space Policy* 21(2), 141–150.
[793] For more information see P Bally, F Boubila, M Viel, S Jutz, S Cheli and S Briggs The International Charter for Space and Major Disasters, ESA Bulletin, August 2010.
[794] See the International Charter web site for the images, http://www.disasterscharter.org/image/journal/article.jpg?img_id=39804&t=1280247059589.
[795] © CRISP, National University of Singapore.
[796] For a technical explanation see T A Warner, M D Nellis and G M Foody eds 2009 *The SAGE Handbook of Remote Sensing*, London: SAGE Publications.
[797] See GEO's Haiti Event Supersite Website http://supersites.unavco.org/haiti.php.
[798] JAXA is the Japan Aerospace Exploration Agency and METI is the Japanese Ministry of Economy, Trade and Industry.
[799] OECD Principles and guidelines for access to research data from public funding, Paris: OECD Publications, 2007.
[800] Policy options for geographic information from Ordnance Survey – Consultation, 2010 London: HMSO.
[801] GEO Data Sharing Principles Implementation. http://www.earthobservations.org/geoss_dsp.shtml.
[802] Ad hoc Strategic Committee on Information and Data, Final Report to the ICSU Committee on Scientific Planning and Review, Paris: ICSU, 2008.
[803] J-L Bessis, A Mahmood, J Bequignon, P Soma and L Lauritson 2002 The International Charter 'Space and Major Disasters' after 20 months of operation, UN Regional Workshop on the Use of Space Technology for Disaster Management, Addis Ababa, Ethiopia, 1-5 July 2002.
[804] See http://www.geoeye.com/CorpSite/corporate/GeoEye_Haiti_Relief_Efforts.aspx, accessed 27 July 2010.
[805] GEO News, issue 7, 19 February 2010. http://www.earthobservations.org/pr_gnl_007.shtml,.
[806] G Brachet and B Deloffre 2006 Space for defence: a European vision, *Space Policy* 22(2), 92–99.
[807] M Rao and K R S Murthi 2006 Keeping up with remote sensing and GI advances – policy and legal perspectives, *Space Policy* 22(4), 262–273.

9. Implications of new trends in small satellite development

Rainer Sandau

9.1. Introduction

Space-borne Earth observation has become a valuable tool for sustainable management and global environmental monitoring because of its unique capability to acquire measurements of various environmental data over large areas of the Earth's surface. The important need for Earth observation missions in order to improve the related data is perhaps most clearly seen in the great number of current initiatives for international co-operation in the field of environment monitoring, in which measurements from Earth observation (EO) satellites are an essential element. This is especially true in cases where we need to acquire, analyse and use data to document the condition of the Earth's resources and environment on a long-term (or permanent) basis.

For example, in 2008 the Group on Earth Observations (GEO), which currently numbers some 74 participating countries, the European Commission (EC) and 51 other organisations, developed concrete plans for its Global Earth Observation System of Systems. Also in 2008, the European Union's Space Council continued to advance Europe's Space Policy, reaffirming the need for the rapid implementation of the Global Monitoring for Environment and Security (GMES) programme.

From a space-based remote sensing point of view, the only way to implement flexible space systems in the service of security and prosperity is to pursue activities aimed at developing and operating cost-effective EO missions to monitor the relevant geophysical phenomena on a global scale.

The following sections deal with general facts and trends in the field of small satellite missions for EO purposes. Special attention is given to the potential spatial, spectral, and temporal resolution of small satellite based systems. The capabilities of small satellites in terms of spatial and spectral resolution are close to what larger satellites can provide. Moreover, satellite constellations give small satellites the unique possibility of providing good daily coverage of the globe and/or allowing observation of various dynamic phenomena through their ability to increase their temporal resolution.

New trends in micro-satellite development include building distributed space systems with a variety of features. The capabilities of both distributed space systems and single micro-satellites have implications for the technical and regulatory aspects of space system development and use. The technical implications concern data rates and volumes, launch services, ground station concepts and space debris-avoiding strategies. Regulatory aspects include registration policy, frequency allocation procedures for the inter-satellite and space-ground communication and space debris problems mitigating policy. Increased awareness of these implications can give competent authorities dealing with regulatory issues the possibility of developing a top-down concept for future requirements.

The current situation and new trends in small satellite development are very much in line with the increasing importance of space-based remote sensing, since we have to face a growing world population and decreasing resources while at the same time asking for an ever increasing number of cost-effective space systems to provide good quality and timely information.

9.2. Small satellite missions: facts and trends

9.2.1. General facts

Small and cost-effective missions are powerful tools to react flexibly to information requirements with space-borne solutions. Small satellite missions can be conducted relatively quickly and inexpensively and provide increased opportunity for access to space. The spacecraft bus and instruments can be based either on optimized off-the-shelf systems with little or no requirement for new technology, or on new high-technology systems. Thus a new class of advanced small satellites, including autonomously operated "intelligent" satellites, may be created to open new fields of applications for scientific purposes as well as operational, public and commercial services. Further milestones in the area of small satellite EO missions are the availability and improvement of small launchers, the development of small ground station networks connected to rapid and cost-effective data distribution methods, and cost-effective management and quality assurance procedures.

For about two decades, small satellites using off-the-shelf technologies for missions focused on specific physical phenomena have been seen as an opportunity for countries with a modest research budget and little or no experience in space technology to enter the field of space-borne EO and its applications. Small satellite technology is a major means of bringing within the reach of every country the opportunity to operate EO missions and to utilise the acquired data effectively and at a low cost, as well as to develop and build application-driven missions. It provides the opportunity to independently conduct or participate in EO missions using small, affordable satellites and associated launches, ground stations, data distributions structures, and space system management approaches.

One possible approach to developing small satellite systems is to take full advantage of ongoing technological developments leading to the further miniaturisation of engineering components and the development of micro-technologies for sensors and instruments that would allow the design of dedicated, well-focused EO missions. At the extreme end of miniaturisation, the integration of micro-electromechanical systems (MEMS) with microelectronics for data processing, signal conditioning, power conditioning and communications leads to the concept of applying specific integrated micro-instruments (ASIM). These micro- and nano-technologies have led to the concept of using nano- and pico-satellites, constructed by stacking wafer-scale ASIMs together with solar cells and antennas on the exterior surface, to create space sensor webs.

The situation in the field of small satellite missions for EO has matured in the last ten years. This can be observed, for example, in the topics and quality of contributions to international conferences in Berlin, Logan, the annual International Astronautical Congress, or those organised by space agencies such as ESA or CNES. The Small Satellite Workshops of Commission I of the ISPRS's provide further evidence of the increased attention being given to this subject.

But what exactly is a small satellite? Table 4 gives some examples of how different entities define or categorise small satellites depending on their products or programmes.[808]

To end this confusion, the International Academy of Astronautics (IAA) has proposed a simplified definition.[809] This definition is reflected in Figure 11 in conjunction with additional features that are essential when discussing small satellite characteristics such as cost and response time. The performance issue is covered in subsequent sections.

Tab. 4: *Confusion of small satellite definitions.*

ESA:	Small Mini Micro	350 kg –700 kg 80 kg –350 kg 50 kg –80 kg
EADS Astrium:	miniXL Mini Micro	1000 kg –1300 kg 400 kg –700 kg 100 kg –200 kg
CNES:	Mini Micro	500 kg + P/L (Proteus) 120 kg + P/L (Myriade)

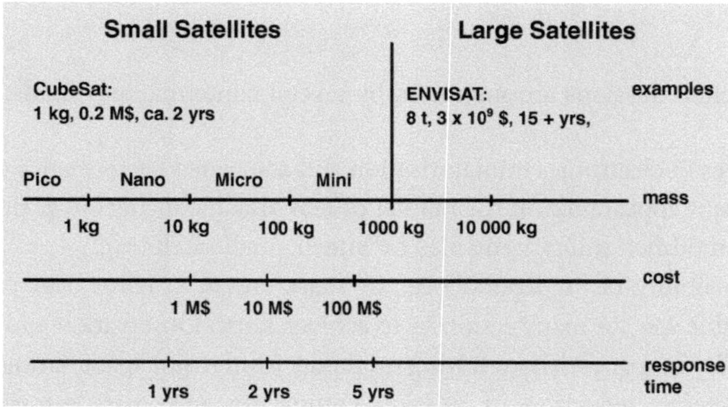

Fig. 11: *Some features of small satellites.*

The cost and response time figures should be considered as ball park figures. They are based on the usage of state-of-the-art technology by professional teams. They may deviate considerably if key technology has to be developed, or if the implementation teams are at the beginning of their learning curve. Figure 11 is complemented by two examples representing the opposite ends of the range. From the customer's point of view, the most important feature of small satellites is their performance. This aspect is covered in the following sections. From Figure 11 we can see that in broad terms, the smaller the satellite is, the less are the cost and the response time. This fact provides a strong incentive to opt for small and especially micro-satellite missions. In this paper we mainly use the term micro-satellite for satellites below a 100 kg mass (including the subsets nano-satellite, pico-satellite etc.) but this definition should also include producer dependent deviations in order to reflect the specific development and mission requirements, as well as cost and response time variables.

The advantages of small satellite missions are:

- more frequent mission opportunities and therefore faster return of science and application data
- larger variety of missions and therefore an equally greater diversification of potential users
- more rapid expansion of the technical and scientific knowledge base
- greater involvement of local and small industry.

Large satellite missions and small satellite missions are considered to be complementary rather than competitive. In some cases large satellite missions can even be a precondition for cost-effective approaches.

9.2.2. General trends

Small satellite missions are supported by several contemporary trends:[810]

- advances in electronics miniaturisation and associated performance capability;
- the recent appearance on the market of new small launchers (e.g. through the use of modified military missiles to launch small satellites);
- the possibility of "independence" in space (small satellites can provide an affordable way for many countries to achieve Earth Observation and a defence capability in space without relying on inputs from major space-faring nations);
- the ongoing reduction of mission complexity and costs associated with management and meeting safety regulations etc.;
- the development of small ground station networks connected to rapid and cost-effective data distribution methods.

In addition, the trend to smaller satellites has been and is still supported by improvements in diverse fields of technology, such as optics, mechanics and materials, electronics, signal processing, communication and navigation, in addition to microelectronics. Mass, volume and power consumption of spacecraft and their instruments have followed the trend to miniaturisation, at the same time allowing for a significant increase in performance. These trends can be observed for passive optical space borne systems as well as for active microwave systems, such as S.A.R. (Synthetic Aperture Radar) systems. They all benefit from overall technology improvements.

9.3. Status and prospects

The focus of this section is mainly on micro-satellites. The capabilities of micro-satellites are shown with respect to their spatial, spectral and temporal features and limitations. The new trend in micro-satellites leads us to the development of distributed space systems and their potential. The knowledge and vision behind this trend is a useful basis for deriving its technical and regulatory implications.

9.3.1. Capabilities of micro-satellites – optical payloads

9.3.1.1. Spatial resolution

The first civil space-borne Earth surface imager was launched in 1972 on the ERTS (Earth Resources Technology Satellite) spacecraft later renamed

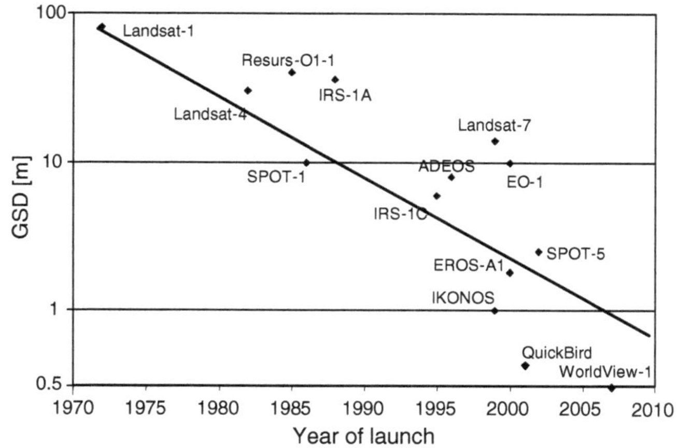

Fig. 12: *Some civil Earth surface Imagers to show the trend of ground resolution (source: GSD).*

Landsat-1. The MMS (Multispectral Scanner System) instrument provided a spatial resolution of 80 m and a swath width of 185 km. With Landsat-4 a more sophisticated multi-spectral imaging sensor was launched in 1982 – the TM (Thematic Mapper) with a spatial resolution of 30 m. There are numerous sensors of different types (mechanical scanners, push-broom scanners, matrix systems) from many countries, including Brazil, China, Argentina, France, India, Thailand, South Africa, Korea, UK, and Germany.[811] Due to the immense improvements in several fields of technology, such as optics, mechanics and materials, electronics, pattern recognition, signal processing, computer technology, communications and navigation, space borne imaging systems have now reached ground sample distances (GSD) of less than one meter. Figure 12 shows the trend of improving resolution (or decreasing GSD) of civil space-based mapping systems that have taken place since Landsat-1 in 1972.[812] The number of space-borne mapping systems indicates the need for high resolution maps to use the best available technologies.[813]

An example of the aforementioned trend is the PIC-2 camera on the small satellite EROS-B from Israel, which provides a GSD of 0.70 m.[814] On 25 April 2006, EROS-B, with a mass of 350 kg, was launched into a 500 km sun synchronous orbit (SSO) by a Russian START-1 launcher. By comparison, the 130 kg micro-satellite TopSat, developed by SSTL/UK and launched into a SSO in November 2005 onboard a KOSMOS-3M from the Pletsesk Cosmodrome, provides a GSD of 2.5 m.[815]

9.3.1.2. Spectral resolution and range

In addition to spatial resolution, spectral resolution is also increasing. An example can be seen in the hyperspectral imager CHRIS on the ESA-funded PROBA satellite.[816] CHRIS, a 14 kg/9 W hyperspectral imager, has a GSD of 18 m and provides up to 19 out of a total of 62 spectral bands in the VIS/NIR spectral range (400 – 1000 nm). PROBA, with a mass below 100 kg (which qualifies it as a micro satellite) was launched into a 600 km sun synchronous orbit (SSO) on 22 October 2001 together with the DLR/Germany micro satellite BIRD for forest fire detection and fire parameter assessment[817] and the main payload TES (India)[818] with the PSLV-C3 launcher from India. The 94 kg micro-satellite BIRD (Bi-spectral InfraRed Detection) is an example of extending the wavelength range of micro-satellite instrumentation to the thermal infrared. BIRD is equipped with two IR cameras in the wavelength ranges of about 4 μm and 9 μm and is used to demonstrate a possible approach to detection and quantitative characterization of high-temperature events like vegetation fires on the Earth's surface. A detailed description of the BIRD system is given in Brieß et al., 2003. Here we only show an example of BIRD performance. Figure 13 gives a comparison of results derived from MODIS (currently the best satellite system to detect fires) and BIRD with

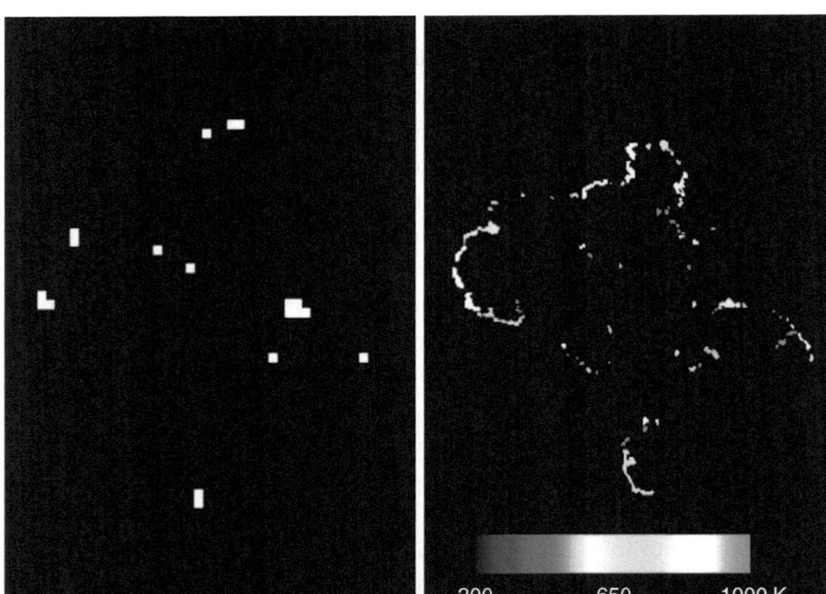

MODIS: Fire map BIRD: Fire map

Fig. 13: *Fire detection by MODIS and BIRD (Australia, 5 January 2002).*

respect to the forest fire in Australia near Sydney on 5 January 2002. The difference is obvious. BIRD data are even capable of providing fire-fighting authorities with important parameters such as fire temperature, front length and front strength in kW/m.

9.3.1.3. Temporal resolution

Small satellites provide a unique opportunity for launching affordable constellations. In this respect, small satellites can do things that are not practical with large satellites. At this point, DMC (Disaster Monitoring Constellation)[819] and RapidEye[820] can serve as examples of constellations of five micro-satellites. More details are given in section 3.4. "Distributed space systems."

9.3.2. Limitations

Small satellite missions usually focus on one specific physical phenomenon to be investigated or monitored. In this context, the restrictions and limitations of small satellite missions by comparison with large complex missions are:

- In orbit lifetime restrictions because of the extended use of advanced technologies (by comparison with conventional satellite missions)
- Limited platform capacity for using instruments with high power consumption or high data rate requirements
- Size limitations and platform stability limitations that do not allow the use of large microwave antennas or long monolithic telescopes
- Restricted options for instrument combinations on a single satellite platform because of the limited size and power capabilities of small satellites.

For these reasons small satellite missions may only be considered as complementary to conventional Earth observation missions.

9.3.3. Distributed space systems

In recent decades, spacecraft mass and power have been reduced to limit costs and risks, leading to shorter development times, as well as to the possibility of flying up-to-date technologies and achieving frequent re-flight. Besides such advantages on a single spacecraft level, it must be considered that cost reduction also opens the possibility of flying multiple spacecraft systems. A system of platforms can replace a monolithic system very effectively with the advantage of substituting the concept of failure with, ideally, one of graceful degradation. In addition, a system of platforms can achieve a performance level unachievable by a monolithic approach. Although in principle the distributed system concept can be applied to both large and small spacecrafts, it is naturally suited for small and especially micro-satellites, for a number of reasons: low system cost, easy replacement of a failed satellite (both in cost and delivery time) and possibility to gradually update technologies in orbit (which is generally an issue for large space systems).

The "Nano-Satellite Constellation Mission Idea Contest"[821] shows that there are already efforts to extensively utilise the low-cost aspect of satellites in constellations. The basic idea of this international competition is to find new remote sensing applications beyond those developed just for academic use. In this way, it is expected to supplement existing remote sensing systems in their performance or application areas, and even to satisfy other applications' needs that have not been met by existing technologies and techniques. For the purposes of this competition, the term constellation covers all distributed space system options.

In thinking about distributed space systems, a distinction can be made between different systems based on the distance between the satellites and the requirements concerning the control of their distance.[822] Using this approach, the following categories can be identified:

- Constellations
- Formations
- Swarms
- Inspection and docking systems

Figure 14 shows local systems with separations of a few meters between the spacecrafts, regional separations of typically a few tens meters to several hundred kilometres, and global systems with separations of more than a thousand kilometres.[823]

The concept of formation flying of satellites is frequently confused with that of satellite constellations. In the following sections we will distinguish between these concepts using the definitions of NASA GSFC:

- A constellation is composed of two or more spacecraft in similar orbits with no active control by either to maintain a relative position.
- Formation flying involves the use of an active control scheme to maintain a relative position.

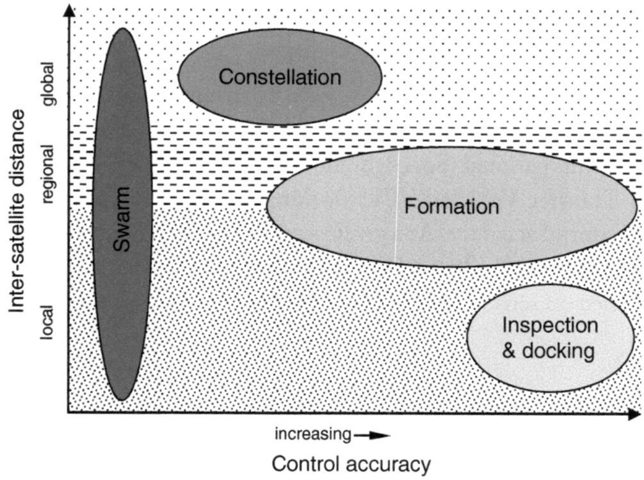

Fig. 14. *Requirements for distributed space systems.*

From the application point of view, the aim of a constellation is generally related to coverage enhancement and, therefore, to the reduction of the repetition time of observing the same ground target and consequently of the required time to achieve global coverage. In practice, a constellation consists of identical spacecrafts whose orbits are designed to adequately cover the globe (or part of it). Constellation products are just the sum of the spacecrafts' products.

On the other hand, formations tackle the issue of achieving a synergic use of payloads onboard different platforms, with no advantage in terms of time resolution. However, they enable additional products with respect to those offered by single spacecraft. In principle, one can imagine a formation implementing a distributed payload, with no product at all delivered by the payloads on board the single spacecrafts. The clear advantage of such an approach is the possibility of flying very large sensor apertures in space.

9.3.3.1. Swarms

While docking, formation flying and constellations are well established implementations of distributed space systems, swarms of spacecraft consisting of several, tens or thousands of satellites have not been deployed yet. Swarms of satellites can characterize for instance the local, regional or global Earth environment, making in situ measurements of the atmosphere or radiation conditions.

9.3.3.2. Inspection and docking systems

Inspection and docking involves two objects in space in close vicinity. This characteristic is typical of an inspector micro- or nano-satellite, orbiting for instance the International Space Station (ISS). Another example is ESA's Automated Transfer Vehicle (ATV) docking at the ISS. This poses very high demands on control accuracy. At a given separation of ten meters for example, the control accuracy should be better by at least a factor of ten (in this case one meter). Control is based on sensors which again need to provide a tenfold better accuracy (or 10 cm).

9.3.3.3. Constellations

To achieve global coverage of the Earth with high time resolution requires a satellite constellation. As already mentioned, small satellites provide a unique opportunity for deploying affordable constellations. In this respect, small satellites can do things that are not practical with large satellites.

DMC may serve as an example of a constellation of five small satellites. The standard spacecraft weighs 88 kg, of which 19 kg is payload. DMC has a GSD of 32 m and a swath width of 600 km (Landsat: GSD = 30 m, Swath width = 185 km). It provides daily coverage of the Earth.[824] The five satellites (AlSat-1, BILSAT-1, NigeriaSat-1, UK-DMC-1, Beijing-1) from five countries have been launched with three COSMOS launchers into the same orbit. DMC-2 is a planned follow up of DMC-1 with improved performances based on new technologies.

Another example is RapidEye, a commercial multispectral Earth observation mission of RapidEye AG of Brandenburg, Germany, that includes a constellation of five micro-satellites (launch August 2008, 150 kg each, GSD = 6.5 m).[825] The mission provides high-resolution multispectral imagery along with an operational GIS (Geographic Information System) service on a commercial basis. The objectives are to provide a range of Earth-observation products and services to the global user community.

9.3.3.4. Satellite formations

Formation flying of satellites is typically associated with a small number of spacecraft flying in a concerted way at regional inter-satellite separation distances. The mission objectives determine the accuracy requirements for their control.

A science mission using interferometry may have high control demands, whereas a formation of two satellites with different instruments can have less stringent control requirements.

Some current remote sensing missions already utilise the formation concept, such as GRACE (NASA/DLR Earth's gravity mapping mission) and NASA's A-TRAIN. A-TRAIN, basically a constellation mission with some formation aspects, integrates different payloads on board different satellites (OCO, Aqua, Aura, PARASOL, Cloudsat, and CALIPSO) whose combined data allow the study of climate change data. The formation flying aspect of the system is particularly related to CALIPSO and Cloudsat, whose time separation is controlled within 15 seconds of each other in order for both instruments to view the same cloud area at nearly the same moment. Such a formation flying requirement emerged from the scientific objective of observing the same clouds (whose lifetime is often less than 15 minutes) at different wavelengths. A-TRAIN consists of large, small (CALIPSO: 635 kg) and micro-satellites (PARASOL: 120 kg).

9.4. Implications

Since the advent of modern technologies, small satellites using off-the-shelf technologies or missions focused on specific physical phenomena have also been perceived as an opportunity for countries with a modest research budget and little or no experience in space technology development to enter the field of space-based Earth observation. Small satellite technology is a way of bringing within the reach of every country the opportunity to operate small satellite Earth observation missions and consequently of utilising relevant data effectively at low cost and developing application-driven missions. It provides the opportunity to independently conduct or participate in multilateral Earth observation missions using small, cost-effective satellites, and associated launches, ground stations, data distribution structures and space system management approaches. The actual performance parameters of small satellites, given in section 3, demonstrate that small satellites are competitive with older larger satellites with respect to their spatial, spectral and temporal resolution. Besides the increase in their number and the improvement of their performance, distributed space systems' use generates an additional operational requirement for reliable inter-satellite communications. In the following section we subdivide the implications arising from the new trends in small and especially micro-satellite development in two different groups: technical implications and regulatory implications.

9.4.1. Technical implications

9.4.1.1. Data rate and volume

Recent technological developments allow for the equipping of even micro-satellites with high spatial and spectral resolution imaging systems, a capability that demands the transmission of huge amounts of data at very high rates. Handling the high data rates is a matter of technology and transmission techniques. Both areas demonstrate significant progress. NigeriaSat-2, a 300 kg small satellite to be launched at the end of 2010 and equipped with a 2.5 m panchromatic band and four 5 m multispectral bands, is able to transmit imagery at a speed of 2×105 Mbps. NigeriaSat-2 is the first space component of the African Resources Management Constellation (ARMC), proposed by South Africa and supported by Nigeria, Algeria and Kenya. As a reference, it should be noted that the related parameters of SPOT-5 are 3000 kg, 5 m panchromatic and 10 m multispectral resolution and 2×50 Mbps downlink rate.

In monitoring EO systems the data volume can be significantly reduced by applying autonomously operating processor systems to extract the information from the sensor data. The BIRD microsatellite, with its neural network capable of producing thematic fire maps, may serve as a technology demonstration example.[826]

9.4.1.2. Access to space

During recent years there have been several small launchers from Brazil, China, Europe, Israel, Japan, Russia and the U.S. available at prices that are quite reasonable compared to the cost of a small satellite.

More commercial launch services are now available on most launch systems, many of which are new vehicles designed or modified to specifically meet international commercial market requirements. The most dramatic shift has been the market entry of Russian and Ukrainian launch systems that are operated as joint ventures with U.S. and European companies. The increasing availability of these low-cost launchers and the development of satellite dispensers have opened up the possibility of launching an entire constellation on a single launch, as individual payloads. The launch of the NASA/DLR GRACE satellites used Eurockot Launch Services, the joint venture owned by Astrium and the Russian company Khrunichev, to place two satellites in a closely controlled formation by using a dispenser. This launch was the first commercial use of the Russian SS-19 ICBM, which provides the two booster stages used by the ROCKOT launch vehicle and has a record of 150 flights. At the other end of the cost and mass spectrum, Ariane 5 has been used to launch six auxiliary payloads along with the primary Helios satellite, using its auxiliary payload adapter ASAP. This included Nanosat, Spain's first small satellite built by the country's INTA national space agency (Instituto Nacional de Técnia Aeroespacial), with a mass of less than 20 kg. In another example, the Cluster mission formed a constellation of four satellites in formation flying, by using two separate launches. Nowadays, most of the large launchers provide payload adapters to accommodate small satellites as secondary payloads.

In the U.S., new private and seed capital is currently being invested in the development of new launchers for small satellites. SpaceX of Al Segundo, California and Air-Launch of Kirkland, Washington are two examples of private enterprises providing launch services. Furthermore, DARPA and the U.S. Air Force have formed a joint venture to work on a small launch vehicle programme.

In addition to the above, space tourism has made its appearance as one of the newest and potentially most vigorous business incentives for the development of commercial small launchers. On 4 October 2004, Burt Rutan and Paul Allen built and flew the world's first private spacecraft to the edge of space, to win the 10 million dollar Ansari X PRIZE. Perhaps by looking back at the history of the early development of commercial aviation, one can have a glimpse at the future of commercial space access for the next twenty years. At the turn of the last century, air travel was relatively risky and quite expensive. As the commercial market for air transport grew, operating costs and investment risk dropped accordingly. Nowadays, air transport is so cost-effective that it is used to ship bulky agricultural goods, such as apples or flowers, half way around the world at prices that are competitive with local transport and production.

Nevertheless, there are some indicators showing that there are still serious efforts to be made in order to achieve comparable results in the space transportation domain. For example:

- The establishment of the annual International Symposium for Personal and Commercial Spaceflight (ISPCS)
- The recognition of a national (U.S.) Centre of Excellence for Commercial Space Transportation by the Federal Aviation Administration (F.A.A.)
- The creation of space tourism companies and the development of related spacecrafts, like Virgin Galactic.

Having these activities in mind, it is probable that in the near future small satellite missions will no longer be strongly constrained by launch costs.

9.4.1.3. Ground systems

The classical approach of ground segments assigns the specific tasks of

- S/C monitoring & control,
- P/L data reception & archives, and
- P/L data products & distribution

to specific ground facilities communicating through complicated protocols. As the number of spacecraft increases in a constellation there would be, without a change in the operations paradigm, a concomitant increase in the costs to operate the constellation. To operate constellations of micro- and nano-satellites, the operation costs have to be low on a per satellite basis especially since some of these constellations are envisioned as consisting of tens or even hundreds of micro- or nano-satellites. Simple downscaling is not sufficient – qualitative changes combining the different tasks and facilities in networks with new features are necessary. Key words for the new ground systems are for instance

- Open systems
- Automation
- "internet" technology
- Multi-session operations
- Ground station networks
- Increasing on-board autonomy.

With respect to the last point, requirements for the space segment also need to be determined. Powerful, cheap microprocessors provide the means for increased autonomy at the individual satellite level and across the constellation. At issue, though, is developing the software to perform these operations and subsequently testing the software so that its operation can be verified before flight. Qualifying these systems for spaceflight will be a challenge that must be addressed.

9.4.2. Regulatory implications

Regulatory measures may become very severe in the future. It is time to think about these now, and not be caught by surprise. Here we just address the reasoning for and the structure of the implications. The solutions need to be provided by the authorities in charge.

9.4.2.1. Space system registration

For the increasing number of micro satellites to come, especially considering distributed space systems based on micro- and nano-satellites that may consist of tens or even hundreds of micro- or nano-satellites, current registration procedures seem to be inadequate. There seems to be a need for new approaches to deal with this development.

9.4.2.2. Transmission frequency allocation

The increasing number of high performance small satellites will be accompanied by an increasing number of data downlink channels to be allocated. At some point we need to think about sophisticated procedures to manage the available frequency slots and/or to extend the usable frequency range using new technologies and transmission techniques to be managed by a competent authority. The problem is even worse if we think about distributed space systems of micro- and nano-satellites, especially since some of these constellations, formations or swarms are envisioned as consisting of tens or even hundreds of micro- or nano-satellites thus adding the inter-satellite, intra- and inter-constellation link problem to the usual downlink requirements.

9.4.2.3. Space debris management

The huge number of future space systems in orbit (satellites and launch system components), again including those distributed space systems consisting of tens or even hundreds of micro- or nano-satellites, has implications for the space debris risk. These need to be addressed now, before we run into a much bigger problem than the one we currently face. The current space debris problem is based on the fact that at the beginning of the space age the space debris problem was simply ignored. Now we think about requesting certain end-of-life procedures for each satellite with the consequence that adequate technical features have to be implemented in the satellite. But this late reaction has more consequences: countries now emerging in space technology are starting with a handicap the space faring nations did not have. This leads to discussions about fairness and responsibility that could have been avoided if we had had vision at the very beginning instead of now being confronted with a vision and a reality of a different kind.

In order to avoid running into problems in the future that seem to suddenly emerge, using mainly technical reasoning this paper has attempted to outline possible future developments in satellite Earth observation and to derive from them the technical and regulatory implications.

[808] Sandau, R., Brieß, K., 2008. Potential for advancements in remote sensing using small satellites. In: IAPRS, **XXXVII**. Part B1. Beijing 2008.

[809] Sandau, Rainer (ed.), 2006. International Study on Cost-Effective Earth Observation Missions. A. A. Balkema Publishers, a member of Taylor & Francis Group plc, Leiden, The Netherlands, ISBN 10: 0-415-39136-9, ISBN 13: 9-78-0-415-39136-8, p. 160.

[810] Ibid.

[811] Kramer, H. J., 2002. Observation of the Earth and its environment – survey of missions and sensors. Springer Verlag Berlin, Heidelberg, New York. 2002, 4 Ed.

[812] Sandau, Rainer, 2009. Satellite Earth Observation and Surveillance Payloads. NATO RTO Lecture Series SCI-209, 1-2 April 2009 at Stanford University, CA, USA, 6-7 April 2009 in Würzburg, Germany, and 8-9 April 2009 in Rome, Italy.

[813] Background information for these needs is given in: Konecny, Gottfried, 2003. Geoinformation: Remote Sensing, Photogrammetry and Geographic Information Systems. London: Taylor & Francis, 2003, p. 248.

[814] EROS-B, 2006. http://www.defense-update.com/directory/erosB.htm (last accessed August 2010).

[815] Skyrocket, 2010. http://space.skyrocket.de/index_frame.htm?http://space.skyrocket.de/doc_sdat/china-dmc.htm (last accessed August 2010).

[816] PROBA, 2002. www.esa.int/esaMI/Proba_web_site/ESAIZ8NSRWC_0.html (last accessed January 11, 2010).

[817] Brieß, K., Jahn, H., Lorenz, E., Oertel, D., Skrbek, W. & Zhukov, B., 2003. Fire Recognition Potential of the Bi-spectral InfraRed Detection (BIRD) Satellite. International Journal of Remote Sensing, 24, 4: 865 – 872. 2003.

[818] TES, 2008. http://directory.eoportal.org/get_announce.php?an_id=15557 (last accessed August 2010).

[819] Da Silva Curiel, 2005. http://www.dlr.de/iaa.symp/Portaldata/49/Resources/dokumente/archiv5/0301_daSilvaCuriel.pdf (last accessed August 2010).

[820] DLR, 2008. http://www.dlr.de/rd/desktopdefault.aspx/tabid-2440/3586_read-5336/ (last accessed August 2010).

[821] Axelspace, 2010. http://www.axelspace.com/missionideacontest/index.html (last accessed August 2010).

[822] Gill, E., 2008. Together in Space, Potentials and Challenges of Distributed Space Systems, 2008. Inaugural speech, TU Delft, Faculty of Aerospace Engineering, September 17.

[823] Sandau, Rainer, 2009. Satellite Earth Observation and Surveillance....

[824] Da Silva Curiel, 2005....

[825] DLR, 2008....

[826] Brieß, K., Jahn, H., Lorenz, E., Oertel, D., Skrbek, W. & Zhukov, B., 2003. Fire Recognition Potential....

PART 3

FACTS AND FIGURES

1. Chronology: June 2009–May 2010

Spyros Pagkratis and Blandina Baranes

1.1. Access to space

Europe	Other countries
LAUNCH LOG	
June 09	
	18 Atlas V 40-LRO LCROSS (S)
	21 Zenit-3SLB-Measat 3a (C)
	27 Delta-4M + GOES 14 (M)
	30 Proton-M-Sirius FM5 (C)
July 09	
01 Ariane V- Terrestar 1 (C)	06 Rokot- Kosmos-2451, Kosmos-06 06 2452, Kosmos-2453 (C)
	14 Falcon 1- Razaksat (C)
	15 Space Shuttle- STS-127 (MF)
	21 Kosmos-3M- Kosmos-2454 (N)
	24 Soyuz-U- Progress M-67 (ISS)
	29 Dnepr- Dubaisat-1 Deimos 1 UK-DMC 2 Nanosat-1B AprizeSat 3 AprizeSat 4 (R) (C)
August 09	
21 Ariane V- JCSAT 12, Optus D3 (C)	11 Proton-M- Asiasat 5 (C)
	17 Delta 7925- GPS 50 (N)
	25 Naro KSLV-1- STSAT-2 (S)
	28 Space Shuttle- STS-128 (MF)
	31 Long March 3B- Palapa-D1 (C)
September 09	
	08 Atlas V 401- USA 207 (C)
	10 H-IIB- HTV-1 (ISS)
	17 Soyuz-2-1b- Meteor-M... (M) (S) (R)
	17 Proton-M- Nimiq-5 (C)
	23 PSLV-CA- Oceansat-2 (C) (R) (S)

	25 Delta 7920- USA 208,209 (EW)
	30 Soyuz-FG- Soyuz TMA-16 (ISS)
October 09	
01 Ariane V- ComsatBw-1, Amazonas-2(C)	08 Delta 7920- WorldView-2 (R)
29 Ariane V- NSS-12, Thor 6 (C)	15 Soyuz-U- Progress M-03M (ISS)
	18 Atlas V 401- DMSP 5D F-18 (M)
November 09	
	02 Rokot- SMOS, PROBA-2 (R)
	10 Soyuz-U- Poisk (ISS)
	12 Long March 2C- Shi Jian XI-1 (D)
	16 Space Shuttle-STS-129 (ISS)
	20 Soyuz-U- Kosmos-2455 (I)
	23 Atlas V 431- Intelsat IS-14 (C)
	24 Proton-M- Eutelsat W7 (C)
	28 H-IIA- IGS-5A (R)
	30 Zenit-3SLB- Intelsat IS-15 (C)
December 09	
18 Ariane V GS- Helios IIB (R)	06 Delta 4 M+- WGS SV3
	09 Long March 2D- Yaohan Weixing VII (R)
	14 Proton-M- Kosmos-2456, Kosmos-2457, Kosmos-2458 (N)
	14 Delta 2 7320- WISE (S)
	15 Long March 4C- Yoagan Weixing VIII Xi Wang 1 (R) (C)
	20 Soyuz-FG- Soyuz TMA-17 (ISS)
	29 Proton-M- DirecTV 12 (C)
January 10	
	28 Long March 3C- Beidou DW3 (N)
February 10	
	03 Proton-M- Raduga-1M (C)
	08 Soyuz-U- Progress M-04M (ISS)
	11 Space Shuttle- STS-130 (MF)
	12 Atlas V 401-SDO (S)
March 10	
	01 Proton-M- Intelsat IS-16 (C)
	04 Proton-M- Kosmos-2459, Kosmos-2460, Kosmos-2461 (N)

	05 Long March 4C- Yaogon Weixing 9, YW-9 subsat 1, YW-9 subsat 2 (I)
	20 Proton-M- Echostar XIV ©
April 10	
	02 Soyuz-FG- Soyuz TMA-18 (ISS)
	05 Space Shuttle- STS-13 (MF)
	08 Dnepr- CryoSat-2 (S)
	15 GSLV Mk II- GSAT-4 (C)
	16 Soyuz-U- Kosmos-2462 (R)
	22 Atlas V 501- USA 212 (D)
	24 Proton-M- SES-1 (C)
	28 Kosmos-3M- Kosmos-2463 (N)
May 10	
21 Ariane V- Astra 3B, ComsatBw-2 (C)	14 Space Shuttle- STS-132 (MF)
	20 H-2A 202- Akatsuki, IKAROS, Unitec-1 Cubesat Negal, Cubesat Waseda-Sat2, Cubesat KSAT (S) (D) (R)
	28 Delta 4M+- GPS 62 (N)

C: Communications – D: Development – I: Intelligence – ISS: International Space Station – M: Meteorological – MF: Manned Flight – N: Navigation – R: Remote Sensing – S: Scientific – EW: Early Warning System

1.2. Space science and exploration

Europe	Other countries
EARTH SCIENCES	
February 10 High-thrust engine demonstrator industrial day in Germany	*February 10* NASA funded research discovers life built with toxic chemical
ASTRONOMY	
	January 10 A new particle of the sun under study
February 10 SDO solar observatory launched successfully	
May 10 Herschel infrared space observatory has discovered the key ingredient for making water in space	*May 10* Primordial Magnetic Fields Discovered Across The Universe
May 10 Final assembling in the CERN of the AMS destined to ISS	
EXPLORATION	
	July 09 New Russian plan decided to launch an interplanetary mission to Venus
	September 10 NASA's Kepler mission discovers two planets transiting the Same Star
	October 10 China's Chang'e 2 probe enters orbit around Moon
October 10 Venus Express new discoveries about Venus's atmosphere	*October 10* NASA and NSF-Funded Research Finds First Potentially Habitable Exoplanet
March 10 Newly Found Exoplanet discovered with Water-laden Clouds	*March 10* The NASA's ion-propelled spacecraft eclipsed the record for velocity change
MANNED SPACEFLIGHT	
	February 10 Atlantis' Final Mission
	April 10 NASA to Launch Human-Like Robot to Join Space Station Crew

1.3. Applications

Europe	Other countries
EARTH OBSERVATION	
November 2 ESA launches the Soil Moisture and Ocean Salinity (SMOS) satellite	*June 21* demonstration of Israel's TecSAR Synthetic Aperture Radar (SAR) satellite
November 19 ESA selects e-Geos to provide EO products for GMES	*August 25* Loss of S. Korean STSAT-2 weather satellite during launch
December 18 Launch of HELIOS 2B satellite	*September 17* Launch of the South African Sumbandila environment monitoring satellite
February 4 France-Germany agreement to jointly develop the CH4 Atmospheric Remote Monitoring Explorer (CHARME)	*September 23* India launches Oceansat-2, its second ocean monitoring satellite
February 3 ESA selects the ThalesAleniaSpace/OHB consortium to built the METEOSAT third generation satellites	*November 23* NOAA QuikScat satellite's scatterometer instrument ceases to function
February 24 CNES selects ThalesAleniaSpace to build Jason-3 Ocean altimetry satellite	*November 28* Japan launches first new generation Information Gathering Satellite (IGS) Optical-3
May 4 OHB contracted to build common Franco-German EO ground segment platform	*February 4* Cancellation of the U.S. National Polar-orbiting Operational Environmental Satellite System (NPOESS) development programme
	May 19 USAF launches TacSat-3, its first hyper-spectral reconnaissance satellite
INTELLIGENCE AND EARLY WARNING	
	June 10 USAF selects Lockheed Martin to build the first three Space Based Infrared System satellites (SBIRS)
	June 11 USAF awards first contracts to develop the ground segment of the future U.S. space surveillance system
	June 15 Northrop Grumman delivers second Space Tracking and Surveillance System (STSS) satellite to USAF
	September 25 U.S. Missile Defence Agency launches two Space Tracking and Surveillance System (STSS) demonstration satellites
	October 26 Raytheon selected to integrate U.S. Missile Defence Agency and USAF space surveillance sensors

NAVIGATION	
October 1 ESA declares the freely accessible service of the European GPS Navigation Overlay Service (Egnos) operational *January 7* European Commission selects OHB to build the first 14 Galileo GNSS satellites	*March 2* Launch of three Glonass GNSS satellites by Russia
TELECOMMUNICATIONS/BROADCASTING	
July 1 Launch of Terrestar-1, largest commercial spacecraft built to date *October 13* ESA-DLR agreement on the management of the European Data Relay System (EDRS) *October 27-29* Decision to create Space Data Association (SDA), a voluntary satellite database during the annual meeting of the Satellite Users Interference Reduction Group (SUIRG) in Cannes, France	*July 23* Indian government approves development of next generation communications satellite GSAT-11
November 3 EDA selects London Satellite Exchange to establish a central European Union database for purchasing commercial communication satellite services *November 30* SES Astra selects EADS Astrium to build four direct broadcast television satellites	*November 23* U.S. Defence Department launches the Internet Routing in Space (IRIS) technology demonstrator, the first dedicated U.S. military payload to orbit on a commercial satellite *December 7* first public presentation of Virgin Galactic's SpaceShipTwo suborbital space-plane
February 9 CNES selects ThalesAleniaSpace to build Athena-Fidus satellite *March 9* UK announces the deployment of a fourth Skynet-5 satellite	*February 3* DARPA selects Inmarsat to provide internet connectivity to LEO satellites services demonstrator *April 15* Loss of the Indian GSAT-1 satellite during launch
TECHNOLOGY DEVELOPMENT	
July 20 ESA selects ThalesAleniaSpace to build the Experimental Re-Entry Test Bed (EXPERT)	*July 11* JAXA and Mitsubishi Heavy Industries complete H-2B launcher ground testing *August 11* Raytheon Space and Airborne Systems presents a new infrared light-wave detector *September 17* DARPA initiates concept studies on LEO orbital debris removal

October 26 ESA launches two experimental maritime Automatic Identification System (AIS) receivers to the ISS	*October 22* DARPA issues RfI to develop an internet connectivity to Low Earth Orbit (LEO) satellites system
November 2 ESA launches Proba-2 future satellite systems demonstrator	*November 23* Cisco Systems Inc. launches the first space-based internet router on board Intelsat 14 commercial SatCom
	December 16 Japanese government cancels GX launcher development
	January 13 Cisco Systems Inc. successfully completes in-orbit testing of first space based internet router
February 24 OHB selected to develop the German Orbital Servicing Mission (DEOS) technology demonstration satellite	*February 3* Iran launches the Kavoshgar-3 carrying small animals to orbit and unveils the full scale model of its future Simorgh rocket
	February 6 NASA Administrator confirms that development of future heavy-lift launcher technologies will continue under the new NASA direction
	April 22 USAF launches its winged unmanned spaceplane demonstrator X-37B Orbital Test Vehicle
	March 3 ISRO successfully tests the Advanced Technology Vehicle (ATV), a new type of sounding rocket
	March 3 MacDonald, Dettwiler and Associates Corp. (MDA) of Canada announces the development of a technology and business model for in-orbit servicing satellites
BUSINESS	
June 9 Signature of an agreement between ThalesAleniaSpace and ESA for the development of the IXV	*June 22 Sea Launch Co LLC declares bankruptcy*
July 1 ASI and Telespazio create e-Geos to commercialise Cosmo-SkyMed radar images	*August 12* ISRO launches indigenous version of Google Earth for India
November 30 SES Astra negotiates sale of its ND SatCom subsidiary to EADS Astrium Services	*November 23* Inmarsat acquires Segovia Inc, a U.S. communications services provider

January 26 OHB selects EADS Astrium as prime subcontractor in manufacturing Galileo satellites *February 23* Dutch Space selected to build solar panels for GMES spacecrafts	*January 26* Intelsat awarded a five year services contract by the U.S. Navy

1.4. Policy and international cooperation

Europe	Other countries
GENERAL POLICY	
October 15 EC President J. M. Barroso delivers the first ever speech dedicated entirely to European Space Policy	
December 1 The Lisbon Treaty on the functioning of the European Union enters into force, inviting under its Article 189 European institutions to implement a long-term European space policy	*December 2* The President of the U.S. Aerospace Industries Association (AIA) urges U.S. President B. Obama to ease ITAR export control restrictions *January 27* U.S. President B. Obama addresses the issue of U.S. export controls reform in his State of the Union speech
February 5 Creation of the GMES Partners Board by the European Commission *March 15* EC announces its intention to remove non-European built components from future Galileo satellites *March 23* Official announcement of the establishment of the UK Space Agency	*February 1* U.S. President B. Obama announces the NASA FY2011 budget, effectively canceling the Constellation programme and outsourcing Human space flight to the private sector *February 23* NASA administrative structure changes announced, emphasizing on R&D activities and expanding the NASA Administrator's capacities *April 14* The Head of China Manned Space Engineering Office confirms plans to build a 30 ton space station by 2022
GENERAL COOPERATION	
	July 20 U.S. And India sign a Technology Safeguards Agreement allowing the latter to launch civil and non-commercial satellites containing U.S. made components *September 14-18* Fourth annual meeting of the International Committee on Global Navigation Satellite Systems (ICG) held in St. Petersburg
October 23 Joint EU-ESA conference on Human Space Exploration held in Prague	*October 19-21* Tenth annual meeting of the United Nations Geographic Information Working Group (UNGIWG) held in Bonn

October 26-27 11th European Inter Parliamentary Space Conference held in London *March 15* EU Member States reach preliminary agreement on the Public Regulated Service user policy for Galileo	November 17 U.S. And China agree to resume annual exchange of Heads of space agencies visits *December 2* The United Nations General Assembly (UNGA) adopts Resolutions 64/28 and 64/49, emphasizing the role of transparency and confidence building measures (TCBM) to avoid an arms race in spacer *December 10* UNGA adopts by consensus resolution 64/86, emphasizing the importance of international cooperation in the peaceful use of outer space *April 9* Kazakh Parliament ratifies agreement with Russia to extent the use of the Baikonur Cosmodrome to 2050
SPACE SCIENCE	
July 30 EC issues the 3rd call for proposals on space related R&D projects within the FP7 framework	
October 8 Brazilian space agency AEB signs a technology exchange agreement with Liege space centre of Belgium	*November 24* U.S. And India sign cooperation agreement on science and technology research, nuclear energy and space
APPLICATIONS	
	May 30 Russia and India agree to establish a joint venture in India to produce Glonass/GPS compatible navigation equipment
October 13 ESA and DLR reach agreement on the management of the future European Data Relay System (EDRS) *October 26* the European Commission officially recognizes maritime surveillance as the next major area of space applications investment	*November 18* NOAA and ISRO agree to share data from India's Oceansat-2 meteorological and oceanographic satellite

2. Country profiles

AUSTRIA

Population[827]	8.35 million
GDP[828]	280 billion euros
Responsibility[829]	The Austrian Space Program is funded by the Federal Ministry for Transport, Innovation and Technology and managed by the Agency for Aeronautics and Space of the Austrian Research Promotion Agency.
Activities[829]	In addition to ESA programmes, two main national programmes: the Austrian Space Applications Programme and the Austrian Radionavigation Technology and Integrated Satnav Services and Products Testbed.
Budget	In 2009, 60.73 million euros (ESA 43.35, national space programs 7.5, Austrian Academy of Sciences 4.02, Eumetsat 4.05, FFG 1.81)
Staff[829]	ALR – 11
Direct employment in space manufacturing industry[830]	318

BELGIUM

Population[827]	10.75 million
GDP[828]	349 billion euros
Responsibility[831]	The Belgian Federal Science Policy Office manages Belgian space activities and the Belgian participation in national and international programmes through it's Department for Space Research and Applications.
Activities[831]	In addition to ESA programmes, there are bilateral cooperation projects with Argentina on SOACOM, France on COROT, SPOT and Pleiades, the U.S. on STEREO and with RUSSIA on MIRAS and SPICAM.
Budget	In 2009, approximately 200 million (ESA 161, EUMETSAT 5, national approximately 34)
Staff	Department for space research and applications: about 20
Direct employment in space manufacturing industry[830]	1,523

CZECH REPUBLIC

CZECH SPACE OFFICE

Population[827]	10.51 million
GDP[828]	145 billion euros
Responsibility[832]	The Ministry of Education, Youth and Sports supervises space activities and cooperation with ESA. The Czech Space Office, a non-profit association, coordinates space activities.
Activities[832]	In addition to the ESA PECS programme, the Czech space activities focus on astronomy, magnetospheric, ionospheric and atmospheric research, microgravity research experiments, scientific instruments and micro-satellites.
Budget	In 2009, 10.4 million euros (national 4.3, ESA 5.5, Other 0.9)
Staff[832]	CSO:13

DENMARK

DTU Space
Institut for Rumforskning og -teknologi

Population[827]	5.54 million
GDP[828]	229 billion euros
Responsibility[833]	The Ministry of Science, Technology and Innovation is responsible for the national space policy and space activities.
Activities[833]	In addition to ESA programmes, bilateral cooperation is undertaken with the U.S., Sweden, Russia and France.
Budget	In 2009, 38.9 million euros (national 7.9, ESA 27.8, EUMETSAT 3.5)
Staff[833]	DTU:136
Direct employment in space manufacturing industry[830]	216

FINLAND

Tekes

Population[827]	5.35 million
GDP[828]	176 billion euros
Responsibility[834]	The Ministry of Trade and Industry, the Funding Agency for Technology and Innovation, and the Academy of Finland are funding space activities in Finland. The Finnish Space Committee consists of representatives of all stakeholders and coordinates all of the activities. Tekes is the executive body for space activities and, together with the Academy of Finland for basic research, manages the Finnish participation within ESA programmes and other international projects.
Activities[834]	In addition to ESA programmes, Finland has bilateral activities with the USA (TWINS, Mars Science Laboratory, Phoenix, ISS) France (Pleiades), Germany (TanDEM-X and TerraSAR-X) and Japan (ISS), as well as a national space technology programme.
Budget	In 2009, 51.66 million euros (national 34, ESA 15, EUMETSAT 2.66)
Staff[834]	Tekes – Environmental Data and Space Applications:14
Direct employment in space manufacturing industry[830]	172

FRANCE

cnes

Population[827]	64.7 million
GDP[828]	1947 billion euros
Responsibility[835]	The Centre National d'Etudes Spatiales (CNES) is responsible for the French space activities. It is under the shared responsibility of the Ministry of Education and Research and of the Ministry of Defense. The Office Nationale d'Etudes et de Recherches Aérospatiales (ONERA) is also responsible for space related research.
Activities[835]	In addition to ESA programmes, civil, military and science programmes are undertaken (Pleiades, Helios, Essaim, e-Gorce), as well as bilateral cooperation with the U.S. (CALIPSO, Jason 2 and 3) and India (Saral, Oceansat 3, Altika-Argos, Megha Tropiques).
Budget	In 2009, CNES had a budget of 1.997 billion euros (including the ESA contribution of 685 million, and a EUMETSAT contribution of 29 million). ONERA had a budget of 210 million euros, of which only a sizable fraction was spent on space-related research.
Staff[835]	CNES: app. 2,400
Direct employment in space manufacturing industry[830]	11,225

GERMANY

DLR

Population[827]	81.8 million
GDP[828]	2432 billion euros
Responsibility[836]	The German Space Agency within the German Aerospace Center (DLR) is responsible for German space activities. It is under the responsibility of the Ministry of Economics and Technology.
Activities[836]	In addition to ESA programmes, Germany has national civil and commercial programmes in Earth observation (RapidEye, TerraSAR-X, TanDEM, EnMAP), Human space flight (ISS, Microgravity experiments), launch services (Eurockot, OHB-Cosmos), associated ground systems and space technologies (such as intersatellite links). Germany is involved in bilateral cooperation with the U.S. (GRACE, Dawn, Sofia) and its military programs include remote sensing satellites (SAR-Lupe radar satellites) and satcoms (Satcom BW).
Budget	In 2009, roughly 920 million euros. DLR contributed 680 million euros to ESA, and 38 million to EUMETSAT.
Staff[836]	DLR for space activities: app. 2,000
Direct employment in space manufacturing industry[830]	5,270

GREECE

GSRT

Population[827]	11.3 million
GDP[828]	237 billion euros
Responsibility[837]	The General Secretariat for Research and Technology (GSRT) of the Ministry of Education is responsible for Greek space activities.
Activities[837]	The Greek space activities cover mainly the fields of space physics, ionospheric physics, Earth observation and telecommunications.
Budget	In 2009, 14.5 million to ESA, and 3.2 million to EUMETSAT.
Staff[837]	For space: <5

HUNGARY

Population[827]	10 million
GDP[828]	96 billion euros
Responsibility[838]	The Hungarian Space Office, under the responsibility of the Ministry of Environment and Water, manages Hungarian space activities.
Activities[838]	Participation in microgravity, Earth observation, life and material sciences and GSTP programs of ESA.
Budget	In 2009, Hungary spent about 2 million euros to satisfy its PECS agreements with ESA and contributed 1.3 million euros to EUMETSAT.
Staff[838]	Total space activities: 250

IRELAND

Population[827]	4.5 million
GDP[828]	159 billion euros
Responsibility[839]	Enterprise Ireland, in association with the Office of Science and Technology of the Department of Enterprise, Trade and Employment, manages and coordinates space activities in Ireland.
Activities[839]	Irish space activities are in the fields of software systems and services, precision mechanical components, advanced materials, electronics/microelectronics and telecommunications systems and service engineering.
Budget	In 2009, Ireland contributed 13.3 million euros to ESA, and 2.3 million euros to EUMETSAT.
Staff[839]	For space: <5
Direct employment in space manufacturing industry[830]	30

ITALY

Population[827]	60.3 million
GDP[828]	1553 billion euros
Responsibility[840]	The Italian Space Agency (ASI), under the Ministry of University and Research, manages Italian space activities.
Activities[840]	Italian civil space activities include ESA programmes and a national programme. National activities include scientific missions (AGILE, PRISMA, MIOSAT), dual-use Earth observation satellites (Cosmos-Skymed) and military satcoms (Sicral). Italy conducts bilateral cooperation with France (Athena, Fidus) and Argentina (SIASGE).
Budget	In 2010, the Italian Space Agency's budget was app. 700 million euros (ESA 369.5 million, EUMETSAT 23.7 million, national app. 307 million)
Staff[840]	ASI: app. 200
Direct employment in space manufacturing industry[830]	4,490

LUXINNOVATION
Agence Nationale pour la Promotion de l'Innovation et de la Recherche

LUXEMBOURG

Population[827]	0.502 million
GDP[828]	39.5 billion euros
Responsibility[841]	Luxinnovation, the National Agency for Innovation and Research, under the responsibility for the Ministry of Culture, Higher Education and Research, coordinates space activities.
Activities[841]	Luxembourg focuses mainly on telecommunications, through SES Astra commercial satellite services provider.
Budget	In 2009, Luxinnovation contributed 12.8 million euros to ESA, and 0.42 million euros to EUMETSAT.
Staff	Luxinnovation has a total of 21 staff members only a few of which are directly involved in space activities.
Direct employment in space manufacturing industry[830]	31

NETHERLANDS

Population[827]	16.6 million
GDP[828]	585 billion euros
Responsibility[842]	The new Netherlands Space Office (2009) is responsible for coordinating international space activities with ESA, NASA and JAXA, as well as for the development of the Dutch national space programme. The Dutch Space Research Organization (SRON) is responsible for national and multilateral space science programmes.
Activities[842]	In addition to it's contributions to ESA programmes, such as Gaia, Exo-Mars, and GMES, NSO conducts bilateral activities with Japan (SPICA-SAFARI) and is responsible for planning the Dutch National Space Programme. SRON conducts space research in the fields of Earth observation, microgravity and planetary exploration.
Budget	In 2009, 133.6 million euros (ESA 99 million, national 25 million, EUMETSAT 8.6 million)
Staff[842]	NSO: 35, SRON: 215
Direct employment in space manufacturing industry[830]	610

NORWAY

Population[827]	4.86 million
GDP[828]	312 billion euros
Responsibility[843]	The Norwegian Space Centre (NSC), under the Ministry of Trade and Industry, manages Norwegian space activities.
Activities[843]	In addition to ESA programmes (in particular Earth observation, telecommunications and launchers), Norway has national support programmes and commercial activities (Telenor). Moreover, Norway operates the Andøya rocket range and the Svalbard ground station. Norway has also a bilateral agreement with Canada on the use of Radarsat 2 data.
Budget	In 2009, Norway contributed 44.6 million euros to ESA, 4 million euros to EUMETSAT and 8 million euros on national space activities.
Staff[843]	NSC: 28
Direct employment in space manufacturing industry[830]	101

Polskie Biuro do spraw Przestrzeni Kosmicznej

POLAND

Population[827]	38.2 million
GDP[828]	348 billion euros
Responsibility[18,19]	Polish space activities are under the joint responsibility of the Ministry for Scientific Research and Information Technology and the Ministry of Economics Affairs and Labour. Within the Academy of Sciences, the Space Research Centre coordinates space activities and hosts the Polish Space Office (PSO).
Activities[18,19]	Polish space activities focus on the fields of space science, navigation and Earth observation applications. Poland is participating in several ESA missions, such as Herschel, Planck, Rosetta, ExoMars and Beppi Columbo.
Budget	In 2009, app. 4 million euros (ESA PECS 1.2 million, national 3 million)
Staff[844]	Space Research Centre: 88

FCT Space Office

PORTUGAL

Population[827]	10.6 million
GDP[828]	170 billion euros
Responsibility[846]	Portuguese space activities are coordinated by the Portuguese Space Office within the Foundation for Science and Technology (FCT), which is accountable to the Ministry of Science, Technology and Higher Education (GRICES).
Activities[846]	Mainly participation in ESA programs (telecommunications systems, technology developments, Earth observation, exploration), as well as space outreach programmes.
Budget	In 2009, Portugal contributed 15.67 million euros to ESA and 2.4 million euros to EUMETSAT.
Staff	App. 10
Direct employment in space manufacturing industry[830]	101

ROMANIA

Population[827]	21.5 million
GDP[828]	123 billion euros
Responsibility[847]	The Romanian Space Agency (ROSA), under the responsibility of the Ministry of Education and Research, is managing Romanian space activities.
Activities[847]	Romania has a PECS agreement with ESA. In addition, Romania has national activities covering space science (space physics and astronomy), space systems (construction of nanosatellites and microgravity experiments) and space applications (telemedicine, Earth observation and navigation space-based services).
Budget	IIn 2009, app.2 million euros was spent to satisfy Romania's PECS agreement with ESA. Romania also spent about 12 million for the "Space and Security" programme within Romania's "National Plan for R&D and Innovation".
Staff[847]	ROSA: 36

SPAIN

Population[827]	46 million
GDP[828]	1052 billion euros
Responsibility[848]	The Centre for the Development of Industrial Technology (CDTI), under the Ministry of Science and Innovation, is funding and coordinating Spanish space activities.
Activities[848]	In addition to ESA programmes, Spain has a national space programme including governmental and commercial programmes, especially in civil and military telecommunications (Hispasat, Spainsat), and Earth observation (SEOSAT/INGENIO, SEOSAR/PAZ, INTAuSAT 1) and it is involved in bilateral cooperation projects with France, Canada, the U.S. and Russia. Also, Spain manages national, ESA and NASA ground facilities.
Budget	In 2009, Spain contributed 184 million euros to ESA and 14.9 million euros to EUMETSAT, while 300 million euros were allocated to national space activities.
Staff[848]	CDTI: app. 150
Direct employment in space manufacturing industry[830]	2,231

SWEDEN

RYMDSTYRELSEN
Swedish National Space Board

Population[827]	9.34 million
GDP[828]	337 billion euros
Responsibility[849]	The Swedish National Space Board (SNSB), under the Ministry of Industry, Employment and Communication, is responsible for space activities in Sweden. Basic research is funded via the Ministry of Education and Research.
Activities[849]	In addition to ESA programmes, Sweden has national programmes (subsystems, satellites and sounding rockets), and bilateral cooperation mainly with France (Proteus, Pleiades, Spot, Vulcain), and Germany. Both countries are partners in the technology demonstration project Prisma. Sweden also operates the test range of Kiruna and the satellite ground station at Esrange.
Budget	In 2009, 83.9 million euros (ESA 56 million, EUMETSAT 5 million, national 22.9 million euros).
Staff[23]	SNSB: 16
Direct employment in space manufacturing industry[830]	664

Schweizerische Eidgenossenschaft
Confédération suisse
Confederazione Svizzera
Confederaziun svizra

SWITZERLAND

Population[827]	7.78 million
GDP[828]	391 billion euros
Responsibility[850]	The Swiss Space Office of the State Secretariat for Education and Research of the Federal Department of Home Affairs is responsible for Swiss space activities and cooperates closely with the Swiss Department of Foreign Affairs on that topic. The Federal Commission for Space Affairs (CFAS) is currently preparing a new Swiss space policy. The interdepartmental coordination committee for space (IKAR) is responsible for coordinating activities.
Activities[850]	Most of the Swiss activities are undertaken within ESA programmes (space science, human spaceflight, launchers, Earth observation, Prodex and navigation).
Budget	In 2009 Switzerland contributed 94.4 million euros to ESA, and 5.4 million euros to EUMETSAT.
Staff[850]	SNSB: 16
Direct employment in space manufacturing industry[830]	783

UK SPACE AGENCY

UNITED KINGDOM

Population[827]	62 million
GDP[828]	1703 billion euros
Responsibility[851]	The UK Space Agency is responsible for all strategic decisions on the UK civil space programme and is responsible for supporting academic research into space technology, as well as raising awareness of UK space activities.
Activities[851]	UK space activities include bilateral cooperation with JAXA, India and the U.S. and within ESA in EO and space exploration (Cassini-Huygens, James Webb Space Telescope, Herschel, and Planck missions).
Budget	In 2010, 312 million euros (ESA 243.4 million, EUMETSAT 27.53 million, national 41.77 million)
Staff[851]	UKSA: 16
Direct employment in space manufacturing industry[830]	3,429

[827] Eurostat – EU Population 501 million at 1 January 2010. http://epp.eurostat.ec.europa.eu/cache/ITY_PUBLIC/3-27072010-AP/EN/3-27072010-AP-EN.PDF.

[828] Eurostat – Gross Domestic Product at Market Prices. http://epp.eurostat.ec.europa.eu/tgm/refreshTableAction.do?tab=table&plugin=1&pcode=tec00001&language=en Accessed 31 Aug. 2010.

[829] FFG Website. http://www.ffg.at/content.php?version=2 Accessed 31 Aug. 2010.

[830] Eurospace – Facts and Figures: The European Space Industry in 2009. http://eurospace.pagesperso-orange.fr/F&F2009/FFdata2009issue2.pdf.

[831] Belgian Federal Science Policy Office Website. http://www.belspo.be/belspo/res/rech/spatres/bilcoop_en.stm Accessed 31 Aug. 2010.

[832] Czech Space Office Website. www.czechspace.cz. Accessed 31 Aug. 2010.

[833] National Space Institute Website. http://www.space.dtu.dk/English.aspx Accessed 31 Aug. 2010.

[834] ESD: European Space Directory 2009 (references in 2008/2009 yearbook on space policy).

[835] CNES website. www.cnes.fr. Accessed 2 Sept. 2010.

[836] DLR website. www.dlr.de Accessed 2 Sept. 2010.

[837] GRST website. www.grst.gr Accessed 2 Sept. 2010.

[838] HSO website. www.hso.hu Accessed 2 Sept. 2010.

[839] Enterprise Ireland website. www.enterprise-ireland.com Accessed 2 Sept. 2010.

[840] ASI – Italian Space Activities 2009. http://www.asi.it/files/COPUOS%202009.pdf Accessed 2 Sept. 2010.

[841] Luxinnovation website. http://www.luxinnovation.lu/site/index.jsp# Accessed 2 Sept. 2010.
[842] Netherlands Space Office website. www.spaceoffice.nl Accessed 3 Sept. 2010.
[843] Norwegian Space Centre website. www.spacecentre.no Accessed 3 Sept. 2010.
[844] Space Research Centre website. http://www2.cbk.waw.pl/ Accessed 3 Sept. 2010.
[845] Polish Space Office Website. http://www.kosmos.gov.pl/ Accessed 3 Sept. 2010.
[846] FCT Space Office Website. http://alfa.fct.mctes.pt/apoios/cooptrans/espaco/Accessed 16 Sept. 2010.
[847] Romanian Space Agency Website. www.rosa.ro Accessed 16 Sept. 2010.
[848] CDTI Website. http://www.cdti.es/index.asp?MP=15&MS=192&MN=3 Accessed 16 Sept. 2010.
[849] Swedish National Space Board Website. http://www.snsb.se/en/Home/Home/ Accessed 16 Sept. 2010.
[850] Swiss Space Office website. http://www.sbf.admin.ch/htm/themen/weltraum_de.html Accessed 16 Sept. 2010.
[851] United Kingdom Space Agency website. http://www.ukspaceagency.bis.gov.uk/default.aspx Accessed 25 Sept. 2010.

3. Bibliography of space policy publications. July 2009–June 2010

Blandina Baranes

3.1. Monographs

Achilleas, Philippe. Droit de l'Espace – Télécommunication, Observation, Navigation, Défense, Exploration. Lyon: Larcier, 2009.

Advanced Professional Education and News Service. 21st Century Chinese Military Issues: Assessment of China's ASAT Anti Satellite and Space Warfare Programs, Policies and Doctrines – Covert Weapons, Attacks, Laser, Plasma. New York: Progressive Management, 2010.

Ansari, Anousheh and Hickham Homer. My Dream of Stars: From Daughter of Iran to Space Pioneer. Hampshire: Palgrave Macmillan, 2010.

Aster, Robert. Missions from JLP – Fifty Years of Amazing Flight Projects. Create Space, 2010.

Badescu, Viorel. Mars: Prospective Energy and Material Resources. New York: Springer, 2010.

Benaroya, Haym, ed. Lunar Settlements. Boca Raton: Taylor & Francis, 2010.

——. Turning Dust to Gold: Building a Future at Moon and Mars. Berlin: Springer Praxis, 2010.

Boyle, Alan. The Case for Pluto: How a Little Planet Made a Big Difference. Hoboken, N.J.: Wiley, 2010.

Buckbee, Ed. Wernher von Braun – The Rocket Man. Huntsville: Ed Buckbee & Associates, 2010.

Burgess, Colin, ed. Footprints in the Dust: the Epic Voyages of Apollo, 1969–1975. Lincoln: University of Nebraska Press, 2010.

Burgess, Colin and French, Francis. In the Shadow of the Moon. A Challenging Journey to Tranquility, 1965–1969. Lincoln: University of Nebraska Press, 2010.

Carmichael, Scott. Moon Men Return – USS Hornet and the Recovery of the Apollo 11 Astronauts. Washington: Naval Institute Press, 2010.

Carroll, Michael. The Seventh Landing: Going back to the Moon, this Time to Stay. New York: Springer, 2009.

Caubarreaux, Eric. For All Mankind: Recipients of the Congressional Space Medal of Honor. Create Space, 2010.

Chuvieco, Emilio, Li, Jonathan and Yang, Xiaojun, eds. Advances in Earth Observation of Global Change. Berlin: Springer, 2010.

Ciancone, Michael L., ed. History of Rocketry & Astronautics. AAS History Series. Vol. 33. San Diego: Univelt, 2010.

Cisco, David. Full Circel: An Incredible Journey of a Lunar Model Spacecraft Technician. His Memoires of his Time at NASA and all the Stories Alo. Washington: DLC Enterprise, 2010.

Coletta, Damon and Pilch, Frances T. Space and Defense Policy. London, New York: Routledge, 2009.

Comiso, Josefino. Polar Oceans from Space. New York: Springer, 2010.

Contant, Jean-Michel and Menschikov, Valeriy A., eds. Space for Security and Prosperity of the Peoples. Moscow: A.A. Maksimov Space Systems Research Institute, 2010.

Damon, Thomas D. Introduction to Space: the Science of Spaceflight. Malabar Fla.: Krieger Publications, 2009.

Daniels, Patricia. The New Solar System: Ice Worlds, Moons, and Planets Redefined. Washington, D.C.: National Geographic Society, 2009.

Davies, Paul. The Eerie Silence: Are We Alone in the Universe. London: Allen Lane, 2010.

Dench, Paul. Carnavon and Apollo: One Giant Leap for a Small Australian Town. Amsterdam: Rozenberg Publisher, 2010.

Denis, John H. and Aldridge, Paul D. eds. Space Exploration Research. Hauppauge, N.Y.: Nova Science Publishers, 2010.
Dickson, Paul. A Dictionary of the Space Age (New Series in NASA History). Baltimore: Johns Hopkins University Press, 2009.
Fernandez, Adolfo J. Military Role in Space Control: A Primer. Kindle Edition, 2010.
Fletcher, Karen. Down to Earth: How Space Technology Improves our Lives. Noordwijk: European Space Agency, 2009.
Freeman, Marsha. Krafft Ehricke's Extraterrestrial Imperative. New York: Apogee Books, 2009.
Friedman, Raymond. A History of Jet Propulsion, Including Rockets. Bloomington: Xlibris Corporation, 2010.
Gangale, Thomas. The Development of Outer Space: Sovereignty and Property Rights in International Space Law. Santa Barbara, Calif.: Praeger, 2009.
Green, Pippa. Chice, Not Fate: Shaping Sustainable Future in the Space Age. Johannesburg: Penguin Books, 2009.
Guo, Huadong and Wu Ji, eds. Space Science and Technology in China: A Roadmap to 2050. New York: Springer, 2010.
Habison, Peter, ed. Himmel@All. Astronomie in Bildung und Kultur. Vienna: Edition Volkshochschule, 2010.
Harland, David. NASA's Moon Program: Paving the Way for Apollo 11. New York: Springer Praxis Books, 2009.
Harvey, Brian, Henk H.F. Smid and Pirard, Theo. Emerging Space Powers: The New Space Programs of Asia, the Middle East, and South America. New York: Springer Praxis Books, 2010.
Howe, Scott A., Sherwood, Brent, and Syd Meade, eds. Out of this World: the New Field of Space Architecture. Reston, VA: American Institute of Aeronautics and Astronautics, 2009.
Jasani, Bhupendra, et al., eds. International Safeguards and Satellite Imagery: Key Features of the Nuclear Fuel Cycle and Computer-Based Analysis. Berlin: Springer, 2009.
Johnson, Les, Matloff, Gregory L. and Bangs C. Paradise Regained: The Regreening of Earth. New York: Springer Praxis Books, 2010.
Kranz, Gene. Failure is not an Option. Mission Control from Mercury to Apollo 13 and Beyond. London: Simon & Schuster, 2009.
Kulacki, Gregory and Lewis, Jeffrey G. A Place for One's Mat: China's Space Program, 1956–2003. Cambridge: American Academy of Arts and Sciences, 2009.
Kunzmann, Klaus R., Willy A. Schmid and Martina Koll-Schretzenmayr. China and Europe: the Implications of the Rise of China for European Space. London, New York: Routledge, 2010.
Lojdahl, Franz, ed. Future U.S. Space Launch Capabilities. Hauppauge, N.Y.: Nova Science Publishers, 2009.
Maguire, Dillon, ed. Exploring the Final Frontier: Issues, Plans and Funding for NASA. Hauppauge, N.Y.: Nova Science Publishers, 2010.
Mailer, Norman. Moon Fire: The Epic Journey of Apollo 11. Los Angeles: Taschen, 2010.
Manber, Jeffrey. Selling Peace: Inside the Soviet Conspiracy that Transformed the U.S. Space Program. New York: Apogee Books, 2009.
Marschall, Laurence A. and Maran, Stephen P. Pluto Confidential: An Insider Account of the Ongoing Battles over the Status of Pluto. Dallas: Benbella Books, 2009.
Morgan, Forrest E. Deterrence and First-Strike Stability in Space: A Preliminary Assessment. Santa Monica: RAND, 2009.
Nansen, Ralph. Energy Crisis: Solution from Space. New York: Apogee Books, 2009.
National Research Council. Defending Planet Earth: Near-Earth Object Surveys and Hazard Mitigation Strategies: Final Report. Washington DC: The National Academies Press, 2010.
O'Brian Frank. The Apollo Guidance Computer: Architecture and Operation. New York: Springer Praxis Books, 2010.
Peabody, Earl M., ed. Sustaining the Global Positioning System. Hauppauge, N.Y.: Nova Science Publishers, 2010.
Pelton, Joseph, Bukley, Angelia P. and Rycroft, Michael, eds. The Farthest Shore: A 21st Century Guide to Space. Burlington Ontario: Apogee, 2010.

Prelinger Shaw, Megan. Another Science Fiction: Advertising the Space Race 1957–1962. New York: Blast Books, 2010.
Rathgeber, Wolfgang, Schrogl, Kai-Uwe and Williamson Ray A., eds. The Fair and Responsible Use of Space. An International Perspective. Vienna: SpringerWienNewYork, 2010.
Ross, Monte. The Search for Extraterrestrials: Intercepting Alien Signals. New York: Springer Praxis Books, 2010.
Rothmund, Christophe, ed. History of Rocketry and Astronautics. AAS History Series, Vol. 32. San Diego: Univelt, 2010.
Sandau, Rainer, Roeser, Hans-Peter and Valenzuela, Arnoldo, eds. Small Satellite Missions for Earth Observation: New Developments and Trends. Berlin: Springer, 2010.
Schilling, Govert. The Hunt for Planet X: New Worlds and the Fate of Pluto. New York: Springer Praxis Books, 2009.
Schoettle, Enid C. Making American Space Policy (1) The Establishment of NASA. New York: General Books LLC, 2010.
Seedhouse, Erik. The New Space Race: China vs. USA. New York: Springer Praxis Books, 2010.
——. Prepare for Launch – The Astronaut Training Process. New York: Springer, 2010.
Shukor, Sheikh Muszaphar. Journey to Space: A Memoir of Malaysia's First Angkasawan. Singapore: MPH Group Publishing Sdn Bhd, 2010.
Siddiqi, Asif. The Rochets' Red Glare: Spaceflight and the Soviet Imagination, 1857–1957. Cambridge: Cambridge University Press, 2010.
Smith, Marcia. Space Launch Vehicles: Government Activities, Commercial Competition, and Satellite Exports. Kindle Edition, 2010.
Sutton, George Paul and Biblarz, Oscar. Rocket Propulsion Elements. 8th edition. Hoboken, N.J.: Wiley, 2010.
Thompson Barbara J., et al., eds. Putting the "I" in IHY. The United Nations Report for the International Heliophysical Year 2007. Vienna: SpringerWienNewYork, 2010.
Treadwell, Terry. Stepping Stones to the Stars: The Story of Manned Spaceflight. Gloucestershire: The History Press, 2010.
Tylor, Frederic W. The Scientific Exploration of Mars. Cambridge, UK, New York: Cambridge University Press, 2010.
Vedda, James A. Choice not Fate: Shaping a Sustainable Future in the Space Age. Bloomington: Xlibris, 2009.
Walsh, Patrick J. Spaceflight: A Historical Encyclopedia. Santa Barbara, Calif.: Greenwood, 2010.
Weiler, Edward. Hubble: A Journey Through Space and Time. New York: Abrams, 2010.
Whitehouse, David. One Small Step: The Inside Story of Space Exploration. London: Quercus Books, 2009.
Wikborg, Elias, ed. Space Tourism Issues. Hauppauge, N.Y.: Nova Science Publishers, 2010.
Wong Wilson W.S. and James Fergusson. Military Space Power: A Guide to the Issues. Santa Barbara, Calif.: Praeger, 2010.

3.2. Articles

Allner, Matthew, et al. "NASA's Explorer School and Spaceward Bound Programs: Insights into Two Education Programs Designed to Heighten Public Support for Space Science Initiatives." Acta Astronautica 66 (2010): 1280–1284.
Arévalo-Yepes, Ciro, et al. "The Need for a United Nations Space Policy." Space Policy 26 (2010): 3–8.
Balogh, Werner R. "Space Activities in the United Nations System – Status and Perspectives of Inter-agency Coordination of Outer Space Activities." Acta Astronautica 65 (2009): 18–26.
Bensoussan, Denis. "Space Tourism Risks: A Space Insurance Perspective." Acta Astronautica 66 (2010): 1633–1638.
Bhaskaranarayana, A., Varadarajan, C. and Hegde, V.S. "Space-based Societal Applications – Relevance in Developing Countries." Acta Astronautica 65 (2009): 1479–1486.

Bonnal, Christophe, Gigou, Jacques and Aubin, Didier. "Space Debris Mitigation Measures Applied to European Launchers." Acta Astronautica 65 (2009): 1679–1688.
Brisibe, Tare C. "Customary International Law, Arms Control and the Environment in Outer Space." Chinese Journal of International Law 8 (2010): 375–393.
Bütfering, Peter. "Regional Convergence Platforms in Europe – Innovation for Space through Technology Partnerships." Acta Astronautica 66 (2010): 1520–1524.
Burzykowska, Anna. "Smaller States and the New Balance of Power in Space." Space Policy 25 (2009): 187–192.
Caisso, Philippe, et al. "A Liquid Propulsion Panorama." Acta Astronautica 65 (2009): 1732–1773.
Coffey, Sarah. "Establishing a Legal Framework for Property Rights to Natural Resources in Outer Space." Case Western Reserve Journal of International Law 119 (2009): 119–147.
Collins, Patrick and Autino, Adriano. "What the Growth of a Space Tourism Industry could Contribute to Employment, Economic Growth, Environmental Protection, Education, Culture and World Peace." Acta Astronautica 66 (2010): 1553–1562.
Conley, Catharine A. and Rummel, John D. "Planetary Protection for Human Exploration of Mars." Acta Astronautica 66 (2010): 792–797.
Cukurtepe, Haydar and Akgun, Ilker. "Towards Space Traffic Management System." Acta Astronautica 65 (2009): 870–878.
Curtis, Jeremy, et al." Reviewing UK Space Exploration." Space Policy 26 (2010): 113–116.
de León, Pablo. "Ricardo Dyrgalla (1910–1970), Pioneer of Rocket Development in Argentina." Acta Astronautica 65 (2009): 1789–1795.
de Montluc, Bertrand. "Russia's Resurgence Prospects for Space Policy and International Cooperation." Space Policy 26 (2010): 15–24.
de O. Bittencourt Neto, Olavo. "Private Launch Activities on Brazilian Territory: Current Legal Framework." ZLW Zeitschrift für Luft- und Weltraumrecht 58 (2009): 429–449.
Dick, Steven J. "Origins and Development of NASA's Exobiology Program, 1958–1976." Acta Astronautica 65 (2009): 1–5.
Doule, Ondrej and Peeters, Walter: "Workforce Policy in the European Sector." Astropolitics 7 (2009): 193–205.
Ehrenfreund, Pascale and Peter, Nicolas. "Toward a Paradigm Shift in Managing Future Global Space Exploration Endeavours." Space Policy 25 (2009): 244–256.
Ehrenfreund, Pascale, et al. "Cross-cultural Management Supporting Global Space Exploration." Acta Astronautica 66 (2010): 245–256.
Eneev, Timur M., et al. "Space Autonomous Navigation System of Soviet Project for Manned Fly By Moon." Acta Astronautica 66 (2010): 341–347.
Ersfeld, Herrmann. "Empfehlungen für eine nationale deutsche Weltraumgesetzgebung." ZLW Zeitschrift für Luft- und Weltraumrecht 59 (2010): 241–251.
Ferencz, Csaba. "Overview of Hungarian Space Activity: Plenty of Potential, not Enough Support." Space Policy 26 (2010): 105–108.
Froehlich, Annette. "Space and the Complexity of European Rules and Policies: The Common Projects Galileo and GMES–Precedence for a new European Legal Approach?" Acta Astronautica 66 (2010): 1262–1265.
Gilbert, Jo-Anne. "'We can Lick Gravity, but…': What Trajectory for Space in Australia?" Space Policy 25 (2009): 174–180.
Grigoriev, Anatoly I., et al. "Space Medicine Policy Development for the International Space Station." Acta Astronautica 65 (2009): 603–612.
Groemer, Gernot, et al. "Human Performance Data in a High Workload Environment During the Simulated Mars Expedition 'AustroMars'." Acta Astronautica 66 (2010): 780–787.
Guglielmi, M., et al. "The Technology Management Process at the European Space Agency." Acta Astronautica 66 (2010): 883–889.
Hansel, Mischa. "The USA and Arms Control in Space: An IR Analysis." Space Policy 26 (2010): 91–98.
Hegde, V.S., Jayaraman, V. and Srivastava, Sanjay K. "India's EO Infrastructure for Disaster Reduction: Lessons and Perspectives." Acta Astronautica 65 (2009): 1471–1478.

Hicks, Colin. "History of UK Contribution to Astronautics: Politics and Government." Acta Astronautica 65 (2009): 1593–1598.
Hiriart, Thomas and Saleh, Joseph H. "Observations on the Evolution of Satellite Launch Volume and Cyclicality in the Space Industry." Space Policy 26 (2010): 53–60.
Hobe, Stephan and Mey, Jan Helge. "UN Space Debris Mitigation Guidelines." ZLW Zeitschrift für Luft- und Weltraumrecht 58 (2009): 388–403.
Hoerber, Thomas. "ESA+EU: Ideology or Pragmatic Task Sharing?" Space Policy 25 (2009): 206–208.
Hoofs, R.M.T., et al. "Venus Express – Science Observations Experience at Venus." Acta Astronautica 65 (2009): 987–1000.
Horneck, Gerda, et al. "Towards a European Vision for Space Exploration: Recommendations of the Space Advisory Group of the European Commission." Space Policy 26 (2010): 109–112.
Huntley, Wade L., Bock, Joseph G. and Weingartner, Miranda. "Planning the Unplannable: Scenarios on the Future of Space." Space Policy 26 (2010): 25–38.
Kolk, Alar and Võõras, Madis. "Estonian Space Policy and Governance in the International Space Community." Space Policy 25 (2009): 218–223.
Johnson, Nicholas L. and Stansbery, Eugene G. "The New NASA Orbital Debris Mitigation Procedural Requirements and Standards." Acta Astronautica 66 (2010): 362–367.
Kraft Newman, Lauri. "The NASA Robotic Conjunction Assessment Process: Overview and Operational Experiences." Acta Astronautica 66 (2010): 1253–1261.
Kristiansen, Raymond and Nicklasson, Per Johan. "Spacecraft Formation Flying: A Review and new Results on State Feedback Control." Acta Astronautica 65 (2009): 1537–1552.
Landis, Rob R., et al. "Piloted Operations at a Near-Earth Object (NEO)." Acta Astronautica 65 (2009): 1689–1697.
Lindenmoyer, Alan and Stone, Dennis. "Status of NASA's Commercial Cargo and Crew Transportation Initiative." Acta Astronautica 66 (2010): 788–791.
Lundquist, Charles A. "A Sputnik IV Saga." Acta Astronautica 65 (2009): 1530–1536.
MacLeish, Marlene Y. and Thomson, William A. "Global Visions for Space Exploration Education." Acta Astronautica 66 (2010): 1285–1290.
Mankins, John C. "Stepping stones to the future: Achieving a Sustainable Lunar Outpost." Acta Astronautica 65 (2009): 1190–1195.
Mantl, Leopold. "The Commission Proposal for a Regulation on the European Earth Observation Programme (GMES) and its Initial Operations (2011–2013). A Major Milestone for GMES." ZLW Zeitschrift für Luft- und Weltraumrecht 58 (2009): 404–422.
Masson-Zwaan, Tanja and Freeland, Steven. "Between Heaven and Earth: The Legal Challenges of Human Space Travel." Acta Astronautica 66 (2010): 1597–1607.
Mathieu, Charlotte: "Assessing Russia's space cooperation with China and India – Opportunities and Challenges for Europe." Acta Astronautica 66 (2010): 355–361.
Millard, Douglas. "A review of UK Space Activity and Historiography, 1957–2007." Acta Astronautica 66 (2010): 1291–1295.
Montebugnoli, Stelio, et al. "The Next Steps in Seti-Italia Science and Technology." Acta Astronautica 66 (2010): 610–616.
Moore, Alan D., et al. "Cardiovascular Exercise in the U.S. Space Program: Past, Present and Future" Acta Astronautica 66 (2010): 974–988.
Morelli, Marianna and Campostrini, Pierpaolo. "Network of European Regions Using Space Technologies an Update on the NEREUS Constitution." Acta Astronautica 66 (2010): 279–284.
Moulin, Hervé. "The International Geophysical Year: Its Influence on the Beginning of the French Space Program." Acta Astronautica 66 (2010): 688–692.
Murthi, K.R. Sridhara, Bhaskaranarayana, A. and Madhusudana, H.N. "New Developments in Indian Space Policies and Programmes – The Next Five Years." Acta Astronautica 66 (2010): 333–340.

Narasaiah, N., et al. "Space Capsule Recovery – Evaluation of Risk Factors, Safety Plans and Procedures and Design of Experiments for Systems Qualification." Acta Astronautica 65 (2009): 1224–1230.
Niebur, Susan M. "Principal Investigators and Mission Leadership." Space Policy 25 (2009): 181–186.
——. "Women and Mission Leadership." Space Policy 25 (2009): 224–235.
Nosanov, Jeffrey P. "International Traffic in Arms Regulations – Controversy and Reform." Astropolitics 7 (2009): 206–227.
Ordyna, Paul. "Insuring Human Space Flight: An Underwriter's Dilemma." Journal of Space Law 36 (2010): 231–251.
Pace, Scott. "Challenges to U.S. Space Sustainability." Space Policy 25 (2009): 156–159.
Paikowsky, Deganit and Israel, Isaac Ben. "Science and Technology for National development: The Case of Israel's Space Program." Acta Astronautica 65 (2009): 1462–1470.
Parameswaran, S. and Shenoy, H.P. "Autonomy in Ground Operations for Geo-missions of ISRO." Acta Astronautica 65 (2009): 1330–1335.
Peeters, Walter. "From Suborbital Space Tourism to Commercial Personal Spaceflight." Acta Astronautica 66 (2010): 1625–1632.
Peter, Nicolas: "Space power and its implications – The case of Europe." Acta Astronautica 66 (2010): 348–354.
Peter, Nicolas and Delmotte, Raphaëlle. "Overview of Global Space Activities in 2007/2008." Acta Astronautica 65 (2009): 295–307.
Petroni, Giorgio, Venturini, Karen and Santini, Stefano. "Space Technology Transfer Policies: Learning from Scientific Satellite Case Studies." Space Policy 26 (2010): 39–52.
Pelton, Joseph N. "A New Space Vision for NASA – And for Space Entrepreneurs too?" Space Policy 26 (2010): 78–80.
Remuss, Nina-Louisa. "Creating a European Internal Security Strategy Involving Space Applications." Space Policy 26 (2010): 9–14.
Robinson, George. "Impact of the U.S. International Traffic in Arms Regulations (ITAR) on International Collaboration Involving Space Research, Exploration and Commercialization." ZLW Zeitschrift für Luft- und Weltraumrecht 58 (2009): 423–428.
Sadeh, Eligar. "National Space Symposium 2009." Astropolitics 7 (2009): 165–170.
Sandal, Gro Mjeldheim and Manzey, Dietrich. "Cross-cultural Issues in Space Operations: A Survey Study among Ground Personnel of the European Space Agency." Acta Astronautica 65 (2009): 1520–1529.
Sandau, Rainer. "Status and Trends of Small Satellite Missions for Earth Observation." Acta Astronautica 66 (2010): 1–12.
Schweickart, Russell L. "Decision Program on Asteroid Threat Mitigation." Acta Astronautica 65 (2009): 1402–1408.
Secara, Teodora and Bruston, Jean. "Current Barriers and Factors of Success in the Diffusion of Satellite Services in Europe." Space Policy 25 (2009): 209–217.
Sheldon, John B. "The Strategic Rationale for Britain in Space. Issues, Opportunities and Challenges." The RUSI Journal 155 (2010): 28–34.
Smith, Lesley Jane and Doldirina, Catherine: "Jurisdiction and Applicable Law in Cases of Damage from Space in Europe – The Advent of the Most Suitable Choice–Rome II." Acta Astronautica 66 (2010): 239–244.
Su, Jinyuan. "The 'Peaceful Purposes' Principle in Outer Space and the Russia – China PPWT Proposal." Space Policy 26 (2010): 81–90.
Suedfeld, Peter, Brcic, Jelena and Legkaia, Katya. "Coping with the Problems of Space Flight: Reports from Astronauts and Cosmonauts." Acta Astronautica 65 (2009): 312–324.
Sumrall, John P. and Creech, Steve. "Update on the Ares V to Support Heavy Lift for U.S. Space Exploration Policy." Acta Astronautica 66 (2010): 1133–1145.
Tarikhi, Parviz. "Iran's Space Programme: Riding High for Peace and Pride." Space Policy 25 (2009): 160–173.
Uhran, Mark L. "Progress toward Establishing a U.S. National Laboratory on the International Space Station." Acta Astronautica 66 (2010): 149–156.

Vasant, Gowarikar and Suresh, B.N. "History of Rocketry in India." Acta Astronautica 65 (2009): 1515–1519.

Webber, Derek. "Point-to-point Sub-orbital Space tourism: Some Initial Considerations" Acta Astronautica 66 (2010): 1645–1651.

Yehia, Julie Abou and Schrogl, Kai-Uwe. "European Regulation for Private Human Spaceflight in the Context of Space Traffic Management." Acta Astronautics 66 (2010): 1618–1624.

Ziliotto, Véronique. "Relevance of the Futron/Zogby Survey Conclusions to the Current Space Tourism Industry." Acta Astronautica 66 (2010): 1547–1552.

Zongpeng, Zhu. "The Current Situation of China Manned Aerospace Technology and the Direction for its Further Development." Acta Astronautica 65 (2009): 308–311.

List of figures and tables

Figures

Part 1: The Year in Space 2009/2010

Figure 1: Global shares in GERD 22
Figure 2: Global Shares in patent applications 23
Figure 3: Public space budgets of major space powers in 2009 (Based on Euroconsult data) 63
Figure 4: Public space budgets (selection) as a share of nom. GDP in 2009 (source: Euroconsult data, IMF). 64
Figure 5: Public space budgets per capita (selection) in 2009 (source: Euroconsult data, UN World Population Prospects) 65
Figure 6: Public space budgets as share of GDP mapped against space budgets per capita in 2009 with the bubble size indicating the absolute space budget (source: Euroconsult data, UN World Population Prospects) 65
Figure 7: Public space budgets as a share of GDP mapped against space budgets per capita in 2009 with the bubble size indicating the absolute space budget, excluding the U.S. and Russia (source: Euroconsult data, UN World Population Prospects) 66
Figure 8: Satellites launched in 2009 by manufacturer and commercial status (source: Futron) 93
Figure 9: GEO satellite orders in 2009 by manufacturer. 94

Part 2: Views and Insights

Figure 1: The G20 Summit on global economic recovery and financial markets (2009) (source: BBC.com). 191
Figure 2: The Lisbon Summit (source: EU Council) 201
Figure 3: The European Court of Justice (source: Wolfgang von Brauchitsch/Bloomberg News). 205
Figure 4: Artist's impression of a Galileo Satellite (source: ESA) 229
Figure 5: The amount spent by the United States on piloted spaceflight from 1959 to 2015 231

Figure 6: The ISS (source: NASA) . . . 249
Figure 7: U.S. President Barack Obama speaking at NASA Kennedy Space Center (source: NASA/Bill Ingalls) . . . 265
Figure 8: Artist's view of Envisat (source: ESA) . . . 276
Figure 9: The Copenhagen Summit (source: The Guardian) . . . 280
Figure 10: Ikonos images of part of Aceh, Sumatra, Indonesia with a pixel size of approximately 2 m (source: CRISP, http://www.crisp.nus.edu.sg/tsunami/tsunami.html, accessed 22 July 2010) . . . 287
Figure 11: Some features of small satellites . . . 299
Figure 12: Some civil Earth surface Imagers to show the trend of ground resolution (source: GSD) . . . 301
Figure 13: Fire detection by MODIS and BIRD (Australia, 5 January 2002) . . . 302
Figure 14: Requirements for distributed space systems . . . 305

Tables

Part 1: The Year in Space 2009/2010

Table 1: Top 10 FSS operators in 2009. . . . 96

Part 2: Views and Insights

Table 1: Costs of US piloted programs by Claude Lafleur, Monday, March 8, 2010 (source: U.S. Office of Management and Budget). . . . 236
Table 2: The members of the International Charter Space and Major Disasters, October 2010. Source: http://www.disasterscharter.org/web/charter/home . . . 284
Table 3: A summary of selected disaster management information systems that provide Earth observation data . . . 290
Table 4: Confusion of small satellite definitions . . . 298

About the authors

Blandina Baranes joined the European Space Policy Institute (ESPI) in Vienna in February 2005 and currently holds the position of ESPI Communications Manager. Prior to that she was the chief librarian of the Jewish Studies' Department of the University of Vienna. During the past years she has also worked as a documentalist and librarian for different institutions, such as the Austrian broadcasting corporation, the Der Spiegel magazine and others. She conducted her studies and research in Austria and Israel and graduated with a Masters Degree from Vienna University, Faculty of Philosophy, Department of Social and Cultural Anthropology.

Isaac Ben-Israel studied Mathematics, Physics and Philosophy at Tel-Aviv University, receiving his Ph.D. in 1988. He joined the Israel Air Force (IAF) after graduating high school and has served continuously up to his retirement in 2002. After his retirement from IDF Isaac Ben Israel joined the University of Tel-Aviv as a professor and headed the Curiel Centre for International Studies (2002–2004), and the Program for Security Studies (2004–2007), while he was also a member of the Jaffe Centre for Strategic Studies (2002–2004). In 2002 he founded and headed the Tel-Aviv University Workshop for Science, Technology and Security. In 2002 he founded RAY-TOP (Technology opportunities) Ltd, a consulting company focused on providing advice to governments and industries in technological and strategic issues. Professor Ben-Israel was a member of the 17th Knesset (Israeli Parliament) between June 2007 and February 2009. He also served as a member of the board of directors of IAI (2000–2002), the board of the Israel Corp. (2004–2007) and the R&D advisory board of TEVA (2003–2007) and as Chairman of the Technion Entrepreneurial Incubator (2007). Professor Ben-Israel has written numerous papers on military and security issues. His book *Dialogues on Science and Military Intelligence* (1989) won the Itzhak-Sade Award for Military Literature.

Simonetta Cheli is Head of the Coordination Office of the Directorate of Earth Observations at ESA ESRIN in Frascati since 2007. She joined ESA HQ International Relations Division in 1988. Consequently she was the Head of the Public and Institutional Relations Office, transferring to ESA ESRIN in 1999 where she was responsible for institutional relations with Italy, Spain and Portugal and communication activities. Prior to this she worked at the European

Commission in the Cabinet of the Commissioner responsible for information, culture and communication.She has made numerous publications on international space policy, aeronautics, and strategic affairs. She was the Chairman of EURISY Programme Committee Working Group and is currently a member of various International committees (ISPRS, IAF) and European Commission Working Groups. She holds a degree in Political Sciences with a specialisation in International Law and obtained a Masters Degree at the Diplomatic and Strategic Institute in Paris, France. She wrote her thesis on International Law of Telecommunication Satellites.

Ray Harris is Emeritus Professor of Geography at the University College of London. Prior to this he was the Executive Dean (2004–2008) and Vice-Dean (2003–2004) of the Faculty of Social and Historical Sciences. Between 1995 and 2008 he taught remote sensing, headed the Faculty's Remote Sensing Unit (RSU) and served as Deputy Head of the Department of Geography. Prior to this he worked as an Earth Observation Manager responsible for consultancy, business development, bid management and project management in Earth observation for Logica UK Ltd (1990–1995) and as a Marketing Manager responsible for strategy development, business planning, financial planning, national and international marketing in the Space Systems, Environment and Civil Systems areas for Software Sciences Limited (1987–1990). He started his academic career as a Lecturer at the Geography University of Durham, where he was responsible for establishing a research group of 12 in remote sensing that led in turn to the university identifying to the UGC remote sensing as a major university strength. He holds a PhD in Geography from the University of Bristol. During his career he has held numerous positions in academic and Earth observation related committees and he has been a member of the Advisory Committee of the Centre for Space Science and Technology Education in Asia and the Pacific (CSSTEAP) in India since 2004.

Spyros Pagkratis currently holds the position of Resident Fellow at the European Space Policy Institute in Vienna, which he joined in 2010. He has been the study leader of two annual ESPI Reports on "Space Policies, Issues and Trends" that provide a comprehensive overview and analysis of developments in the field of international space policies on a yearly basis and are considered among the Institute's flagship publications. Prior to that, he briefly worked as a space policy analyst with the Western European Union Assembly and EADS Astrium Space Transportation in Paris. He holds a degree in history from Athens University and Master's degrees in international relations' history from Sorbonne, as well as in defence policy and armament procurement planning from Panthéon/Assas Universities in Paris.

Deganit Paikowsky holds the position of Research Fellow at the School of Government and Policy, and project manager of the "Science, Technology & Society" unit and "Tel Aviv Workshop for Science, Technology and Security". She completed her Master's Degree in international relations with distinction (magna cum laude) in Tel Aviv's political science department. She holds a Ph.D. in Political Science. Her research thesis focused on the impact of space technologies on warfare and force build-up in the US army and the IDF. She is the holder of numerous scholarships and academic awards.

Ian Pryke retired from the European Space Agency at the end of September 2003. He is currently a Senior Fellow/Assistant Professor at the Center for Aerospace Policy Research in the School of Public Policy of George Mason University and also operates as an independent consultant. He joined the European Space Research Organisation [later ESA] in 1969 working in the areas of data processing and satellite communications. In 1976 he transferred to the Agency's Earth Observation Programme Office, where he was involved in the formulation of ESA's Remote Sensing programme. In August 1979 he moved to the ESA Washington Office, where he was engaged in liaison work with both government and industry in the United States and Canada, taking over as Head of the Office in November 1983. He holds a B.Sc. Degree in Physics from the University of London and a M.Sc. Degree in Space Electronics and Communications from the University of Kent. He has been involved with the International Space University since its founding. He served as Chairman of the Board of Directors from September 1988 to September 1990, was created an Associate Founder in April 1995, and has been a member of the Board of Trustees from May 1997 until the present. He currently Chairs the Strategic Planning Committee. Mr. Pryke is a Fellow of the AAS [Current Vice President Public Policy], a Fellow of the AIAA, a Member of the IAA, and a Fellow of the BIS. He is the recipient of an AAS President's Recognition Award, the AIAA International Cooperation Award, the NASA Public Service Medal, and an Aviation Week and Space Technology 2002 Laurels Award.

Rainer Sandau retired from DLR after over 30 years of experience in airborne and spaceborne remote sensing activities. He was involved in instrumentations of space missions to Venus, Mars and Earth, and also in numerous concepts for instruments and small satellites for or with different countries and space agencies, e.g. Argentina, GB, Russia, Taiwan, Tunisia, CNES, ESA, NASA, ranging from the concept of a German stereo camera on-board the French SPOT 5 mission to a lander concept jointly done with NASA/JPL for ESA's cometary mission ROSETTA. He is member of various national and international associations, for instance member of the International Academy of Astronautics (acting as the

Technical Director, Satellites and Space Applications) and chairman of the International Policy Advisory Committee (IPAC) of the International Society of Photogrammetry and Remote Sensing (ISPRS). He authored or co-authored over 250 publications, holds over 30 patents, is member of the Editorial Advisory Board of the ISPRS Journal "Photogrammetry and Remote Sensing", and is editor or co-editor of 19 books or conference proceedings. Specialist Subjects: International security, North-East Asia, the international politics of space, balance of power theory, comparative liberal politics.

Kai-Uwe Schrogl is the Director of the European Space Policy Institute (ESPI) in Vienna, Austria since 1 September 2007. Before, he was Head Corporate Development and External Relations Department in the German Aerospace Center (DLR). In his previous career he worked with the German Ministry for Post and Telecommunications and the German Space Agency (DARA). He has been delegate to numerous international forums and recently served as the chairman of various European and global committees (ESA International Relations Committee, was chairman at UNCOPUOS working groups "Launching State" and "Registration Practice"). Kai-Uwe Schrogl has published nine books and more than 100 articles, reports and papers in the fields of space policy and law as well as telecommunications policy. He is Member of the Board of Directors of the International Institute of Space Law, Member of the International Academy of Astronautics (chairing its Commission on policy, economics and law) and the Russian Academy for Cosmonautics as well as member in editorial boards of international journals in the field of space policy and law (Acta Astronautica, Space Policy, Zeitschrift für Luft- und Weltraumrecht, Studies in Space Law/Nijhoff). He holds a doctorate degree in political science, lectures international relations at Tübingen University, Germany (as a Honorarprofessor) and has been a regular guest lecturer i.a. at the International Space University and the Summer Courses of the European Centre for Space Law.

Michael Sheehan has been a Professor of politics and international relations at the Department of Political and Cultural Studies of the University of Swansea since 2004. His current research focuses on the military use of outer space, particularly the arms control issues surrounding anti-satellite systems, and on the military space policies of the European Union. He is continuing his research into the meaning of the concept of security in the contemporary world. He has published in a variety of journals including Defense and Security Analysis, Mediterranean Quarterly, Contemporary Security Policy, Jane's Intelligence Review, NATO Defence College Monograph Series, Korean Journal of Defense Analysis, Diplomacy and Statecraft, Defense Analysis, Contemporary South Asia, Pacific Review, Review of International Studies, History, and Arms Control.

He previously held the position of Director at the Scottish Centre for International Security of the University of Aberdeen and worked at the International Institute for Strategic Studies in London. He has published several books on the field of international security, space policy and space security.

Lesley Jane Smith is a professor of comparative law at the Leuphana University of Lüneburg, Germany. She has long-standing academic experience as professor of International and European Economic Law, including comparative law, including as a guest lecturer within various international law programmes in Europe. She has worked for both public and private legal services. Her legal expertise extends to complex international contracts and conflicts, and latterly included law reform work in transformation states. Her research interests are in European competition and intellectual property law, European private law and in particular, the interface between business and space law. She is an active member of the International Institute of Space Law.

Christophe Venet is a Ph.D. candidate at the Institute for Political Science at Tübingen University, Germany and an Associate Fellow of the European Space Policy Institute (ESPI) since January 2010. In 2009, he was Research Assistant at the Institute. In this capacity, he contributed to the Report "Space Policies, Issues and Trends in 2008/2009" and collaborates as a co-editor and co-author to the "Yearbook on Space Policy 2008/2009". He also worked on the issue of space commerce and space entrepreneurship, preparing several presentations on these issues. He was invited to become a peer-reviewer for the journal "Acta Astronautica" in August 2009. He graduated from the Institut d'Etudes Politiques de Strasbourg, France, and studied international relations at the Moscow State Institute of International Relations (MGIMO), Russia. He also holds a Masters degree in Peace Studies and International Politics from Tübingen University. His dissertation deals with EU policy in the field of space security, focusing on the actorness of Europe and on interests and norms underlying the policy processes within the European Space Policy.

Heike Wieland currently holds the position of Principal Director of legal services at the European Patent Office. Prior to this she served as Head of the GSA Legal Office (2006–2010) and as the Agency's Acting Director in 2010. Her fields of specialisation include international technology projects and transfers, venture capital for new technology projects, aviation law and legal risk management. She has worked as Senior Legal Counsel at EADS Deutschland, GmbH, Defence and Security Systems from 2002 to 2006. Prior to that, Ms. Wieland was a self-employed attorney for over a decade, focusing in particular on German-Italian activities. She holds a law degree from the University of Passau, Germany and has been a member of the German bar (Munich) since 1990.

Contributors to the Yearbook on Space Policy 2009/2010 from left: Spyros Pagkratis (ESPI), Rainer Sandau (ISPRS), Ian Pryke (Geroge Mason University, Washington DC), Lesley-Jane Smith (Leuphana University), Christophe Venet (University of Tübingen), Blandina Baranes (ESPI), Michael Sheehan (University of Swansea), Simonetta Cheli (ESA), Ray Harris (University College, London), Heike Wieland (GSA) and Kai-Uwe Schrogl (ESPI).

Index

A

Afghanistan 7, 8, 75, 99, 106
Agenzia Spaziale Italiana (ASI) 42, 43, 75, 191, 248, 256, 320, 329
Algeria 60, 284, 290, 308
Anti-Satellite test (ASAT) 59, 82, 104, 113, 115, 263, 264, 336
Applications 7, 9, 11, 14, 17, 23, 24, 26, 27, 30, 33, 35, 37, 40, 42, 43, 46, 55, 57–60, 71, 72, 100, 102, 113, 114, 136, 159, 167, 176, 187, 189, 192–194, 196, 235, 236, 260, 262, 268, 269, 272, 297, 304, 318, 323, 324, 326, 331, 332, 343, 349
Ares 1 254
Ares 5 139
Ariane 31, 34, 38, 39, 43, 69, 74, 78, 82, 85, 87, 89–91, 134, 135, 165, 203, 256, 309, 314–316
Arianespace 72, 77–79, 81, 89–92, 139
Automated Transfer Vehicle (ATV) 75, 89, 145, 249, 255, 306, 320

B

Baikonur 78, 86, 90, 112, 139, 140, 161, 323
Belgium 38, 61, 64, 65, 153, 262, 323, 324
Brazil 19, 61, 96, 143, 262, 274, 301, 308
British National Space Centre (BNSC) 43, 256, 284
Budgets 4, 20, 30, 43, 58, 59, 62–66, 72, 99, 102, 139, 187, 190, 191, 261, 343

C

Canada 63–66, 96, 106, 116, 148, 248, 284, 287, 288, 320, 330, 332, 348
Cassini 150, 334
Centre National des Études Spatiales (CNES) 37–39, 42, 77, 78, 99, 151, 171, 172, 256, 284, 286, 298, 318, 319, 326, 348
China 5, 13, 14, 19, 23, 24, 26, 45, 58, 59, 63–66, 79, 81, 87–92, 94, 95, 114, 141, 142, 147, 150, 159, 160, 186, 236, 260, 262, 263, 274, 284, 288, 290, 301, 308, 317, 322, 323
Climate Change 9, 10, 27, 29, 39, 48, 52, 149, 173, 209, 264, 272–279, 293, 307
Cold War 187, 193
Columbus 39, 89, 249
Commercial Orbital Transportation Services (COTS) 82
Committee on Global Navigation Satellite Systems (ICG) 26–28, 322
Committee on the Peaceful Uses of Outer Space (COPUOS) 26
Competitiveness 22, 23, 32, 37, 44–46, 50, 54, 140, 160, 194, 200, 209, 247, 262, 269, 279
Conference on Disarmament (CD) 25, 26, 28
Constellation 34, 37, 42, 48, 50, 53, 56, 60, 72, 75–77, 80, 81, 86, 99, 100, 101, 103, 104, 106, 112, 114, 116, 135, 137–139, 161, 168, 175, 222, 235, 237, 244, 245, 254, 255, 266, 284, 290, 296, 303–311, 322
COROT (COnvection, ROtation and planetary Transits) 155, 324
COSMO-Skymed 42, 75, 106, 288, 320
Czech Republic 325

D

Data policy 162, 277, 285, 290, 291, 294
Debris 26, 41, 47, 50, 86, 98, 101, 104, 117, 118, 152, 169, 174, 239, 240, 263, 297, 311, 319

Department of Defense (DoD) 62
Deutsches Zentrum für Luft- und Raumfahrt (DLR) 40, 41, 148, 153, 157, 158, 256, 284, 290, 302, 307, 308, 319, 323, 327, 348, 349
Direct Broadcast Services (DBS) 67, 76
Direct-to-home (DTH) television 67, 68, 73, 188
Disarmament and International Security Committee (DISEC) 26
Disaster monitoring 281, 303
Dragon 82
Dual-use 97–101, 112–115, 264, 329

E

Earth Observation 16, 29, 31, 35, 37, 39, 40, 42, 44, 46, 48, 57, 59, 60, 62, 75, 78, 82, 86, 89, 99, 101, 102, 110–114, 116, 117, 136, 157, 162, 163, 175, 187, 239, 242, 267, 275–278, 281, 282, 285, 287–296, 300, 303, 306, 307, 312, 318, 327–333, 345–348
Energy 8, 11–13, 29, 56, 103, 110, 147, 156, 166, 175, 186, 234, 235, 264, 278, 279, 323, 336, 337
Environment 4, 9, 11, 13–16, 41, 45–54, 57, 60, 68, 75, 101, 104, 106, 117, 144, 149, 150, 152, 157, 165, 168, 171, 173, 193, 194, 220, 232, 237, 239, 242, 264, 267, 270, 272, 274–278, 282, 283, 296, 305, 318, 326, 347, 328
European Defence Agency (EDA) 16, 32, 35, 72, 100, 101, 209, 210, 319
European Space Policy (ESP) 32, 33, 35, 193, 194, 200, 246–248, 322, 350
European Space Policy Institute (ESPI) 217, 346, 347, 349–351
EU Code of Conduct (CoC) 118
EU Council 32, 218
Eumetsat 33, 36, 78, 162, 163, 277, 324–334
Europe 6, 12, 13, 16, 18, 19, 21, 23, 24, 29–32, 35–37, 40, 51, 63, 67–69, 73, 74, 76, 78, 88–93, 95, 98, 99, 101, 102, 134, 140, 145, 146, 150, 152, 154, 157, 159, 160, 162–164, 168, 169, 171, 173, 184, 186, 190, 192, 193, 195, 196, 199–202, 206, 209, 211, 217, 222, 224–227, 229, 242, 243, 245–250, 252, 255, 257, 258, 260, 261, 275, 278, 279, 282, 288, 292, 296, 308, 314, 317, 318, 322, 350
European Aeronautic Defence and Space Company (EADS) 34, 36, 38, 67, 76, 92–95, 99, 106, 146, 167, 189, 298, 319–321, 347, 350
European Aeronautic Defence and Space Company (EADS)-Astrium 34, 36, 38, 67, 76, 92–95, 99, 146, 189, 298, 319–321, 347
European Commission (EC) 16, 22, 23, 31–36, 69, 76, 100, 134, 135, 159, 161–163, 193, 217–220, 246, 277, 279, 292, 296, 319, 322, 323, 347
European Council 186, 194, 217
European Geostationary Navigation Overlay Service (EGNOS) 16, 33, 76, 160, 161, 319
European Parliament (EP) 217, 279
European Space Agency (ESA) 16, 29–44, 63, 75, 78, 100, 101, 116, 134, 135, 145, 146, 148–152, 154, 157, 158, 161–164, 166–168, 171–176, 191, 194, 196, 200, 202–205, 207–209, 217–219, 222, 225, 229, 242, 246–248, 250–252, 255, 256, 259, 260, 263, 267, 275–277, 283, 284, 286, 291, 298, 299, 302, 306, 318–320, 322–334, 344, 346, 348, 349, 351
European Space Policy 32, 33, 35, 193, 200, 246, 247, 248, 322
European Space Programme 32, 217, 246
European Union (EU) 20, 22, 32, 35, 36, 69, 100, 199, 200, 205–211, 217–223, 225, 227, 236, 246, 263, 277–279, 293, 296, 319, 322, 347, 349
ExoMars 148, 157, 331
Exploration 15, 26, 30–32, 39, 40, 43, 46, 48, 57, 59, 82, 88, 116, 117, 136, 139, 140, 143–147, 149, 150, 152, 154, 155, 157, 158, 163, 173, 191, 200, 231–235,

238–240, 242–249, 251–258, 261, 263, 284, 317, 322, 330, 331, 334, 336–341, 344
Export control 46, 49, 53–55, 71, 92, 257, 322

F

Fixed Satellite Services (FSS) 68, 96, 344
Framework Programme (FP7) for Research, Technological Development 31, 35, 195, 323
France 21, 23, 35–40, 42–44, 61, 63–66, 69, 77, 96, 98–100, 145, 151, 158–160, 174, 186, 190, 192, 203, 283, 284, 288, 301, 318, 319, 324–326, 329, 332, 333, 347, 350
Full Operational Capability (FOC) 134

G

Galileo 16, 31–34, 42, 76, 91, 94, 95, 101, 135, 140, 160, 161, 194, 199, 200, 202, 203, 211, 217, 218, 220, 222–224, 226, 227, 229, 237, 282, 319, 321, 322, 323
GEOSS (Group on Earth Observation System of Systems) 29, 291, 292
Geostationary orbit 89
Geostationary orbit (GEO) 29, 31, 41, 42, 77, 89–91, 93, 94, 174, 189, 275, 289, 291, 294, 296, 343
Geosynchronous Satellite Launch Vehicle (GSLV) 59, 142, 316
Germany 21, 23, 24, 28, 34, 36, 38–41, 43, 63–66, 75, 98, 99, 145, 146, 153, 164, 167, 173, 225, 288, 301, 302, 306, 317, 318, 326, 327, 333, 349, 350
GIOVE-A 202
GIOVE-B 202
Global Exploration Strategy (GES) 242, 243, 255, 256
Global Monitoring for Environment and Security (GMES) 16, 32, 33, 35, 36, 41, 42, 75, 78, 101, 162, 163, 194, 199, 211, 222, 276, 277, 290, 294, 296, 318, 321, 322, 330
Global Navigation Satellite System (GLONASS) 26, 27, 47, 56, 86, 91, 112, 160, 161, 222, 223, 319, 322, 323
Global Positioning System (GPS) 47, 57, 73, 76, 80, 114, 154, 160, 161, 189, 221, 223, 230, 282, 314, 316, 319, 323
Globalisation 49, 209, 278
GNSS Supervisory Authority (GSA) 160, 223, 350, 351
Governance 57, 162, 163, 207, 208, 219, 228, 235
Ground equipment 62, 73, 189
Group on Earth Observation (GEO) 29, 31, 41, 77, 90, 91, 93, 94, 174, 189, 275, 289, 291, 294, 296, 343

H

Hispasat 91, 94, 95–97, 332
Human exploration 139, 253, 255, 257, 339
Human spaceflight 33, 49, 135–137, 143, 144, 242, 243, 245, 251, 256, 333, 342
Hungary 328

I

India 5, 13, 19, 24, 45, 59, 61, 63–68, 81, 88–90, 95, 115, 142, 158, 159, 161, 186, 189, 191, 260–264, 284–286, 301, 302, 318, 320, 322, 323, 326, 334, 342, 347
Indian Space Research Organisation (ISRO) 59, 60, 68, 93, 115, 142, 191, 284, 286, 320, 323, 341
Inmarsat 69, 72, 83, 91, 105, 108, 159, 161, 187, 282, 319, 320
Innovation 15–17, 21, 23, 24, 32, 43, 44, 54, 136, 165, 167, 174, 194, 195, 209, 235, 247, 261, 267, 268, 279, 324–326, 329, 332, 339
Intelsat 79, 81, 83, 90–92, 95, 96, 107, 108, 159, 187, 315, 320, 321
Intergovernmental Panel on Climate Change (IPCC) 272, 276

International cooperation 2, 5, 10, 14, 18, 25, 45–48, 50, 52, 53, 57, 109, 118, 136, 139, 140, 146, 157, 173, 195, 224, 228, 232, 242–247, 252, 257, 258, 262, 263, 266, 267, 269, 270, 278, 283, 322, 323, 348
International Heliophysical Year (IHY) 27, 338
International Launch Services (ILS) 79, 91, 81, 92
International Space Station (ISS) 33, 39, 40, 43, 48, 75, 82, 86, 89, 94, 136, 140, 143–146, 154, 158, 168–172, 232–234, 242–254, 258, 266, 306, 314–316, 319, 326, 327, 339, 341, 344
International Telecommunications Union (ITU) 26, 27, 61, 118, 227, 263
International Traffic in Arms Regulations (ITAR) 46, 53, 87, 50, 54, 81, 322, 341
Iran 8, 9, 13, 82, 88,–90, 116, 118, 142, 143, 262, 263, 320, 336, 341
Israel 81, 82, 117, 157, 260, 264, 268, 269, 301, 308, 318, 341, 346

J

Japan 6, 7, 12, 21–24, 40, 56–58, 63–66, 86–90, 95, 96, 113, 114, 140, 145, 146, 151, 163, 166, 168, 169, 173, 186, 190, 248, 251, 263, 284, 288, 289, 308, 318, 326, 330
Japan Aerospace Exploration Agency (JAXA) 57, 58, 113, 140, 145, 155, 169, 170, 284, 289, 319, 330, 334

L

Laser Communication 41
Launch industry 88, 92, 140
Launch sector 87, 88, 90, 92, 189
Launch site 203
Launcher 34, 38, 39, 42, 43, 81, 82, 90, 102, 106, 110, 112, 113, 115–117, 134, 135, 139–141, 143, 144, 163, 164, 170, 232, 233, 235, 253, 297, 300–302, 306, 308, 309, 319, 320, 330, 333, 339
Law 27, 56, 169, 201–203, 205, 206, 245, 257, 337, 339, 341, 347, 349, 350
Lockheed Martin 80, 83, 90, 93, 94, 102, 104–106, 138, 173, 318
Low Earth Orbit (LEO) 38, 39, 41, 83, 88, 90, 104–106, 118, 136, 165, 169, 170, 174, 232, 244, 253, 255–257, 269, 319
Lunar CRater Observing and Sensing Satellite (LCROSS) 314
Lunar exploration 146, 147, 158, 191, 256
Lunar reconnaissance Orbiter (LRO) 314

M

Malaysia 286
Mars 48, 85, 139, 147–151, 157, 163, 166, 175, 232–235, 240, 241, 245, 246, 253–256, 282, 326, 330, 331, 336, 338, 348
Mars Express 148
Mars Reconnaissance Orbiter (MRO) 149
Meteosat Third Generation (MTG) 36, 318
Military 7, 8, 17, 23, 24, 35, 37, 42, 44, 55, 57, 58, 67, 72, 76, 77, 80, 83, 86, 89, 97–107, 111–117, 136, 138, 160, 161, 175, 187, 188, 209, 210, 227, 230, 231, 234, 237, 238, 261, 263, 264, 294, 300, 319, 326, 327, 329, 332
Missile Defence Agency (MDA) 80, 103, 108, 109, 110, 138, 318, 320
Mobile Satellite Services (MSS) 71, 72
Moon 136, 139, 146, 147, 150, 158, 175, 177, 232, 233, 235, 245, 253, 254, 255, 262, 317
Multinational Space-based Imagery System (MUSIS) 32

N

National Aeronautics and Space Administration (NASA) 50, 53, 57, 59, 82, 102, 104, 111, 135–140, 144, 145, 148–150, 152–158, 165, 170–173, 175, 227, 231–236, 239, 240, 242–245, 248–250, 253–257, 265, 266, 275, 286, 304, 307, 308, 317, 320, 322, 330, 332, 344, 348

National Oceanic and Atmospheric Administration (NOAA) 78, 154, 173, 277, 284, 318, 323
National Reconnaisance Office (NRO) 62, 102, 104, 111
National Space Policy 44, 48, 221, 244, 250, 266, 325, 347
Natural resources 11, 14, 19, 272
Navigation 7, 16, 17, 26–28, 33, 34, 47, 56, 60, 73, 76, 86, 91, 101, 102, 112, 114, 115, 134, 140, 155, 159–161, 164, 168, 174, 187, 188, 194, 217, 221, 256, 267, 282, 300, 301, 316, 319, 322–324, 331–333
Near Earth Objects (NEOs) 26, 48, 254
Netherlands 24, 35, 64, 65, 78, 134, 174, 330
Nigeria 60, 191, 284, 290, 308
North Atlantic Treaty Organization (NATO) 75, 100, 349
North Korea 88–90, 143

O

OHB 33, 34, 36, 41, 76, 94, 95, 99, 318–321, 327
Operations 7, 8, 30, 40, 46, 48, 52, 61, 68, 70, 105, 107, 111, 136, 149, 157, 162, 166–168, 208, 229, 238, 239, 243, 244, 249–252, 258, 263, 285, 310
Operators 20, 35, 53, 67, 70, 71, 75, 79, 81, 87, 91, 92, 95, 96, 103, 106, 107, 154, 159, 169, 170, 211, 344
Orbital Sciences Corp 90
Organisation for Economic Co-operation and Development (OECD) 12, 188, 291, 294
Orion 135, 138, 233, 234, 244, 254

P

Patent 23, 24, 176, 343, 349, 350
Plesetsk 112, 163
Poland 21, 35, 109, 331
Polar Satellite Launch Vehicle (PSLV) 302, 314
Prevention of an Arms Race in Outer Space (PAROS) 25, 26, 28
Propulsion 141, 144, 155, 163, 164, 166–168, 175, 234, 235
Public funding 20, 211, 294
Public-Private Partnership 46, 100, 175, 195, 264

R

Reconnaissance satellites 40, 70, 87, 99
Remote sensing 48, 71, 116, 158, 239, 281, 282, 283, 286, 289, 296, 297, 304, 307, 316, 327, 347–349
Research and Development (R&D) 15, 21–24, 32, 35, 37, 40, 46, 49, 52, 54, 57, 94, 109, 112, 115, 134, 136, 137, 139, 162, 166, 167, 175, 188, 194, 199, 244, 322, 323, 332, 346
Robotic exploration 48, 232, 234, 239, 244, 246
Romania 6, 21, 332
Roscosmos 139, 145, 158, 166, 175
Russia 5, 14, 19, 21, 22, 24, 26, 28, 36, 38, 40, 45, 55, 56, 63–66, 73, 85, 86, 88, 89, 90–94, 96, 111, 112, 114, 117, 118, 135, 139, 140, 145, 147, 157, 158–161, 163, 168, 174, 175, 186, 191, 192, 222, 224, 248, 252, 260, 261–263, 293, 308, 319, 323–325, 332, 343, 348–350

S

Satellite services 62, 66, 68–72, 75, 77, 88, 92, 95, 96, 189, 264, 319, 329
Saturn 150
Science and Technology (S&T) 20, 21, 23, 27, 43, 58, 60, 95, 111, 137, 244, 291, 323, 328, 331, 347
Sea Launch 73, 79–81, 88–92, 320
Security 7, 8, 13, 16, 17, 25, 26, 28, 32, 35, 38, 41, 45, 47–55, 57, 74, 75, 81, 82, 97, 100–102, 113, 114, 160, 187, 193, 206, 207, 209, 210, 221, 227, 235–239, 253, 261, 262–264, 266, 276, 279, 285, 290, 296, 332, 346, 348–350

Shuttle 111, 138, 144, 145, 165, 231–234, 243, 248, 249, 253, 266, 286, 288, 294, 314–316, 344
Solar and Heliospheric Observatory (SOHO) mission 173
Solar observation 158
Solar Terrestrial Relations Observatory (STEREO) 173
South Africa 60, 191, 262, 274, 279, 280, 301, 308
South Korea 22–24, 63–66, 88–90, 95, 116, 117, 142, 191, 262
Soyuz 34, 72, 78, 89, 91, 134, 135, 139, 140, 203, 234, 249, 255, 314–316
Space Council 32, 194, 205, 207, 211, 236, 246, 296
Space debris 26, 41, 47, 50, 98, 101, 104, 117, 169, 174, 240, 263, 297, 311
Space environment 45–47, 50, 51, 104, 117, 157, 232, 237, 239, 264, 270
Space expenditures 62, 63, 190
Space markets 66
Space powers 60, 63, 161, 186, 199, 207, 209
Space Situational Awareness (SSA) 16, 26, 31, 32, 47, 57, 98, 101, 109, 114, 173, 209, 240, 264
Space surveillance 80, 82, 98, 103, 109, 118, 159, 240, 318
Space Surveillance Network (SSN) 82, 159
Space transportation 34, 46, 67, 90, 134–137, 139, 140, 141, 245, 247, 253, 255, 256, 309, 347
Spacecraft 33, 34, 36, 38–42, 47, 58–62, 68, 69, 75–78, 81–85, 87–89, 91–93, 95, 99, 100, 104–106, 108–112, 114–116, 118, 134, 135, 137, 140, 145, 148–151, 155, 156, 158–161, 163, 165, 166, 168–175, 231, 233–235, 255, 297, 300, 303–306, 309, 310, 317, 319, 321
Spaceport 34, 78, 85, 91, 139, 140
SpaceX 88, 90, 309
Spain 38, 63–66, 75, 96, 167, 210, 290, 309, 332
Starsem 78
Sweden 21, 23, 64, 65, 325, 333
Switzerland 24, 64, 65, 171, 333

T

TecSAR 71, 81, 103, 117, 318
Telecommunications satellites 41, 42
Telesat 81, 96
Thailand 61, 286, 301
Thales Alenia Space 93
Transparency and Confidence-Building Measures (CBMs) 25, 26, 46, 47, 51
Treaty of Lisbon 199, 200, 246
Turkey 75, 117, 191, 290

U

Ukraine 94, 158, 161, 262
United Kingdom (UK) 3, 23, 35, 43, 44, 63–67, 85–88, 99, 100, 105, 158, 160, 167, 186, 190, 261, 262, 267, 284, 290, 291, 301, 306, 314, 319, 322, 334
United Nations (UN) 7, 9, 18, 25–28, 47, 65, 66, 202, 203, 263, 273, 274, 276, 282, 283, 287, 290, 292, 322, 323
United Nations Platform for Space-based Information for Disaster Management and Emergency Response (UNSPIDER) 27, 290
United States (U.S.) 3, 4, 6–8, 12, 13, 22–24, 26, 36, 40, 45–59, 62–67, 69–71, 73–93, 95, 97, 102–112, 114, 118, 135–140, 143, 144, 148, 154, 155, 157, 159–161, 163, 165, 168–170, 174–176, 185–196, 230, 231, 233, 235–240, 242–245, 248–258, 260, 261, 264, 265, 267, 277, 278, 284, 308, 309, 318–327, 332, 334
U.S. Wideband Global Satcom (WGS) 315

V

Vega 39, 42, 78, 135, 164, 203
Venezuela 191
Venus 150, 151, 317, 348
Venus Express 150, 151, 317
Virgin Galactic 83, 169, 170, 309, 319

Vision for Space Exploration (VSE) 233, 235, 249
Vostochny 139, 140

W

Weaponisation 28, 113, 237, 238, 240
White House 54, 136, 137, 232, 236, 238
World economy 273

X

X Prize 309

MIX
Papier aus verantwortungsvollen Quellen
Paper from responsible sources
FSC® C105338

If you have any concerns about our products,
you can contact us on
ProductSafety@springernature.com

In case Publisher is established outside the EU,
the EU authorized representative is:
Springer Nature Customer Service Center GmbH
Europaplatz 3, 69115 Heidelberg, Germany

Printed by Libri Plureos GmbH
in Hamburg, Germany